Electronics for Sensors

Electronics for Sensors

Editors

Giuseppe Ferri
Gianluca Barile
Alfiero Leoni

MDPI • Basel • Beijing • Wuhan • Barcelona • Belgrade • Manchester • Tokyo • Cluj • Tianjin

Editors
Giuseppe Ferri
University of L'Aquila
Italy

Gianluca Barile
University of L'Aquila
Italy

Alfiero Leoni
University of L'Aquila
Italy

Editorial Office
MDPI
St. Alban-Anlage 66
4052 Basel, Switzerland

This is a reprint of articles from the Special Issue published online in the open access journal *Sensors* (ISSN 1424-8220) (available at: https://www.mdpi.com/journal/sensors/special_issues/electronic_sensor).

For citation purposes, cite each article independently as indicated on the article page online and as indicated below:

LastName, A.A.; LastName, B.B.; LastName, C.C. Article Title. *Journal Name* **Year**, *Volume Number*, Page Range.

ISBN 978-3-0365-1240-2 (Hbk)
ISBN 978-3-0365-1241-9 (PDF)

© 2021 by the authors. Articles in this book are Open Access and distributed under the Creative Commons Attribution (CC BY) license, which allows users to download, copy and build upon published articles, as long as the author and publisher are properly credited, which ensures maximum dissemination and a wider impact of our publications.

The book as a whole is distributed by MDPI under the terms and conditions of the Creative Commons license CC BY-NC-ND.

Contents

About the Editors . . . . **vii**

Giuseppe Ferri, Gianluca Barile and Alfiero Leoni
Electronics for Sensors
Reprinted from: *Sensors* **2021**, *21*, 2226, doi:10.3390/s21062226 . . . . **1**

Ra úl Esteve Bosch, José F. Toledo Alarcón, Vicente Herrero Bosch, Ander Simón Estévez, Francesc Monrabal Capilla, VicenteÁlvarez Puerta, Javier Rodríguez Samaniego, Marc Querol Segura and Francisco Ballester Merelo
The Event Detection System in the NEXT-White Detector
Reprinted from: *Sensors* **2021**, *21*, 673, doi:10.3390/s21020673 . . . . **5**

Pietro Antonio Paolo Calò, Savino Petrignani, Michele Di Gioia and Cristoforo Marzocca
Analytical Study of Front-End Circuits Coupled to Silicon Photomultipliers for Timing Performance Estimation under the Influence of Parasitic Components
Reprinted from: *Sensors* **2020**, *20*, 4428, doi:10.3390/s20164428 . . . . **23**

Vincenzo Stornelli, Leonardo Pantoli, Gianluca Barile, Alfiero Leoni and Emanuele D'Amico
Silicon Photomultiplier Sensor Interface Based on a Discrete Second Generation Voltage Conveyor
Reprinted from: *Sensors* **2020**, *20*, 2042, doi:10.3390/s20072042 . . . . **41**

Hyuntak Jeon, Injun Choi, Soon-Jae Kweon and Minkyu Je
A Power-Efficient Radiation Sensor Interface with a Peak-Triggered Sampling Scheme for Mobile Dosimeters
Reprinted from: *Sensors* **2020**, *20*, 3255, doi:10.3390/s20113255 . . . . **55**

Andriy Holovatyy, Vasyl Teslyuk, Natalia Kryvinska and Artem Kazarian
Development of Microcontroller-Based System for Background Radiation Monitoring
Reprinted from: *Sensors* **2020**, *20*, 7322, doi:10.3390/s20247322 . . . . **65**

Nelson Castro, Margarida M. Fernandes, Clarisse Ribeiro, Vítor Correia, Rikardo Minguez and Senentxu Lanceros-Méndez
Magnetic Bioreactor for Magneto-, Mechano- and Electroactive Tissue Engineering Strategies
Reprinted from: *Sensors* **2020**, *20*, 3340, doi:10.3390/s20123340 . . . . **79**

Canjian Zhan, Jiafeng He, Mingjin Pan and Dehan Luo
Component Analysis of Gas Mixture Based on One-Dimensional Convolutional Neural Network
Reprinted from: *Sensors* **2021**, *21*, 347, doi:10.3390/s21020347 . . . . **93**

Montree Kumngern, Nattharinee Aupithak, Fabian Khateb and Tomasz Kulej
0.5 V Fifth-Order Butterworth Low-Pass Filter Using Multiple-Input OTA for ECG Applications
Reprinted from: *Sensors* **2020**, *20*, 7343, doi:10.3390/s20247343 . . . . **105**

Jianfeng Chen, Kai Zhang, Leiyang Wang and Mingyue Yang
Design of a High Precision Ultrasonic Gas Flowmeter
Reprinted from: *Sensors* **2020**, *20*, 4804, doi:10.3390/s20174804 . . . . **121**

Zhipeng Song, Zhixiang Zhao, Hongsen Yu, Jingwu Yang, Xi Zhang, Tengjie Sui, Jianfeng Xu, Siwei Xie, Qiu Huang and Qiyu Peng
An 8.8 ps RMS Resolution Time-To-Digital Converter Implemented in a 60 nm FPGA with Real-Time Temperature Correction
Reprinted from: *Sensors* **2020**, *20*, 2172, doi:10.3390/s20082172 . . . . **137**

Sounghun Shin, Yoontae Jung, Soon-Jae Kweon, Eunseok Lee, Jeong-Ho Park, Jinuk Kim, Hyung-Joun Yoo and Minkyu Je
Design of Reconfigurable Time-to-Digital Converter Based on Cascaded Time Interpolators for Electrical Impedance Spectroscopy
Reprinted from: *Sensors* **2020**, *20*, 1889, doi:10.3390/s20071889 . . . . **159**

Iván Zamora, Eyglis Ledesma, Arantxa Uranga and Núria Barniol
Miniaturized 0.13-μm CMOS Front-End Analog for AlN PMUT Arrays
Reprinted from: *Sensors* **2020**, *20*, 1205, doi:10.3390/s20041205 . . . . **177**

Chuangze Li, Benguang Han, Jie He, Zhongjie Guo and Longsheng Wu
A Highly Linear CMOS Image Sensor Design Based on an Adaptive Nonlinear Ramp Generator and Fully Differential Pipeline Sampling Quantization with a Double Auto-Zeroing Technique
Reprinted from: *Sensors* **2020**, *20*, 1046, doi:10.3390/s20041046 . . . . **195**

Chunxi Zhang, Xin Wang, Lailiang Song and Longjun Ran
Temperature Hysteresis Mechanism and Compensation of Quartz Flexible Accelerometer in Aerial Inertial Navigation System
Reprinted from: *Sensors* **2021**, *21*, 294, doi:10.3390/s2101029 . . . . **211**

Zbigniew Marszalek and Krzysztof Duda
Multifrequency Vector Measurement System for Reliable Vehicle Magnetic Profile Assessment
Reprinted from: *Sensors* **2020**, *20*, 4933, doi:10.3390/s20174933 . . . . **229**

Sara Pettinato, Andrea Orsini and Stefano Salvatori
Compact Current Reference Circuits with Low Temperature Drift and High Compliance Voltage
Reprinted from: *Sensors* **2020**, *20*, 4180, doi:10.3390/s20154180 . . . . **245**

About the Editors

Giuseppe Ferri (Professor) was born in L'Aquila, Italy, in 1965. He received a Laurea degree (cum laude) in electronic engineering in 1988. In 1991, he joined the Department of Electronic Engineering, University of L'Aquila, L'Aquila, Italy, where he is a full professor of Electronics and Microelectronics. His research activity mainly concerns the design of analog electronic circuits for integrated sensor applications both in voltage and current modes. In this field of research, he is author or coauthor of 7 patents, 3 international books, 1 book chapter, and more than 400 publications in international journals and conferences. He is the Ph.D. coordinator for his department and responsible for the sensors, microsystems, and instrumentation area in Italian Electronic Group. He is also an IEEE senior member and Editor of some international journals and special issues.

Gianluca Barile (Ph.D.) was born in Avezzano, Italy. He received a master's degree (cum laude) in Electronics Engineering from the University of L'Aquila (Italy) in 2016 and a Ph.D. degree from the same university in 2020, where he is currently a researcher in electronics. His research activities include voltage and current-mode sensor interfaces, integrated circuits design and systems for industrial electronics, and the IoT.

Alfiero Leoni (Ph.D.) was born in L'Aquila, Italy. He received a Laurea (M.S.) degree (cum laude) in electronic engineering, and a Ph.D. degree in electrical and information engineering from the University of L'Aquila, L'Aquila, Italy, in 2016 and 2020, respectively. Several months before his master's Laurea degree, he joined the Department of Industrial and Electronic Engineering at the University of L'Aquila as an external collaborator, where he is currently a researcher in electronics. His research activity mainly consists of the design of analog electronic circuits and systems for energy harvesting, and industrial and microwave applications.

Editorial

Electronics for Sensors

Giuseppe Ferri *, Gianluca Barile and Alfiero Leoni

Department of Industrial and Information Engineering and Economics, University of L'Aquila, 67100 L'Aquila, Italy; gianluca.barile@univaq.it (G.B.); alfiero.leoni@univaq.it (A.L.)
* Correspondence: giuseppe.ferri@univaq.it

Citation: Ferri, G.; Barile, G.; Leoni, A. Electronics for Sensors. *Sensors* **2021**, *21*, 2226. https://doi.org/10.3390/s21062226

Received: 11 March 2021
Accepted: 18 March 2021
Published: 23 March 2021

Publisher's Note: MDPI stays neutral with regard to jurisdictional claims in published maps and institutional affiliations.

Copyright: © 2021 by the authors. Licensee MDPI, Basel, Switzerland. This article is an open access article distributed under the terms and conditions of the Creative Commons Attribution (CC BY) license (https://creativecommons.org/licenses/by/4.0/).

Research on systems and circuits for interfacing sensors has always been, and will surely be, a highly prioritized, widespread, and lively topic. This is because each application that has usefulness in the real world inevitably needs to perceive and elaborate some kind of magnitudes coming from it. For this reason, however, advances in technology in many fields continuously pose new challenges in designing electronic interfaces. In particular, they have to satisfy parameters strictly related to the magnitude under evaluation, such as sensitivity and resolution, in addition to fulfilling new constraints coming from the particular macro-system into which the sensor interface is embedded, such as low power consumption and low cost. To make it even more challenging, with the first interfacing stage typically analog, it does not gain much benefit from the technology scaling in terms of chip area reduction and power consumption, whereas it has typically to deal with parasitic elements that are of the same magnitude as the sensing element. Therefore, the aim of this Special Issue has been to gather works that present new possible solutions regarding electronics for sensors, both related to the analog processing stage as well as including the digital processing section. Published manuscripts well embody the advances and the latest novel and emergent results on the Special Issue topic, showing best practices, implementations, and applications.

This Special Issue contains a selection of 16 papers covering photomultiplier tubes and silicon photomultipliers (PMTs and SiPMs) interfaces and applications [1–3], techniques for monitoring radiation levels [4,5], electronics for biomedical applications [6–8], design and applications of time-to-digital converters [9–11], interfaces for image sensors [12,13] and general-purpose theory and topologies for electronic interfaces [14–16].

In [1], Bosch et al. describe how it is possible to use together photomultiplier tubes and silicon photomultipliers in a high-pressured Xenon particle detection chamber as part of the Neutrino Experiment with a Xenon Time projection chamber (NEXT)–White experiment. In [2], Calò et al. address how it is possible to take full advantage of SiPMs intrinsic fast response time for characterizing and modeling the same SiPM when parasitic elements are associated with the interconnection between the sensor and the interface degrade its performances. In [3], Stornelli et al. propose a discrete-transistors level electronic interface for SiPMs. The interface is based on the novel active block called second-generation voltage conveyor (VCII) and has therefore the capability of working both in voltage and current mode taking advantage of both approaches.

In [4], Jeon et al. propose a power-efficient interface for battery-powered dosimeters. The manuscript presents both the analog and the analog-to-digital conversion stages and proposes a sampling scheme that allows the analog to digital converter (ADC) to only operate near its peak value. In [5], Holovatyy et al. design and implement a microcontroller-based radiation monitoring system that is able to detect if excessive radiation exposure is experienced in order to prevent long-term damages.

In [6], Castro et al. propose a novel bioreactor for biological tissue engineering that is able to remotely stimulate magnetoelectric scaffolds through a magnetic field, and therefore, via the scaffolds, provide a mechanical or electrical stimulus to the cells contained in the reactor. In [7], Zhan et al. propose a novel methodology for indoor gas analysis based

1

on a one-dimensional convolutional neural network, capable of detecting harmful gas components by extracting nonlinear features of the analyzed samples. In [8], Kumngern et al. propose the implementation of a fifth-order low pass filter for electrocardiogram (ECG) acquisition purposes.

In [9], Chen et al. propose a novel gas flowmeter based on a microcontroller and a time-to-digital converter aimed at solving common gas flowmeter drawbacks such as pressure loss, reduced range, and contact measurement. In [10], Song et al. proposed the implementation of a nonuniform multiphase time to digital converter, by means of a Cyclone 10 FPGA as a working platform, achieving a resolution of 8.8 ps and showing insensitivity to temperature variations. Another time-to-digital converter implementation is shown by Shin et al. in [11], with the feature to be reconfigurable and the aim to be used in electrical impedance spectroscopy.

Zamora et al. propose in [12] a complementary metal oxide semiconductor (CMOS)-based analog front-end transceiver for ultra-sonic imaging. The piezoelectric ultrasonic transducer was fabricated on top of the interface substrate. In [13], Li et al. tackle some of the limitations that affect typical CMOS sensors such as linearity degradation caused by in-pixel circuitry and propose the introduction of a pre-distortion circuit based on a nonlinear ramp generator to replace linear ones commonly adopted.

In [14], Zhang et al. propose a modeling of the temperature hysteresis of inertial navigation systems used in aerial navigation and a novel compensation methodology. Marszalek and Duda, in [15], present the design and performances of a simultaneous and multifrequency impedance measurement system used with four inductive loop sensors for detecting vehicles' magnetic profile. Finally, in [16], Pettinato et al. propose a current reference circuit based on the LT199x amplifier, which has increased accuracy and linearity with respect to commonly adopted op-amp-based solutions.

Guest Editors would like to take this opportunity to thank all the authors for having submitted their papers to this special issue and also want to thank all the reviewers for having dedicated their time and helping to improve the quality of the submitted papers.

Author Contributions: Conceptualization, G.B. and A.L.; methodology, G.F.; validation, G.F., G.B. and A.L.; investigation, G.F., G.B. and A.L.; writing—original draft preparation, G.B. and A.L.; writing—review and editing, G.F.; supervision, G.F. All authors have read and agreed to the published version of the manuscript.

Funding: This research received no external funding.

Conflicts of Interest: The authors declare no conflict of interest.

References

1. Esteve Bosch, R.; Toledo Alarcón, J.F.; Herrero Bosch, V.; Simón Estévez, A.; Monrabal Capilla, F.; Álvarez Puerta, V.; Rodríguez Samaniego, J.; Querol Segura, M.; Ballester Merelo, F. The Event Detection System in the NEXT-White Detector. *Sensors* **2021**, *21*, 673. [CrossRef] [PubMed]
2. Calò, P.A.P.; Petrignani, S.; Di Gioia, M.; Marzocca, C. Analytical Study of Front-End Circuits Coupled to Silicon Photomultipliers for Timing Performance Estimation under the Influence of Parasitic Components. *Sensors* **2020**, *20*, 4428. [CrossRef] [PubMed]
3. Stornelli, V.; Pantoli, L.; Barile, G.; Leoni, A.; D'Amico, E. Silicon Photomultiplier Sensor Interface Based on a Discrete Second Generation Voltage Conveyor. *Sensors* **2020**, *20*, 2042. [CrossRef] [PubMed]
4. Jeon, H.; Choi, I.; Kweon, S.-J.; Je, M. A Power-Efficient Radiation Sensor Interface with a Peak-Triggered Sampling Scheme for Mobile Dosimeters. *Sensors* **2020**, *20*, 3255. [CrossRef] [PubMed]
5. Holovatyy, A.; Teslyuk, V.; Kryvinska, N.; Kazarian, A. Development of Microcontroller-Based System for Background Radiation Monitoring. *Sensors* **2020**, *20*, 7322. [CrossRef] [PubMed]
6. Castro, N.; Fernandes, M.M.; Ribeiro, C.; Correia, V.; Minguez, R.; Lanceros-Méndez, S. Magnetic Bioreactor for Magneto-Mechano- and Electroactive Tissue Engineering Strategies. *Sensors* **2020**, *20*, 3340. [CrossRef] [PubMed]
7. Zhan, C.; He, J.; Pan, M.; Luo, D. Component Analysis of Gas Mixture Based on One-Dimensional Convolutional Neural Network. *Sensors* **2021**, *21*, 347. [CrossRef] [PubMed]
8. Kumngern, M.; Aupithak, N.; Khateb, F.; Kulej, T. 0.5 V Fifth-Order Butterworth Low-Pass Filter Using Multiple-Input OTA for ECG Applications. *Sensors* **2020**, *20*, 7343. [CrossRef] [PubMed]

9. Chen, J.; Zhang, K.; Wang, L.; Yang, M. Design of a High Precision Ultrasonic Gas Flowmeter. *Sensors* **2020**, *20*, 4804. [CrossRef] [PubMed]

10. Song, Z.; Zhao, Z.; Yu, H.; Yang, J.; Zhang, X.; Sui, T.; Xu, J.; Xie, S.; Huang, Q.; Peng, Q. An 8.8 ps RMS Resolution Time-To-Digital Converter Implemented in a 60 Nm FPGA with Real-Time Temperature Correction. *Sensors* **2020**, *20*, 2172. [CrossRef] [PubMed]

11. Shin, S.; Jung, Y.; Kweon, S.-J.; Lee, E.; Park, J.-H.; Kim, J.; Yoo, H.-J.; Je, M. Design of Reconfigurable Time-to-Digital Converter Based on Cascaded Time Interpolators for Electrical Impedance Spectroscopy. *Sensors* **2020**, *20*, 1889. [CrossRef] [PubMed]

12. Zamora, I.; Ledesma, E.; Uranga, A.; Barniol, N. Miniaturized 0.13-Mm CMOS Front-End Analog for AlN PMUT Arrays. *Sensors* **2020**, *20*, 1205. [CrossRef] [PubMed]

13. Li, C.; Han, B.; He, J.; Guo, Z.; Wu, L. A Highly Linear CMOS Image Sensor Design Based on an Adaptive Nonlinear Ramp Generator and Fully Differential Pipeline Sampling Quantization with a Double Auto-Zeroing Technique. *Sensors* **2020**, *20*, 1046. [CrossRef] [PubMed]

14. Zhang, C.; Wang, X.; Song, L.; Ran, L. Temperature Hysteresis Mechanism and Compensation of Quartz Flexible Accelerometer in Aerial Inertial Navigation System. *Sensors* **2021**, *21*, 294. [CrossRef] [PubMed]

15. Marszalek, Z.; Duda, K. Multifrequency Vector Measurement System for Reliable Vehicle Magnetic Profile Assessment. *Sensors* **2020**, *20*, 4933. [CrossRef] [PubMed]

16. Pettinato, S.; Orsini, A.; Salvatori, S. Compact Current Reference Circuits with Low Temperature Drift and High Compliance Voltage. *Sensors* **2020**, *20*, 4180. [CrossRef] [PubMed]

sensors

The Event Detection System in the NEXT-White Detector

Raúl Esteve Bosch [1,*], José F. Toledo Alarcón [1], Vicente Herrero Bosch [1], Ander Simón Estévez [2], Francesc Monrabal Capilla [3,4], Vicente Álvarez Puerta [1], Javier Rodríguez Samaniego [1], Marc Querol Segura [5] and Francisco Ballester Merelo [1]

[1] Instituto de Instrumentación para Imagen Molecular (I3M), Centro Mixto CSIC—Universitat Politècnica de València, Camino de Vera s/n, 46022 Valencia, Spain; jtoledo@eln.upv.es (J.F.T.A.); viherbos@eln.upv.es (V.H.B.); vialpue@upv.es (V.Á.P.); jarodsa@upv.es (J.R.S.); fballest@eln.upv.es (F.B.M.)
[2] Nuclear Engineering Unit, Faculty of Engineering Sciences, Ben-Gurion University of the Negev, P.O. Box 653, Beer-Sheva 8410501, Israel; ander@post.bgu.ac.il
[3] Donostia International Physics Center (DIPC), Paseo Manuel Lardizabal 4, 20018 Donostia-San Sebastian, Spain; francesc.monrabal@dipc.org
[4] IKERBASQUE, Basque Foundation for Science, 48013 Bilbao, Spain
[5] Instituto de Física Corpuscular (IFIC), CSIC & Universitat de València, Calle Catedrático José Beltrán 2, 46980 Paterna, Spain; marc.querol@ific.uv.es
* Correspondence: rauesbos@eln.upv.es

Abstract: This article describes the event detection system of the NEXT-White detector, a 5 kg high pressure xenon TPC with electroluminescent amplification, located in the Laboratorio Subterráneo de Canfranc (LSC), Spain. The detector is based on a plane of photomultipliers (PMTs) for energy measurements and a silicon photomultiplier (SiPM) tracking plane for offline topological event filtering. The event detection system, based on the SRS-ATCA data acquisition system developed in the framework of the CERN RD51 collaboration, has been designed to detect multiple events based on online PMT signal energy measurements and a coincidence-detection algorithm. Implemented on FPGA, the system has been successfully running and evolving during NEXT-White operation. The event detection system brings some relevant and new functionalities in the field. A distributed double event processor has been implemented to detect simultaneously two different types of events thus allowing simultaneous calibration and physics runs. This special feature provides constant monitoring of the detector conditions, being especially relevant to the lifetime and geometrical map computations which are needed to correct high-energy physics events. Other features, like primary scintillation event rejection, or a double buffer associated with the type of event being searched, help reduce the unnecessary data throughput thus minimizing dead time and improving trigger efficiency.

Keywords: xenon TPC; trigger concepts; data acquisition circuits; FPGA

Citation: Esteve Bosch, R.; Toledo Alarcón, J.F.; Herrero Bosch, V.; Simón Estévez, A.; Monrabal Capilla, F.; Álvarez Puerta, V.; Rodríguez Samaniego, J.; Querol Segura, M.; Ballester Merelo, F. The Event Detection System in the NEXT-White Detector. *Sensors* **2021**, *21*, 673. https://doi.org/10.3390/s21020673

Academic Editor: Giuseppe Ferri
Received: 2 December 2020
Accepted: 15 January 2021
Published: 19 January 2021

Publisher's Note: MDPI stays neutral with regard to jurisdictional claims in published maps and institutional affiliations.

Copyright: © 2021 by the authors. Licensee MDPI, Basel, Switzerland. This article is an open access article distributed under the terms and conditions of the Creative Commons Attribution (CC BY) license (https://creativecommons.org/licenses/by/4.0/).

1. Introduction

NEXT-White [1] is the phase one of NEXT-100 [2], an experiment to measure the neutrino double beta decay. The main objectives of NEXT-White are to demonstrate the High-Pressure Xenon Gas (HPGXe) TPC technology and to measure and validate the background model for the next generation of NEXT (Neutrino Experiment with a Xenon TPC) detectors. NEXT is a staged experimental program aiming at the detection of $\beta\beta0\nu$ in 136Xe, using successive generations of high-pressure gaseous xenon electroluminescent Time Projection Chambers (HPXe EL-TPC). While other xenon experiments (EXO [3], KamLAND-Zen [4]) have an approach of large exposures at the cost of losing energy resolution (and therefore background rejection potential), the HPGXe TPC technology has demonstrated an excellent energy resolution (1% FWHM at the Q value of the double beta decay [5]) and an impressive topological rejection factor thanks to the relatively low density of the xenon gas and to the tracking system with SiPMs [6]. These two tools provide a very low background index in the region of interest near $Q_{\beta\beta}$, a fundamental parameter for next generation detectors.

NEXT-White uses a 5 kg high-pressure gaseous xenon TPC with electroluminescence (EL) charge amplification and optical readout. The detector, a prototype of NEXT-100, is based on a PMT plane for energy measurements and a SiPM tracking plane for topological event filtering. The experiment has been operating since the end of 2016 at the Laboratorio Subterráneo de Canfranc (LSC) in Spain and will end its data-taking and be replaced by NEXT-100, during 2021. The NEXT-White detector, being a demonstrator of the 100 kg scale NEXT-100 detector, is needed to prove all the technical solutions that will be implemented in the later detector. Specially, the background budget of the different materials and components was strictly controlled to minimize it.

NEXT-White is an optical TPC. Ionizing radiation excites and ionizes the gas producing primary scintillation (S_1) and ionized electrons. The presence of a moderate electric field in the active volume prevents their recombination, drifting them towards the amplification region near the anode (EL region). In the anode, a more intense electric field accelerates the electrons producing a secondary scintillation (S_2) as a result of an amplification process. The light is detected with two sensor planes in opposite sides of the detector vessel (see Figure 1). The PMT plane measures event energy by integrating the amplified S_2 signal with 12 PMTs (Hamamatsu R11410-10) with a total area coverage of 31%. This plane is also responsible for the detection of the primary scintillation light that provides the event t0 and allows to position the event in the axial axis of the detector [7,8]. The SiPM Plane, instrumented with 1792 SiPMs (SensL series-C) in an array at a pitch of 10 mm, detects the light produced in the EL region allowing topological reconstruction that helps to discriminate double beta decay events from the background, being a unique characteristic of pressure gas detectors.

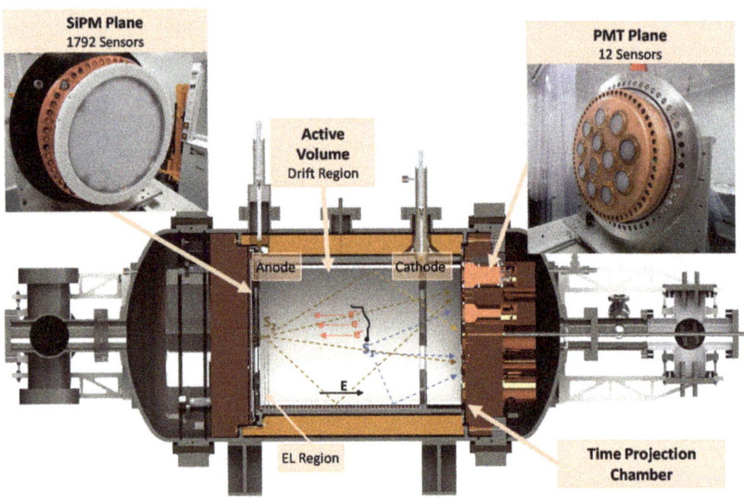

Figure 1. Schematic view of the NEXT-White Detector. Top right: Picture of the PMT sensors plane. Top left: Picture of the SiPM sensors plane. In the active volume of the TPC: Drawing with the principle of operation of the detector.

The Data Acquisition System (DAQ) in NEXT-White is an enhancement of the system developed for the NEXT-DEMO detector [9], the first prototype of NEXT-100. A double data circular buffer has also been implemented to reduce dead time. DAQ also implements data compression to reduce data throughput, which also has a remarkable impact on decreasing overall system dead time.

Event detection systems for argon or xenon TPCs such as those from LUX, LZ, DARKSIDE-50 and PANDA [10–13], as well as NEXT-White TPC, are based on fast detection events (S_1, S_2, or both) either per PMT, or using a signal sum of several PMTs. The detection operation is carried out by means of temporal, amplitude or energy parameters. Moreover, it is common to have a second level of detection that requires the coincidence of hits in a time window. In addition to these, the event detection system described in this article brings some relevant and new functionalities. First, it adds more complex event selection flexibility since the S_2 expected signals generated by MeV electrons in NEXT-White are very diverse due to the large variety of possible topologies of the events. Second, the system allows to carry out calibrations while taking physics data, ensuring a dynamic calibration. And third, the double circular buffer can be prioritized for a type of event thus reducing dead time for this selected event, generally, physics data versus calibration data.

2. Event Detection in NEXT

NEXT-White can provide a measurement of the two-neutrino double beta decay mode. Furthermore, to fully validate the detection technique, the NEXT collaboration must perform several studies in three different areas. First, the comparison of the background model extracted from in situ data-taking with the expected contribution to the radioactive budget of the different components. Second, to determine the energy resolution that must be in the order of 1% FWHM for energy depositions near the Q-value of the decay ($Q_{\beta\beta}$), that is 2.458 MeV. And third, the background rejection efficiency of the topological signature characteristic of this type of TPC.

Detector calibration and definition of the background model requires the detection of events of different characteristics. Alpha, electron, gamma, x-ray events associated with diverse sources (^{222}Rn, ^{137}Cs, ^{232}Th, ^{22}Na, and ^{83m}Kr) and muons, together with distinct field settings, have been studied to understand the detector in a wide range of energies, as well as to measure its energy resolution [5,14–16]. Equalization of the PMT gain is also foreseen in dedicated runs by means of regular LEDs that generate pulses periodically.

The event detection is performed by reading the signal of the PMT sensors. As commented in Section 1, one of the strengths of the NEXT technology is the capability to reconstruct the topology of the electrons in the gas. In a TPC, different topologies imply different temporal distributions of the charge reaching the amplification region and thus different amplitude and time duration pulses in the PMT sensors.

In order to identify diverse types of events, different event detection approaches have been explored. Figure 2 shows the typical type of searches applied in the experiment. In the most common approach, DAQ must identify online S_2 signals in a set of PMT sensors synchronously. The S_2 signal is centered in the middle of the data acquisition time window, while the pre-trigger data shows the S_1 signal, indicating the time drifted by the electrons that have a linear relation with their initial position in the detector, thus allowing for a 3D reconstruction of the event. In this approach, a peak-finding offline algorithm searches for the S_1 that can be as small as one photoelectron per PMT. In other approach, the same identification procedure is used to identify synchronously S_1 signals online in a set of PMT sensors, searching also for the corresponding S_2 offline. Both approaches are valid for detector calibration and physics data.

External trigger and a set of LEDs arranged in the chamber planes are also needed for test purposes and gain equalization of the different types of sensors.

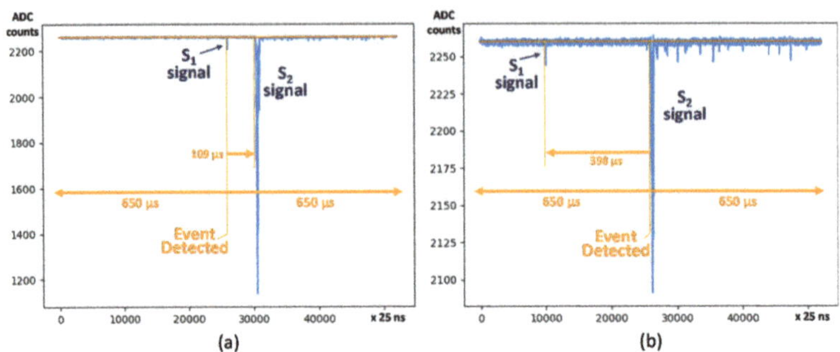

Figure 2. NEXT-White most common signal searches: (**a**) Online S_1 signal search, with offline S_2 signal search; (**b**) Online S_2 signal search with offline S_1 signal search. In both cases, a data acquisition window of 1300 µs and pre-trigger of 650 µs is applied.

3. NEXT-White Data Acquisition Hardware Architecture

3.1. Hardware Architecture

NEXT-White DAQ System, described in [17], is based on the SRS-ATCA (Scalable Readout System ported to the Advanced Telecommunications Computing Architecture standard) jointly developed by the NEXT Collaboration, CERN-PH and IFIN-HH Bucarest in the framework of the CERN RD51 collaboration [18,19]. Based on FPGA, it provides a customizable interface between a wide range of front-end and DAQ. Originally developed for micropattern gaseous detectors (MPGDs), it has also been used with other sensors such as PMTs and SiPMs. This flexibility is achieved via generic plug-in cards to a common module for all applications, the SRS DAQ module. The SRS default online system for RD51 users is DATE (ALICE Data Acquisition and Test Environment) [20], though other software options are possible, like RCDAQ [21,22].

The SRS DAQ module (referred to as "ATCA blade" in Figure 3) provides real-time digital processing through two Xilinx Virtex-6 FPGAs (XC6VLX240T-1ff1156). Each FPGA is connected to a DDR3 SO-DIMM memory module that can be used as a data memory buffer. Two on-board custom mezzanine connectors provide I/O flexibility for a wide range of front ends. We use two different mezzanines: the EAD-M1 unit (ADC Card), with 24 ADC channels (12 bit, up to 60 MSa/s) and the DTC card, with 12 DTC links [23] on HDMI connectors to interface digital front ends at a maximum speed of 200 Mb/s. DTC stands for Data Trigger and Control and was specified in the framework of RD51's SRS as a 4-pair differential interface between digital front-ends and the SRS DAQ module. It can be used for configuration and data transfer, and trigger purposes. The SRS DAQ module also includes connection to a Rear Transition Module (RTM) for multiple GbE, 10 GbE and other I/O connectivity, like two HDMI connectors for DTC connection and external input/output NIM connection.

In the case of NEXT-White, on the one side, the SiPM plane is readout through the SiPM Front-End Board (FEB) [24] and a DTC link to the SRS DAQ module. FEBs provide the analogue front-end electronics for 64 SiPM sensors and digital processing through a Xilinx Virtex-6 FPGA (XC6VLX130T-1ff784). On the other side, the PMT plane analogue front-end is interfaced directly through the EAD-M1 unit, that is used to digitize the analogue signals of the PMT sensors. In both cases, sensor data are sent individually to the DAQ PCs for offline processing.

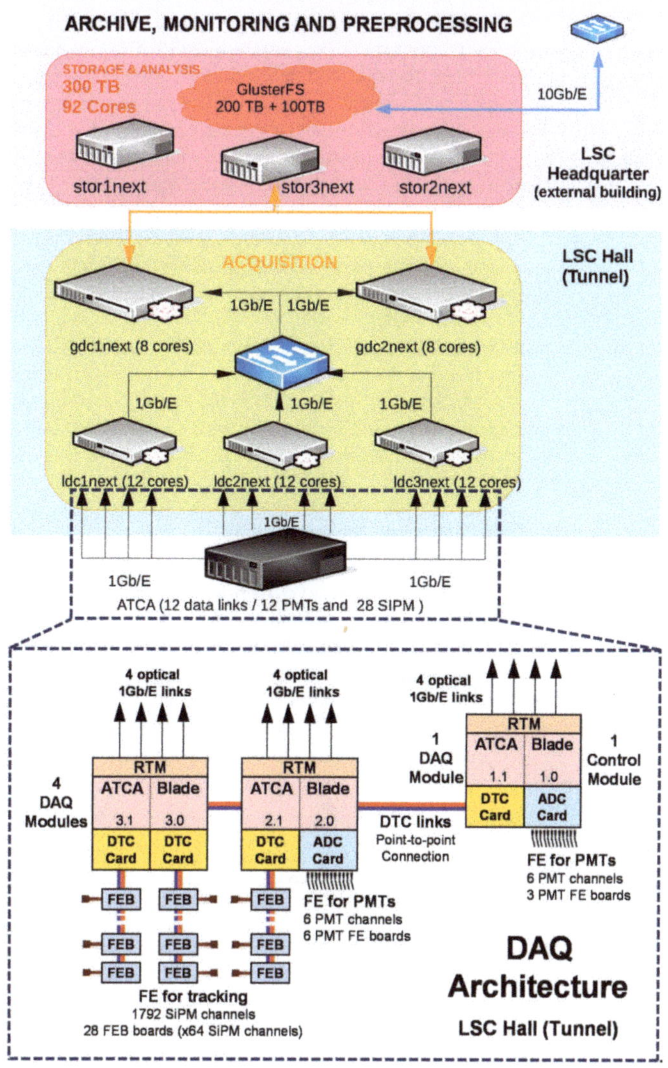

Figure 3. NEXT-White Data Acquisition Hardware Architecture.

3.2. System Architecture

As shown in Figure 3, three SRS DAQ modules (in the figure ATCA blades) are needed to read out the data from PMT and SiPM sensors, as well as to perform the event detection and system control. In NEXT, each SRS DAQ module constitutes two independent modules, five of them are used mainly for data acquisition purposes (DAQ Modules) and one for system control purposes (Control Module). DAQ modules devoted to readout PMT sensors can readout up to 12 PMTs each, while DAQ modules devoted to read SiPM sensor can readout data from 12 FEB boards each. Since each FEB board digitizes, formats and sends data form 64 SiPM sensors, 28 boards are enough to readout the total amount of sensors needed, 1792 SiPM sensors.

For system control duties, the Control Module is connected with the other modules through four LVDS-pair point-to-point connections (DTC links) devoted to clock distribution, synchronization, system configuration and event command handling. DTC links are also used to connect DAQ Modules to FEBs through DTC cards.

The readout works in push mode. DAQ Modules read out, time stamp and store data coming from the front-end in a reconfigurable-length circular buffer, whose maximum size corresponds to approximately six times the maximum detector drift time (up to 3200 µs). The circular buffer is indeed a double circular buffer implemented on DDR3 memory.

Data are sent from the DAQ Modules to the DAQ PCs using optical 1 Gb/E links. The online system, DATE, defines a hierarchy in the DAQ computers in which Local Data Concentrator (LDC) PCs receive data that is merged in sub-events and sent to a set of Global Data Concentrator (GDC) PCs. GDCs send complete events to the final storage.

DAQ Modules can be configured to limit data throughput. This feature prevents data loss and also limits the number of LDC and GDC PCs needed. In NEXT-White three LDCs and two GDCs suffice to handle the event data load of 120 MB/s initially set. Data acquired near the detector is sent outside the LSC Hall to an external building where data are analyzed and processed in a set of servers before being sent to final storage on tape.

A JAVA GUI implemented in the System Configuration PC is used to configure and control the complete system through the Control Module. All the modules also send slow control information that is processed by the GUI and sent to the detector Slow Control System.

4. Event Detection System

4.1. Event Detection Architecture

In NEXT-White, event detection is based on two processing levels, distributed among different modules, as shown in Figure 4. In the first level, from the early energy measured in the PMT sensor, the Double Event Processor (DEP) generates event candidates, which are sent to the second level, located in the Control Module. In this module, event candidates are stored in the Multi-Hit Memory to be later processed by the Coincidence Event Processor (CEP). This processor generates an Event Accept signal that produces data upload from each DAQ Module buffer to the online system for later offline analysis.

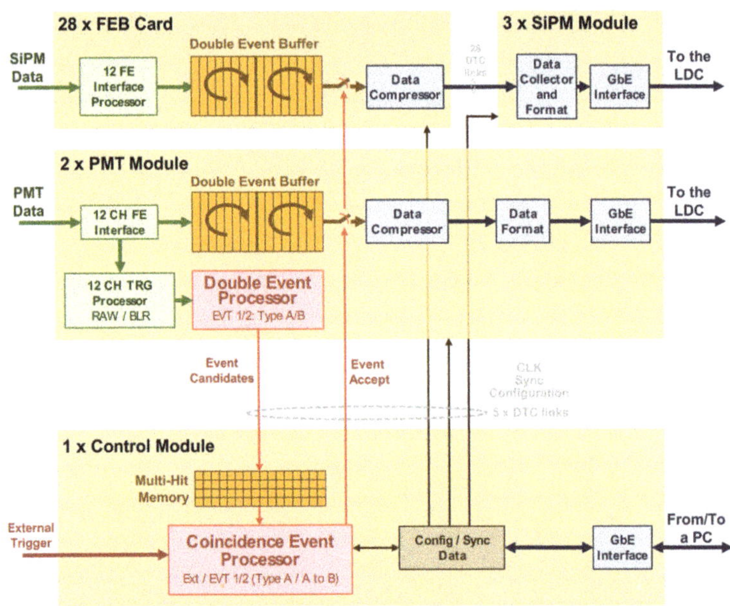

Figure 4. NEXT-DEMO trigger scheme.

Both DEP and CEP have a two-processor based parallel architecture. Each pair of processors can be configured with a set of parameters to search for a different type of event marked as EVT1 or EVT2 type. As mentioned in Section 2, the most obvious application of these double event detection feature is to permit calibrations to be carried out while taking physics data, ensuring high-quality and properly calibrated physics data. In this context, EVT1 type is generally set to detect calibration events, while EVT2 type is devoted to physics events.

As described in Section 3.2, DAQ data storage is partitioned into two circular buffers. A central memory state machine manages the assignment and disposal of the buffers and controls the memory readout. Sampled data are stored in the first available buffer. Moreover, the use of the double circular buffer can be associated to the type of event, allowing two modes of operation:

- Buffer Mode 1: Every type of event detected, EVT1 or EVT2 type, uses the double buffer indistinctly. No priority is given no any event type.
- Buffer Mode 2: EVT1 type events only use one buffer. If at least one buffer is full, the free remaining buffer is always reserved for the accommodation of EVT2 type events. That means that priority is given to EVT2 type events. In this mode of operation, EVT2 type is devoted to detect physics data since it guarantees the minimum dead time for this type of event.

4.2. External Trigger

The DAQ has a third mode of operation for external trigger and test purposes called Test Mode. When this mode is set, event type configuration is not relevant and event processors are set off. In this mode, when an external NIM pulse is detected, the Event Accept is generated automatically. The trigger can be also be internally generated and configured in frequency, so the external NIM trigger source is not needed.

In NEXT, the external trigger is used to calibrate the detector sensors. The aim of the calibration is to equalize precisely the gain of the sensors. NEXT-White sensors are calibrated using LEDs located in the PMT and the SiPM planes. In both type of sensor calibration, the light of the LEDs is adjusted to have a response in the sensors on the order of the single photo-electron.

In the case of the SiPM sensors, the LED, placed in the opposite plane, is connected to a pulse generator outside the detector which generates synchronously the NIM external trigger signal. This configuration allows acquiring data only in a window where the LED is active. PMT sensors are calibrated with a similar procedure but, in this case, the LED is placed in the SiPM plane and is controlled through the FEBs. No external NIM input is needed since the Event Accept generation is internally synchronized with the signal needed for the LEDs.

In both cases, since the signal under study is related to the single photo-electron signal, the buffer needed can be of the order of hundred microseconds. On the one side, this configuration strongly accelerates the acquisition rate as the data throughput is highly reduced, compared to the usual data-taking. On the other side, it ensures that the signal induced by the LED is in a fixed time position defined by the acquisition system facilitating the data processing performed offline.

4.3. The Double Event Processor
4.3.1. Event Detection

As shown in Figure 5, each of the Double Event Detection processors generates event candidates per selected PMT channel by means of configurable thresholds on amplitude (baseline deviation and maximum amplitude relative to the self-calculated baseline), on energy (maximum and minimum energy), and time (minimum and maximum length of the event). Different configuration can be applied per PMT channel. Start and end of the signal under study is defined by the baseline deviation threshold.

Event timing is defined by the 40 MHz DAQ master clock generated in the Control Module and it is distributed among DAQ Modules. This frequency matches with that of the PMT signal sampling rate.

Due to capacitive coupling of the PMTs signal, the baseline needs to be restored by a baseline restoration (BLR) algorithm. The effect of these DC rejecting capacitors on the obtained analogue signals is similar to that of a high pass filter with a very low cutoff frequency. Thus, the algorithm being implemented on the Event Detection block inside the FPGA, aims at applying the inverse function of that high pass filter to the digitally acquired signal from the PMT front end, described in [25]. A careful selection of the numerical precision required for the digital implementation and a proper individual calibration of every PMT channel allow a reconstruction of the original PMT output signal with a neglectable error.

A configurable moving average filter applied to the output of the baseline restoration block calculates its baseline value. Energy estimation is computed from the accumulation of the difference between the signal and the calculated baseline whenever the signal is beyond the baseline deviation threshold (noise is below 1 ADC count). The signal length is also estimated. Comparing the estimated values with the aforementioned thresholds in time, energy and amplitude, the event processor establishes whether the current signal is a valid event candidate or not.

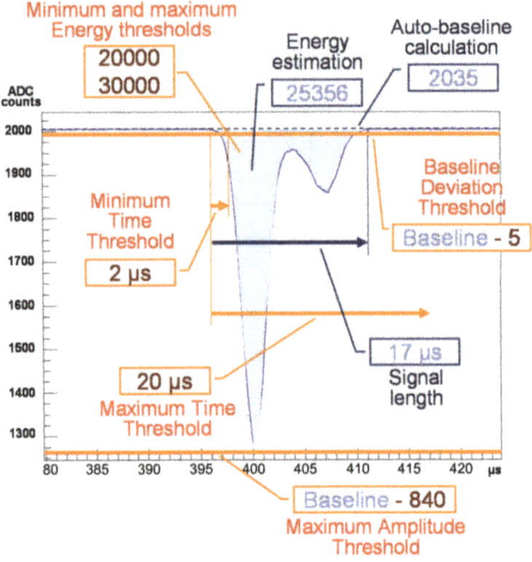

Figure 5. Example of signal candidate generation. In red, the complete set of configuration parameters to generate an event candidate from a PMT signal. In blue, data estimated by the event processor over the reconstructed signal by the BLR algorithm.

Table 1 shows the parameters applied to detect different calibration sources, muons and beta decay physics. This set of parameters allows physic studies for low energy in the range of tens of keV, while for high energy, from hundreds of keV to over 2.65 MeV.

Figure 6 shows a set of different event candidates for a RUN with a different range of energies. In this case, EVT1 and EVT2 types are set for low and high energy event detection respectively. The Kr event corresponds to an EVT1 type event, and the high energy electron, the muon and the alpha events correspond to EVT2 type events.

Table 1. Event detection parameters applied to detect some of the calibration sources, muons and beta decay physics during RUN V. Same configuration applied in the event processor for all the PMT sensors. Circular Buffer size of 1600 µs and pre-trigger of 800 µs.

Event Type	Event Processor Parameters						Coincidence Event Processor Parameters	
	Energy		Amplitude		Time		Time Co-incidence Window	Coincident Events
	Accumulation of ADC Counts		Relative to the Baseline		µs		µs	Number of PMTs
	Min	Max	Min	Max	Min	Max		
$\beta\beta$	100,000	16,777,215 [1]	10	-	2	600	1.2	2
^{83m}Kr	5000	50,000	10	1000	2	40	1.2	2
^{232}Th	220,000	16,777,215	10	2000	10	250	0.5	2
Muons	100,000	16,777,215	10	-	2	600	1.2	2

[1] Maximum possible value (24 bits parameter).

4.3.2. S_1 Event Rejection System

As mentioned in Section 2, one of the aims of the NEXT DAQ is to be able to identify S_1 signals online in a set of PMT sensor synchronously and, in this case, to search for the corresponding S_2 offline afterwards. This is challenging due to the low amplitude of the S_1 signal, especially for low energy events.

When searching for S_1 signals, due to the topology of the signal generated by some radioactive sources used for calibration, a high number of non S_1 hits are detected as signal candidates. Some of the runs studied have up to 25% of "false" S_1 signals. This happens mostly at the beginning or at the end of S_2 signals, as shown in Figure 7. S_1 signal hits can be tagged as false due to their topology or other effects that produce small signals too close to an S_2 signal to be considered a true S_1.

The Double Event Processor has an additional module to reject the above mentioned false S_1 signals. When this system is activated, if an S_1 signal is a valid event candidate, it automatically searches for possible S_2 signals in a time window around the time of the detected event candidate. In this special case, if the S_2 signal is detected, the previously selected event candidate is rejected. The rejection time window is configurable in length, as well as the time around the detected event candidate to start the search.

4.4. The Coincidence Event Processor

4.4.1. Event Accept Generation

The global Coincidence Event Processor (CEP) centralizes the searching of events in the Control Module. This extra discrimination level is necessary to reject spurious events in individual channel. Two more levels of discrimination are available. First, it is possible to configure searches over a minimum number of event candidates per type in a time window (Coincidence Window). Searches can involve from one to all of the PMTs hits, and the Coincidence Window can be extended up to 1.6 µs. And second, two type of search can be configured since the processor can search for only one signal, or for two following signals in a defined time. These two different types of search give rise to two modes of operation:

1. Single Searching Mode.

This mode is oriented to the search of S_1 or S_2 signals. A set of S_1 or S_2 signals in a coincidence window generate an Event Accept. Signals must be of the same type (EVT1 or EVT2), but both can be searched in parallel. Priority is given to EVT2 events if both are detected simultaneously.

2. Double Searching Mode.

This mode is oriented to the search of events, that is, an S_1 signal followed by the corresponding S_2 signal. A double search implies that the system must be able to define which event must be detected first and which one after. In the case of this searching mode, in the Double Event Processor, event candidates can also be defined as Type A or B, having to be detected in this order. Generally, Type A represents an S_1-like signal and Type B is an S_2-like signal. The Event Accept is defined as a combination of the two-consecutive type of signal detected. Strictly, the Event Accept is generated when a set of Type A detected signals is followed, inside a window time, by a set of Type B detected signals, which can be different. In other case, the system looks for another set of type A signals, starting the Event Accept searching process again.

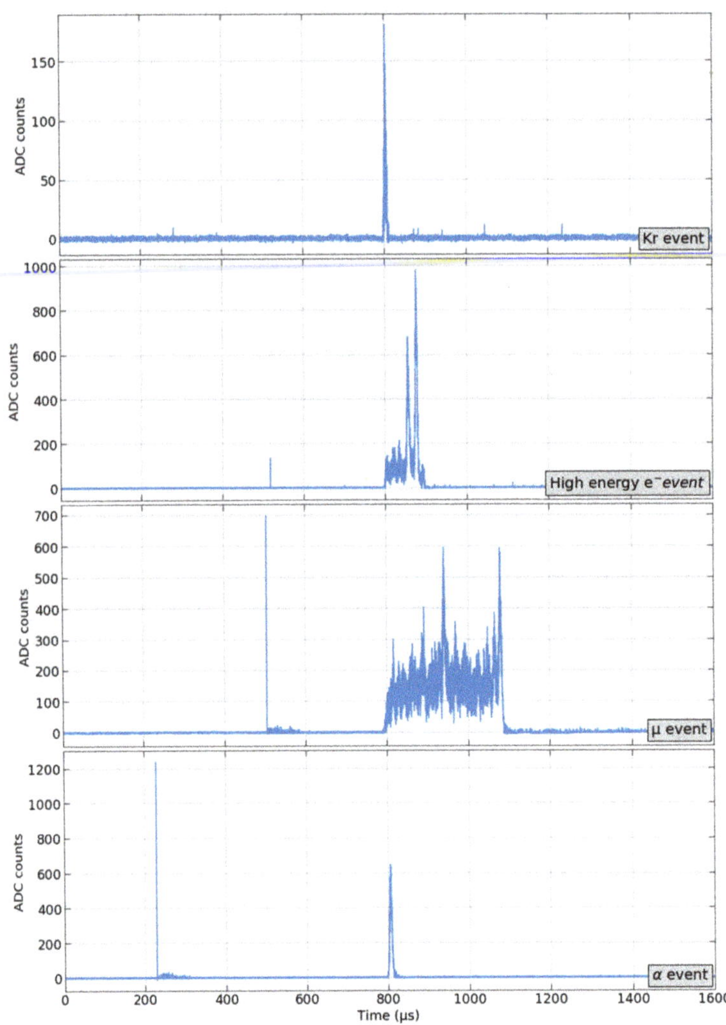

Figure 6. Set of events with a different range of energies from RUN 8250. General configuration: Circular Buffer size of 1600 μs and pre-trigger of 800 μs. EVT1 type set for low energy: Maximum amplitude of 1000 ADC counts, minimum and maximum amplitude thresholds of 5000 and 50,000 sum of ADC counts, and minimum and maximum time thresholds of 2 and 40 μs. EVT2 type set for high energy: Maximum amplitude of 4095 ADC counts (maximum possible value), minimum and maximum amplitude thresholds of 50,000 and 16,777,215 (maximum possible value) sum of ADC counts, and minimum and maximum time thresholds of 2 and 600 μs.

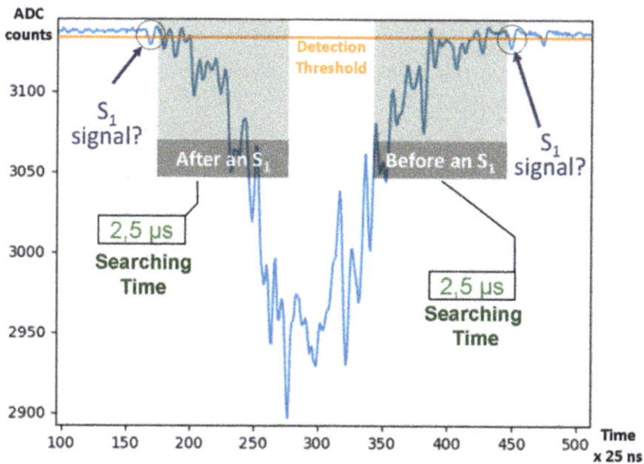

Figure 7. Run 7979 ^{83m}Kr energy deposition signal (S_2) in PMT0 with possible false S_1 signals after and before the S_2 signal that could be set as possible event candidates.

In the case of Single Searching Mode, only Type A signals defined are used by the CEP to generate an Event Accept. Type B signal configuration is ignored by the processor.

Figure 8 shows an example of double search that generates an Event Accept. In blue, number of events and time measured; in brown, configuration required for an Event Accept. For each type of signal, if more than one PMT is involved, the starting time is set by the first detected hit among them. Once a set of signals satisfies Type A conditions (at least 3 signals in the same time bin in the example), the processor looks for Type B signal candidates. As seen in the figure, in less than 625 µs (maximum difference allowed between Type A and B signals), at least 3 Type B signals are detected in 850 ns, satisfying Type B proposed conditions and, therefore generating an Event Accept.

Since PMT sensors can be readout by different DAQ Modules, event timing is defined by the 40 MHz DAQ master clock generated in the Control Module as previously stated in Section 4.3.1. This clock is distributed through the DTC links and synchronized at the level of a clock period among modules.

Statistics per event type of the Event Accept finally processed and lost are continuously sent to the offline system for later analysis of the background rate and half-live of the two neutrino double beta decays over the time for a run.

Events lost occurs when an Event Accept is rejected in some conditions. Independently of the searching mode, when an Event Accept is generated, processing continues and no other events are accepted until the buffer is complete and ready to be downloaded. In this special case, new Event Accept are rejected but do not contribute to the event lost statistics. However, when buffers are full, Event Accepts are also rejected and counted as lost events. In the case of the buffer full condition, the system will reject Event Accepts depending on the Buffer Mode selected and event type detected, as described in Section 4.1.

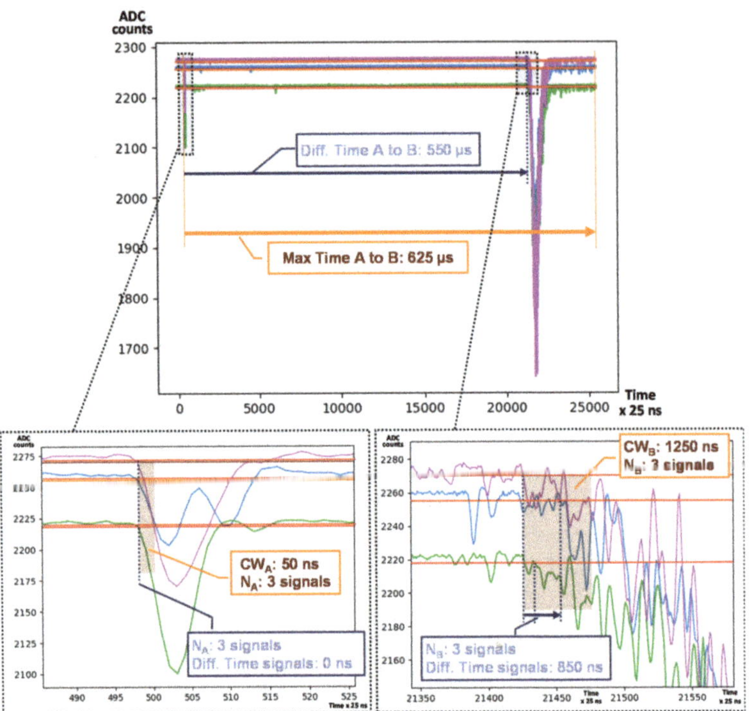

Figure 8. Event Accept example in Double Searching Mode: Run 4405 electron like Type A signal followed by a Type B signal for PMT0, PMT1 and PMT2. Double Search configuration set: 625 µs Maximum Time Event A to B. Type A signal configuration set: 50 ns Coincidence Window (CW$_A$) and 3 minimum number of PMT hits (N$_A$). Type B signal configuration set: 1250 ns Coincidence Window (CW$_B$) and 3 minimum number of PMT hits (N$_B$).

4.4.2. Multi-Hit Memory Concept

DAQ Modules are continuously sending event candidates to the Control Module. Event candidate related data, time-stamp and channel number, are stored in a Multi-Hit Memory that is implemented in a true dual port RAM. Each of the available Multi-Hit Memories (one per type of event) has the size, or an exact divisor, of the programmed Circular Buffer, allowing multiple configurations in size of the Circular Buffer. The Multi-Hit Memory stores, per cell, the number of possible hits, type A or B for double search, and channel hit, with the minimum number of bits to reduce the data memory needed.

Each memory data cell of a Multi-Hit Memory points to a timer related value. This timer is the Fine Timer (FT). The FT is connected to the 40 MHz DAQ master clock. The FT synchronizes event time-stamps and event accept timing among modules.

As shown in Figure 9, when a signal is detected, the system performs a set of steps to generate an Event Accept:

1. When a signal is below the threshold set (Baseline Deviation Threshold), the event time-stamp and channel are stored: FT$_N$ and CH$_N$ respectively.
2. When the signal is above the Baseline Deviation Threshold, estimated values are checked. If the signal satisfies the configuration set, the event is considered as detected, if not, event candidate data stored are discarded.
3. The event candidate data are sent to the Control Module. Event data transmission and memory writing take around 2 µs.

4. Event candidate's data received in the Control Module are stored in the Multi-Hit Memory in the correspondent position to FT_N in the memory. As mentioned above, DAQ and Control Timers are synchronized by the DAQ system master clock.

5. In order to guarantee that the complete signal under study is stored, an amount of time must elapse prior to starting processing the hits. The time elapsed must be at least the time corresponding to the configured buffer pre-samples and maximum time threshold.

6. It is then when processing starts from the beginning of the memory. To speed up processing, the Coincident Event Processor runs at 200 MHz. Moreover, groups of 16 memory cells are read out and stored in a shift register performing a flash detection in one clock cycle. Moreover, only one clock cycle is needed to read and clear the cells. In the case a hit is found, it takes 16 cycles to process the shift register. Processing implies searching for other hits and getting the total number of hits in the coincidence window set. The fast processing guarantees that the event detected in FT_K time is sent completely including the configured number of pre-event samples.

7. Depending on the configuration applied (searching mode), only Type A, or also Type B defined signals, are considered. If the number of hits and the time between them satisfies the configuration, Event Accept is generated and sent to the DAQ Modules. The Control Module sends to the PC farm data about the configuration set and the sensors involved in the Event Accept generated.

8. Upon arrival of the Event Accept in the DAQ Modules, the end of the buffer is calculated with the Event Accept data sent and the configured buffer size and pre-events. Once the buffer is filled, buffer data is readout and formatted, and sent to the PC farm.

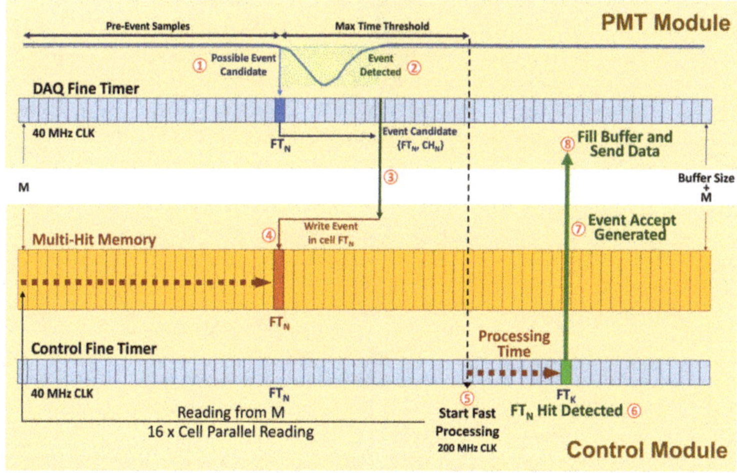

Figure 9. System event detection and Multi-Hit Memory scheme and functionality example.

As seen above, the event detection system processors introduce latency that does not affect its functionality due to the large buffer provided and the possibility to send a configurable number of pre- and post-event data.

5. System Performance

5.1. Data Acquistion

The intensive calibration campaigns, especially with ^{83m}Kr sources and their event topology associated, have prioritized the S_2 signal detection over S_1 using the Single Searching Mode. This decision was taken based on sheer simplicity as S_2 signals are noticeably more prominent than S_1 signals further requiring a less precise configuration. This is especially the case with ^{83m}Kr sources as their weak S_1 signal is hard to distinguish from a dark count-induced response at a given PMT, as can be seen on Figure 6. This, of course, is not the case of the high energy sources where S_1 triggering would not be as challenging. However, for offline processing simplicity, because the signal that causes the Event Accept is always in the same position in the acquisition window, it is desirable to configure both processors for S_2 signal detection. Despite this, all the described features, among them the Double Searching Mode, have been successfully tested and validated.

An advantage of the event detection system is the possibility to simultaneously perform very long calibration runs with low energy sources (tens of keV) combined with searches for high energy physics (few MeV) events. The simultaneous data-taking scheme is of vital importance for the experiment and one of the highlights of the acquisition system as it permits to have deep knowledge at all times of the electron lifetime and the geometrical corrections maps, necessary ingredients to correct high energy events and achieve the objective energy resolution as described in [5,16]. Moreover, this scheme is not only important to maximize the detector potential but also to monitor the detector status during the data-taking as the event rate of the high energy events, aside from the calibration periods, is considerably small, of the order of few hundreds of mHz. Therefore, any deviation from the standard operation is considerably harder to notice in the high energy events, and if relaying only on it, the reaction time would be much worse. Once this feature was included and tested properly, NEXT-White has carried out several runs during two years, combining calibration with low and high energy, using ^{83m}Kr and ^{228}Th sources, low background or double beta decay searches. Around 900 million events have been processed along this period of time. In each run, the low-energy events acquired are used to compute the lifetime and geometric correction maps that are used to correct the high-energy events.

In addition to the simultaneous data-taking, another tool that is proving itself to be greatly beneficial for the experiment, is the capability to keep count of rejected Event Accepts that are considered lost events by the DAQ (see Section 4.4.1). This can be used as an estimate of the dead time of the detector on a run-by-run basis, allowing to correct each run individually by the number of known lost events. Proper understanding of the lost events is required in order to compare the measured background with the expected one and thus correctly extract the half-live of the two neutrino double beta decays over the time for a run.

5.2. Dead Time

Another innovative feature reported is the association of the type being searched with the use of a double buffer. This feature has an impact in the dead time for the events selected as EVT1 or EVT2 type, as described in Section 4.1. Table 2 shows three runs taken with the same configuration where dead time increases with data rate, which has been fixed by adjusting the quantity of ^{83m}Kr in the chamber. Data rate can be lowered by adjusting the event detection parameters though runs are configured with a wide margin in them so that no bias is expected in the data analysis even at the cost of an increase of the dead time. Moreover, the table shows that almost the total data rate is due to the ^{83m}Kr source events, with the beta decay events representing less than 0.5% of the total rate.

Table 2. Three run statistics with general configuration: Single Searching Mode, two type of S_2 events (^{83m}Kr and beta decay events), 24 h, buffer priority for beta decay events, at least 2 PMT hits in a Coincidence Window of 1.2 μs, data compression applied.

	Total Event Rate	Data Rate	EVT Rate	Dead Time Rate	
			EVT2	EVT1	EVT2
7520	20.78 Hz	31.78 MB/s	1.95 Hz	9.4%	3.8%
7512	27.27 Hz	41.66 MB/s	1.01 Hz	16.4%	5.3%
7502	76.53 Hz	77.28 MB/s	0.84 Hz	39.2%	15.15%

Data throughput bottleneck can be easily solved adding more Local and Global Data Concentrator PCs (LDCs and GDCs) to the online system. For the time being, due to the current needs of the detector, a maximum throughput of around 120 MB/s has been fixed in the online hardware configuration described in Section 3.2.

Dead Time improves when only one event processor is used. This is mainly because the double processor shares the double buffer and the GbE links to readout the data. Table 3 shows that dead time for a high energy run can be almost negligible if only one processor is used. In the runs shown, for beta decay events, dead time decreases from 0.46% to 0.03%.

Table 3. Two run statistics with general configuration: Single Searching Mode, 24 h, buffer priority for beta decay events, at least 2 PMT hits in a Coincidence Window of 1.2 μs, no data compression applied. RUN 8087 is set for two type of S_2 events (^{83m}Kr and beta decay events) and RUN 8091 is set for one type of S_2 events (beta decay events).

RUN	Total Event Rate	Data Rate	EVT Rate	Dead Time Rate	
			EVT2	EVT1	EVT2
8087	8.56 Hz	26.25 MB/s	0.089 Hz	1.06%	0.46%
8091	0.097 Hz	0.31 MB/s	0.097 Hz	-	0.03%

5.3. Trigger Efficiency

Trigger efficiency has been also studied, for low and high energy searches, specifically for ^{83m}Kr and ^{228}Th sources. In both cases, independent events have been obtained from coincidences with other events. This has been done by looking at events that have an additional S_2 candidate aside from the S_2 which caused the Event Accept. The additional S_2 signal has been characterized and only signals which are fully compatible with the event under study have been kept, filtering out any incompatible S_2. In particular, events have been chosen based on their well-known energy and their position within the buffer window, leaving enough margin from the event accept position (at the middle of the buffer) to avoid interferences. These events are completely uncorrelated with other events in the buffer window and, therefore, are an unbiased sample which can be used to evaluate the event detection efficiency. This strategy has been used for low and high energy events. Selected data have been processed by a Python model of the event detection system. For each type of event under study, individual event search and possible coincidences are set with the same conditions of the runs under study. Event Processor has shown an efficiency of 100%, but as it is shown in Table 4, the Coincidence Event Processor reduces the event detection efficiency for both type of events to a value between 97–98%. Table 4 also shows that there is a margin of improvement to the 98–99% widening the coincidence window to its maximum possible value. In any case, the number of events lost due to event detection efficiency is below the one shown due to dead time.

Table 4. Coincidence Event Processor efficiency results. Configuration under study: Two minimum PMT hits in a Coincidence Window (CW) of 1.2 µs (run parameter) and 1.6 µs (maximum possible value).

RUN	Event Type Searched	Number of Events	Event Detection Efficiency	
			CW = 1.2 µs	CW = 1.6 µs
7492	^{83m}Kr	41,101	97.33%	98.25%
7738	^{228}Th	3511	98.15%	99.09%

The efficiency reported integrates all selected events for the study without taking any additional considerations. Full geometrical and energetic characterization will be performed in the future.

6. Conclusions

This paper has presented the NEXT-White Event Detection System, adapted from the latter prototype NEXT-DEMO, fully commissioned and deployed in NEXT-White, and ready to be dimensioned (thanks to its modularity) and installed in NEXT-100. The system has evolved during the four years of operation of the detector, being an important tool to prove its event identification capabilities. Its performance, reliability and robustness in the simultaneous calibration and double-beta physics searches have been thoroughly tested.

Among many other, the key feature of NEXT-White, and future NEXT detectors, is the capability of detecting a wide variety of events with the highest efficiency. This fact helps to avoid an excess of data throughput and reduces the amount of offline data to be analyzed. The system described is able to search for events in a very wide dynamic range, triggering in the S_1 or S_2 signals of different types of sources, from Kr events that are very low in amplitude and short, to other events with very high amplitude, like S_2 alpha events, or very long like muons, thanks to the high versatility and configurability of the processors implemented.

Moreover, the number of lost events shows a remarkable improvement when only one event processor is used, as stated in Section 5.2. Anyway, in the case the double processor is used, the number of lost events is low for one type of event, considered of interest. In this sense, the modular hardware used that allows to expand easily the data throughput, as well as the programmability of the implemented system, have both demonstrated its adaptability to the new detector demands.

7. Future Work

NEXT final detector is more demanding in terms of readout channels and chamber drift time, which implies more DAQ buffering and processing capacity. The event detection system will need to cope with the processing of more channels, up to 60 PMT signal sensors compared to the 12 of NEXT-White. From this point of view, the advantage of the current modular DAQ and event detection system is that it has been dimensioned from the beginning to cope with the number of sensors that will allocate NEXT-100. Due to the modularity of the RD51 electronics and DATE, DAQ hardware and online system can be easily scaled.

Next step in the NEXT collaboration will be to extrapolate the technology to a 1-ton scale detector. This will imply a deep revision of the detector that will affect completely the DAQ and the event detection system like the replacement of the PMT sensors of the Energy Plane by SiPM sensors. This will imply the study of new detection techniques since the sensors show a completely different behavior. Studies of new detections algorithms based on the concepts learned in the previous detectors must be carried on.

Another line of improvement is the reduction of the dead time in double searching mode (see Sections 4.4.1 and 5.2). In addition to scaling the online system, a double buffer per type of event could be foreseen. Both could reduce dead time drastically for every type of event. In the same direction, to provide 10 Gb Ethernet connection would speed up data download, reducing dead time as well. DAQ modules have foreseen this type of connection, but the current online system, DATE, does not support any more UDP transactions. For this reason, the use of other online systems, like RCDAQ (PHENIX experiments), is under study. In the short term, next version of DAQ will provide a configurable size of buffer per event type. This will reduce the data throughput for low energy events in the simultaneous calibration and double-beta physics searches.

Author Contributions: Conceptualization, R.E.B., J.F.T.A., V.H.B., A.S.E., F.M.C., V.Á.P., J.R.S., M.Q.S. and F.B.M.; methodology, R.E.B. and A.S.E.; software, R.E.B. and A.S.E.; validation, R.E.B., A.S.E. and F.M.C.; formal analysis, R.E.B. and A.S.E.; investigation, R.E.B., J.F.T.A., V.H.B., A.S.E., F.M.C., V.Á.P., J.R.S. and M.Q.S.; writing—original draft preparation, R.E.B.; writing—review and editing, R.E.B., J.F.T.A., V.H.B., A.S.E., F.M.C. and J.R.S.; visualization, R.E.B. and A.S.E.; supervision, R.E.B.; project administration, R.E.B., J.F.T.A.,V.H.B., A.S.E. and F.M.C.; funding acquisition, R.E.B., J.F.T.A., V.H.B. and F.M.C. All authors have read and agreed to the published version of the manuscript.

Funding: The NEXT collaboration acknowledges support from the following agencies and institutions: the European Research Council (ERC) under the Advanced Grant 339787-NEXT, the Ministerio de Economía y Competitividad and the Ministerio de Ciencia, Innovación y Universidades of Spain under grants FIS2014-53371-C04, RTI2018-095979, the Severo Ochoa Program SEV-2014-0398 and the María de Maetzu Program MDM-2016-0692; the GVA of Spain under grants PROMETEO/2016/120 and SEJI/2017/011; the Portuguese FCT under project PTDC/FIS-NUC/2525/2014, under project UID/FIS/04559/2013 to fund the activities of LIBPhys, and under grants PD/BD/105921/2014, SFRH/BPD/109180/2015 and SFRH/BPD/76842/2011; the U.S. Department of Energy under contracts number DE-AC02-06CH11357 (Argonne National Laboratory), DE-AC02-07CH11359 (Fermi National Accelerator Laboratory), DE-FG02-13ER42020 (Texas A&M) and DE-SC0019223/DE-SC0019054 (University of Texas at Arlington); and the University of Texas at Arlington. DGD acknowledges Ramon y Cajal program (Spain) under contract number RYC-2015-18820. We also warmly acknowledge the Laboratori Nazionali del Gran Sasso (LNGS) and the Dark Side collaboration for their help with TPB coating of various parts of the NEXT-White TPC. Finally, we are grateful to the Laboratorio Subterráneo de Canfranc for hosting and supporting the NEXT experiment.

Informed Consent Statement: Not applicable.

Data Availability Statement: Data collected through research presented in the paper are available on request from the corresponding authors.

Conflicts of Interest: The authors declare no conflict of interest.

References

1. Monrabal, F.; Gómez-Cadenas, J.J.; Toledo, J.; Laing, A.; Álvarez, V.; Benlloch-Rodríguez, J.; Cárcel, S.; Carrión, J.; Esteve, R.; Felkai, R.; et al. The NEXT White (NEW) detector. *J. Instrum.* **2018**, *13*, P12010. [CrossRef]
2. Álvarez, V.; Borges, F.I.G.M.; Cárcel, S.; Carmona, J.M.; Castel, J.; Catalá, J.M.; Cebrián, S.; Cervera, A.; Chan, D.; Conde, C.; et al. NEXT-100 Technical Design Report (TDR). Executive summary. *J. Instrum.* **2012**, *7*, T06001. [CrossRef]
3. EXO-200 Collaboration. Search for neutrinoless double-beta decay with the upgraded EXO-200 detector. *Phys. Rev. Lett.* **2018**, *120*, 072701. [CrossRef] [PubMed]
4. KamLAND-Zen Collaboration. Search for Majorana neutrinos near the inverted mass hierarchy region with KamLAND-Zen. *Phys. Rev. Lett.* **2016**, *117*, 082503. [CrossRef]
5. The NEXT Collaboration; Renner, J.; López, G.D.; Ferrario, P.; Morata, J.H.; Kekic, M.; Martínez-Lema, G.; Monrabal, F.; Gómez-Cadenas, J.J.; Adams, C.; et al. Energy calibration of the NEXT-White detector with 1% resolution near $Q_{\beta\beta}$ of ^{136}Xe. *JHEP* **2019**, *2019*, 230. [CrossRef]
6. The NEXT Collaboration; Ferrario, P.; Benlloch-Rodríguez, J.M.; López, G.D.; Morata, J.H.; Kekic, M.; Renner, J.; Usón, A.; Gómez-Cadenas, J.J.; Adams, C.; et al. Demonstration of the event identification capabilities of the NEXT-White detector. *JHEP* **2019**, *2019*, 52. [CrossRef]
7. Gómez-Cadenas, J.J.; Martín-Albo, J. NEXT, a HPXe TPC for neutrinoless double beta decay searches. *J. Physics Conf. Ser.* **2008**, *136*, 042048. [CrossRef]
8. Fernandes, L.M.P.; Freitas, E.D.C.; Ball, M.; Gómez-Cadenas, J.J.; Monteiro, C.; Yahlali, N.; Nygren, D.; Dos Santos, J.M.F. Primary and secondary scintillation measurements in a Xenon Gas Proportional Scintillation Counter. *JINST* **2010**, *5*, P09006. [CrossRef]

9. Esteve, R.; Toledo, J.; Monrabal, F.; Lorca, D.; Serra, L.; Marí, A.; Gómez-Cadenas, J.; Liubarsky, I.; Mora, F. The trigger system in the NEXT-DEMO detector. *JINST* **2012**, *7*, C12001. [CrossRef]

10. Akerib, D.S.; Araújo, H.; Bai, X.; Bailey, A.J.; Balajthy, J.; Beltrame, P.; Bernard, E.; Bernstein, A.; Biesiadzinski, T.; Boulton, E.; et al. FPGA-based trigger system for the LUX dark matter experiment. *NIM A* **2016**, *818*, 57–67. [CrossRef]

11. Druszkiewicz, E. The Data Acquisition System for LZ. *JINST* **2016**, *11*, C02072. [CrossRef]

12. Agnes, P.; Albuquerque, I.; Alexander, T.; Alton, A.; Arisaka, K.; Asner, D.; Ave, M.; Back, H.; Baldin, B.; Biery, K.; et al. The Electronics, Trigger and Data Acquisition System for the Liquid Argon Time Projection Chamber of the DarkSide-50 Search for Dark Matter. *JINST* **2017**, *12*, P12011. [CrossRef]

13. Wu, Q.; Ren, X.; Chen, X.; Ji, X.; Liu, J.; Lei, S.; Wang, M.; Xiao, M.; Xie, P.; Yan, B.; et al. Update of the trigger system of the PandaX-II experiment. *JINST* **2017**, *12*, T08004. [CrossRef]

14. The NEXT Collaboration; Simón, A.; Felkai, R.; Martínez-Lema, G.; Monrabal, F.; González-Díaz, D.; Sorel, M.; Hernando, J.A.; Gómez-Cadenas, J.; Adams, C.; et al. Electron drift properties in high pressure gaseous xenon. *JINST* **2018**, *13*, P07013. [CrossRef]

15. Martínez-Lema, G.; Hernando, J.A.; Palmeiro, B.; Botas, A.; Ferrario, P.; Monrabal, F.; Laing, A.; Renner, J.; Simón, A.; Para, A.; et al. Calibration of the NEXT-White detector using ^{83m}Kr decays. *JINST* **2018**, *13*, P10014. [CrossRef]

16. The NEXT Collaboration; Novella, P.; Palmeiro, B.; Sorel, M.; Usón, A.; Ferrario, P.; Gómez-Cadenas, J.J.; Adams, C.; Álvarez, V.; Arazi, L.; et al. Radiogenic backgrounds in the NEXT double beta decay experiment. *JHEP* **2019**, *2019*, 51. [CrossRef]

17. Esteve, R.; Toledo, J.; Rodríguez, J.; Querol, M.; Álvarez, V. Readout and data acquisition in the NEXT-NEW Detector based on SRS-ATCA. *JINST* **2016**, *11*, C01008. [CrossRef]

18. Martoiu, S.; Muller, H.; Costa, F.; Tarazona, A.; Toledo, J.; Zang, F. The SRS scalable readout system for micropattern gas detectors and other applications. In Proceedings of the TWEPP 2012, Oxford, UK, 17–21 September 2012.

19. Toledo, J.; Muller, H.; Esteve, R.; Monzó, J.M.; Tarazona, A.; Martoiu, S. The front-end concentrator card for the RD51 scalable readout system. *JINST* **2011**, *6*, C11028. [CrossRef]

20. RD51 Collaboration. Development of Micro-Pattern Gas Detectors Technologies. Available online: http://rd51-public.web.cern.ch/rd51-public/S (accessed on 15 January 2021).

21. Martin, L. Purschke, RCDAQ, a Lightweight yet Powerful Data Acquisition System. 2020. Available online: https://www.phenix.bnl.gov/~{}purschke/rcdaq/rcdaq_doc.pdf (accessed on 15 January 2021).

22. Purschke, M.L. Readout of GEM stacks with the CERN SRS system. In Proceedings of the 2012 18th IEEE-NPSS Real Time Conference, Berkeley, CA, USA, 9–15 June 2012. [CrossRef]

23. Tarazona, A.; Gnanvo, K.; Martoiu, S.; Muller, H.; Toledo, J. A point-to-point link for data, trigger, clock and control over copper or fibre. *JINST* **2014**, *9*, T06004. [CrossRef]

24. Rodriguez, J.; Toledo, J.; Esteve, R.; Lorca, D.; Monrabal, F.; Alarcón, J.F.T. The front-end electronics for the 1.8-kchannel SiPM tracking plane in the NEW detector. *JINST* **2015**, *10*, C01025. [CrossRef]

25. Álvarez, V.; Herrero-Bosch, V.; Esteve, R.; Laing, A.; Rodríguez, J.; Querol, M.; Monrabal, F.; Toledo, J.; Gómez-Cadenas, J. The electronics of the energy plane of the NEXT-White detector. *NIM A* **2019**, *917*, 68–76. [CrossRef]

Article

Analytical Study of Front-End Circuits Coupled to Silicon Photomultipliers for Timing Performance Estimation under the Influence of Parasitic Components

Pietro Antonio Paolo Calò *, Savino Petrignani, Michele Di Gioia and Cristoforo Marzocca *

Department of Electrical and Information Engineering, Politecnico di Bari, via E. Orabona 4, I-70125 Bari, Italy; savino.petrignani@poliba.it (S.P.); m.digioia2@studenti.poliba.it (M.D.G.)
* Correspondence: pietroantoniopaolo.calo@poliba.it (P.A.P.C.); cristoforo.marzocca@poliba.it (C.M.)

Received: 6 July 2020; Accepted: 5 August 2020; Published: 8 August 2020

Abstract: Full exploitation of the intrinsic fast timing capabilities of analog silicon photomultipliers (SiPMs) requires suitable front-end electronics. Even a parasitic inductance of a few nH, associated to the interconnections between the SiPM and the preamplifier, can significantly degrade the steepness of the detector response, thus compromising the timing accuracy. In this work, we propose a simple analytic expression for the single-photon response of a SiPM coupled to the front-end electronics, as a function of the main parameters of the detector and the preamplifier, taking into account the parasitic inductance. The model is useful to evaluate the influence of each parameter of the system on the slope of its response and to guide the designer in the definition of the architecture and the specifications for the front-end electronics. The results provided by the model have been successfully compared with experimental measurements from a front-end circuit with variable configuration based on a bipolar junction transistor (BJT), coupled to a 3×3 mm^2 SiPM stimulated by a fast-pulsed laser source.

Keywords: silicon photomultiplier; front-end electronics; single-photon response; timing accuracy

---

1. Introduction

Accuracy in time measurements represents a challenging task for a growing number of photo-detection systems in applications such as time-of-flight positron emission tomography (ToF-PET) [1], γ-ray spectroscopy [2], time-correlated single photon counting (TCSPC) [3,4], and distance measurements (LiDAR) [5,6]. In this kind of applications, time resolution as low as 100 ps FWHM, or even less, are often required.

Silicon photomultipliers (SiPMs) are becoming the detectors of choice for such applications, owing to their intrinsically fast response, characterized by sub-nanosecond risetime, and single-photon sensitivity. The design of effective SiPM-based detection systems aiming at good single-photon time resolution calls for the development of high-speed and low-noise front-end electronics. When the detector equivalent capacitance is quite large, as in the case of SiPMs, the best achievable time resolution is often constrained by the characteristics of the front-end electronics [7]. Indeed, electronic noise is one of the main causes of jitter in time measurements and often dominates other sources of error, such as the intrinsic jitter due to avalanche build-up statistics [8,9]. The effects of electronic noise are emphasized when the maximum slope of the output pulse produced by the front-end electronics is reduced because of either non-optimal coupling between detector and preamplifier or bandwidth limitations.

In fact, the most common time pick-off technique is leading-edge discrimination. A fast comparator, cascaded to the preamplifier, fires when the signal overcomes a suitable threshold V_{TH}. The resulting rms time jitter σ_t is expressed by the following well-known equation:

$$\sigma_t = \frac{\sigma_n}{\left.\dfrac{dV_{out}(t)}{dt}\right|_{V_{out} = V_{TH}}}, \tag{1}$$

where σ_n is the *rms* electronic noise at the preamplifier output, and the denominator represents the slope of the output pulse $V_{out}(t)$, evaluated around the level of the discriminator threshold [10]. Hence, whenever good timing accuracy is required, it is important to figure out how the parameters of the front-end electronics, such as the input resistance and the bandwidth, affect the leading edge of the output pulse and its slope.

For this purpose, even though powerful simulation tools and accurate models of the system can be used, analytical expressions of the response $V_{out}(t)$ and its slope as a function of the main parameters involved can be very helpful. For instance, if these expressions are available, the initial choice of the specifications for the front-end electronics can be easily made. Moreover, the designer can understand with little effort in which direction the circuit must be modified to optimize the time performance of the detection system. From this perspective, useful analytical expressions of $V_{out}(t)$ are already available in the literature [11–13], but their accuracy can be improved if relevant parameters, not yet considered, are added to the model used.

In fact, it is well known that also the small parasitic inductance L_{par}, associated to the interconnection between the SiPM and the front-end electronics, plays a relevant role in shaping the very fast waveform of the SiPM signal [14,15]. Consequently, an effective model of the whole detection system, besides including the parameters of both the SiPM and the preamplifier, must also take account of this inductance. This results in increased complexity of the transfer function of the system, which makes it very difficult to obtain the desired analytical expression for both the system response and its slope as a function of time. Previous attempts have been made in this regard, but they have been based on an oversimplified detector model and are not supported by experimental validation [16].

We propose a new approximate analysis which results in few simple mathematical equations that relate some features of the single-photon time response of a SiPM readout circuit to the most important circuit parameters, including the parasitic inductance L_{par}. In particular, the approximate expression obtained for the slope of the single-photon response correctly reproduces the behavior of the complete model of the system.

A factorized expression of the overall system transfer function is derived, such that the individual contributions of the parameters involved to the formation of the leading edge of the output response can be easily identified and evaluated.

The proposed analysis allows us to draw some conclusions about the way parasitic components interact with the detector and circuit parameters and influence the timing performance of a SiPM-based detection system. The resulting analytic expressions provide the designer of the front-end with practical guidelines for the selection of the most suitable architecture and useful indications about its specifications.

Experiments carried out with a 3×3 mm^2 SiPM coupled to a BJT-based front-end circuit have also been carried out to validate the effectiveness of the proposed analysis. The achieved results show that our analytical model reproduces the slope of the single-photon circuit response as a function of the main parameters involved.

2. Transfer Function of a Typical SiPM-Based Detection System

The complete electrical model of a SiPM detector with N micro-cells coupled to a front-end circuit based on the classic current-mode approach [17,18] is shown in Figure 1. The preamplifier is a current buffer with gain A_i, which exhibits a low input resistance R_{in}. The load resistance R_L converts the output current of the buffer into the voltage V_{out}, whereas the bandwidth of the preamplifier is dominated by the single output time constant $\tau_L = R_L C_L$.

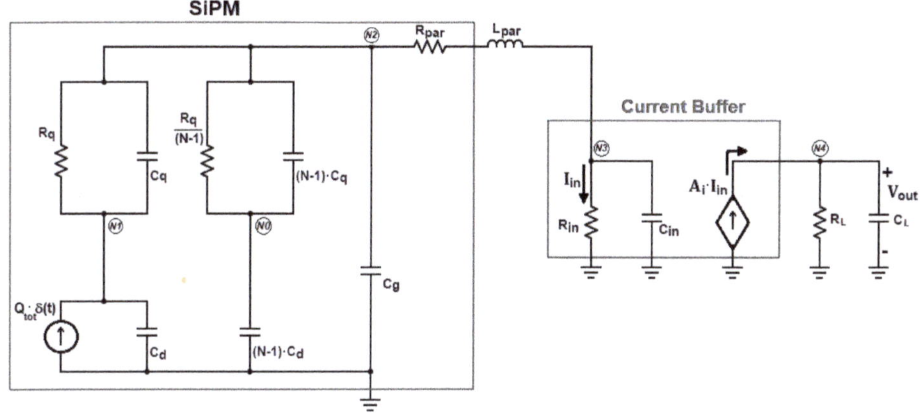

Figure 1. Equivalent electrical model of the silicon photomultiplier (SiPM) coupled to a current-mode front-end.

In Figure 1, only one micro-cell of the detector is supposed to undergo a Geiger discharge, since our analysis is restricted to single-photon events.

The model of the SiPM includes the delta-like current source $I_{\mu cell}(t) = Q_{tot} \cdot \delta(t)$, which represents the very fast Geiger discharge of the fired micro-cell, the quenching resistor R_q, the parasitic capacitance across it C_q, the photodiode capacitance C_d, and the capacitance C_g, due to the routing metal grid used to connect in parallel all the micro-cells. L_{par} represents the parasitic inductance associated with the interconnection wire between the SiPM and the front-end.

The same electrical model of the SiPM has been already used to derive the transfer function from the input Dirac's delta to the current I_{in} flowing through R_{in} in Figure 1 [19–21]. Compared with the analysis proposed in Ciciriello et al. [19], limited to the Laplace domain, the only additional element introduced in the model is the resistor R_{par}. This component, as described in Licciulli et al. [22], includes both the series substrate resistance of the SiPM, R_{sub}, that improves the accuracy of the SiPM model, and other series parasitic resistances associated with the interconnections between the detector and the electronics. In all the practical cases the contribution of R_{sub} is dominant in R_{par}, thus $R_{sub} \cong R_{par}$. To simplify the analysis in the s-domain, the input section of the circuit in Figure 1 has been redrawn, after applying the Norton equivalent to the SiPM model, resulting in the schematic of Figure 2.

The expression of the Norton equivalent current $I_N(s)$ in Figure 2 is the following:

$$I_N(s) = Q_{tot} \frac{1 + \tau_q s}{1 + \tau_r s},\tag{2}$$

where the time constants $\tau_q = R_q C_q$ and $\tau_r = R_q (C_d + C_q)$ appear.

After expressing the parallel admittances $Y_{det}(s)$ and $Y_{par,in}(s)$ respectively as

$$Y_{det}(s) = N C_d \frac{s(1 + \tau_q s)}{1 + \tau_r s} + s C_g,\tag{3}$$

$$Y_{par,in}(s) = \frac{1}{R_s} \cdot \frac{\omega_n^2}{s^2 + 2\zeta \omega_n s + \omega_n^2}(1 + \tau_{in} s),\tag{4}$$

where $\tau_{in} = R_{in}C_{in}$, $R_s = R_{in} + R_{par}$, $\omega_n^2 = \frac{1}{L_{par}C_{in}} \cdot \frac{R_s}{R_{in}} = \frac{1}{\tau_{in}} \cdot \frac{R_s}{L_{par}}$, and $2\,\zeta\,\omega_n = \frac{1}{\tau_{in}} + \frac{R_{par}}{L_{par}}$, application of the current divider rule leads to the following expression for the current $I_{in}(s)$:

$$I_{in}(s) = \frac{1}{1 + \tau_{in}s} \cdot \frac{Y_{par,in}(s)}{Y_{par,in}(s) + Y_{det}(s)} \cdot I_N(s). \tag{5}$$

Replacing expressions from (2) to (4) in (5), the Laplace transform of the current flowing into the input resistance of the preamplifier R_{in} can be rearranged as follows:

$$I_{in}(s) = \frac{Q_{tot}\omega_n^2(1 + \tau_q s)}{\omega_n^2(1 + \tau_{in}s)(1 + \tau_r s) + R_{ss}\left(s^2 + 2\,\zeta\,\omega_n\,s + \omega_n^2\right)\left[C_g(1 + \tau_r s) + NC_d(1 + \tau_q s)\right]} \tag{6}$$

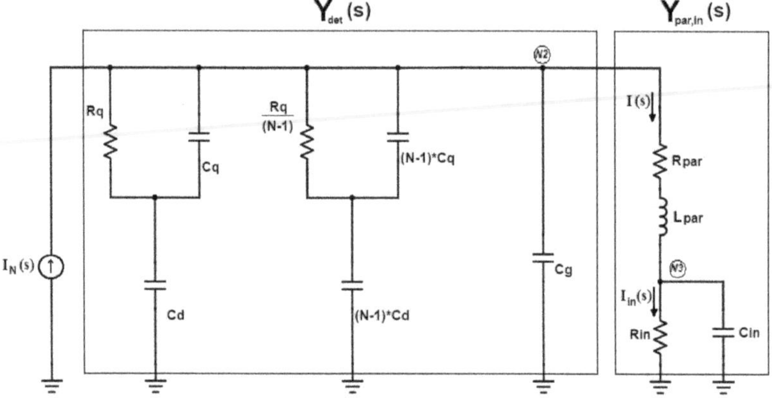

Figure 2. The input section of the circuit in Figure 1, redrawn as the parallel of two admittances, Y_{det} and $Y_{par,in}$.

In Figure 1, the output current of the current buffer is converted into the output voltage

$$V_{out}(s) = \frac{K_R}{(1 + \tau_L s)} \cdot I_{in}(s), \tag{7}$$

where $K_R = A_i \cdot R_L$ represents the overall transimpedance gain of the preamplifier. Equations (6) and (7) can be expressed in the following way:

$$I_{in}(s) = Q_{tot} \cdot \left(\frac{1 + \tau_q s}{1 + a_1 s + a_2 s^2 + a_3 s^3 + a_4 s^4}\right), \tag{8}$$

$$V_{out}(s) = Q_{tot} \cdot K_R \cdot \left[\frac{1 + \tau_q s}{(1 + \tau_L s)(1 + a_1 s + a_2 s^2 + a_3 s^3 + a_4 s^4)}\right], \tag{9}$$

where

$$a_1 = \tau_{in} + \tau_r + R_s\left(C_g + NC_d\right),$$
$$a_2 = \tau_{in}\tau_r + C_g\left(R_s\tau_r + L_{par} + R_{par}\tau_{in}\right) + NC_d\left(R_s\tau_q + L_{par} + R_{par}\tau_{in}\right),$$
$$a_3 = L_{par}\left(C_g + NC_d\right)\tau_{in} + \left(L_{par} + R_{par}\tau_{in}\right)\left(C_g\tau_r + NC_d\tau_q\right),$$
$$a_4 = L_{par}\left(C_g\tau_r + NC_d\tau_q\right)\tau_{in}.$$

Equation (9) is the relation between the total charge Q_{tot} released by a single fired micro-cell and the voltage at the output of the current-mode preamplifier in Figure 1.

In case a voltage-mode approach is used for the front-end, as, for instance, in Di Lorenzo et al. [23] and Fisher et al. [24], the current pulse of the detector is first converted into a voltage by means of the input resistor R_{in}. The voltage across R_{in} is amplified by means of a voltage amplifier with gain A_V, as depicted in Figure 3.

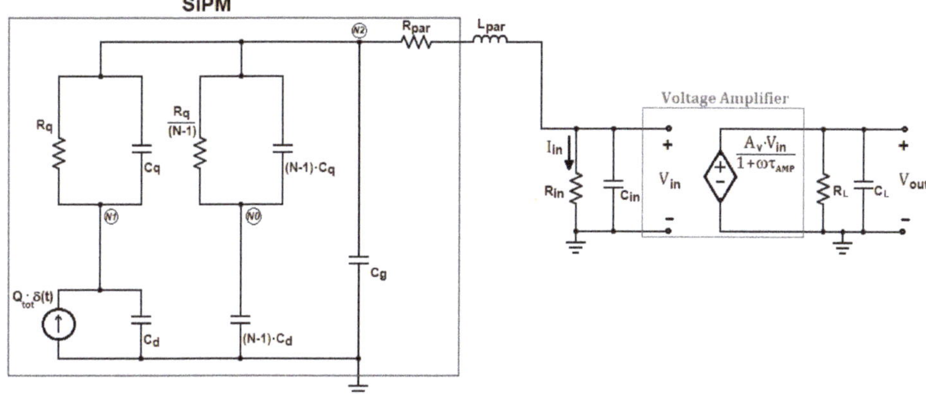

Figure 3. Model of the SiPM coupled to a voltage-mode front-end circuit.

In this case, the output voltage $V_{out}(s)$ can still be expressed by Equation (9). The only formal difference is that the overall transimpedance gain $K_R = A_i \cdot R_L$ of the current-mode approach must be replaced with the factor $K_R' = A_V \cdot R_{in}$, and the dominant time constant of the voltage amplifier τ_{AMP} must be considered instead of τ_L.

The transfer function in Equation (9) is characterized by one zero and five poles, thus the corresponding expression of the pulse $I_{in}(t)$ in the time-domain is rather complex. This expression can be either calculated as the inverse Laplace transform of Equation (9), for instance using a MATLAB® script, or plotted by means of SPICE simulations of the circuit in Figure 1. In both cases, to distinguish the influence of each model parameter on the slope of the leading edge of the pulse, sets of parametric simulations would be required. In order to effectively support the choice of the architecture and the design of the preamplifier, it is more convenient to simplify the complex analytic expressions derived so far, considering suitable approximations.

3. Analytic Approximation of the Model for the Study of the Fast Transient

We are interested in the fast initial transient of the output pulse, which dominates the timing performance of the detection system. Thus, an analytic approximation of the model that does not affect the high frequency components of the response is needed. In this perspective, the slow second order term $(1 + \tau_{in}s)(1 + \tau_r s)$ can be neglected in Equation (6), since its contribution is irrelevant as compared to the remaining fourth order polynomial of the denominator.

The resulting expression of the current $I_{in}(s)$, valid for the early fast transient of the response, can be written and factorized as follows:

$$I_{in}(s) \cong V_0 \cdot \frac{\left(1 + \tau_q s\right)}{s\left(1 + \tau_p s\right)} \cdot \frac{1}{R_s} \cdot \frac{\omega_n^2}{\left(s^2 + 2\,\zeta\,\omega_n\,s + \omega_n^2\right)}, \tag{10}$$

where $\tau_p = \frac{C_g \tau_r + N C_d \tau_q}{C_g + N C_d}$ and $V_0 = \frac{Q_{tot}}{C_g + N C_d}$.

By substituting expression (10) in (7), the following approximation for the output pulse $V_{out}(s)$ is obtained:

$$V_{out}(s) \cong V_0 \cdot \frac{(1+\tau_q s)}{s(1+\tau_p s)} \cdot \frac{1}{R_s} \cdot \frac{\omega_n^2}{\left(s^2 + 2\,\zeta\,\omega_n\,s + \omega_n^2\right)} \cdot \frac{K_R}{(1+\tau_L s)}. \tag{11}$$

In Equation (11), the whole system is represented in the s-domain as the cascade of three submodules, as shown in Figure 4. The blocks of Figure 4 correspond to the factors of Equation (11). $G_1(s)$ accounts for the SiPM model with its electrical parameters; $G_2(s)$ models the interaction between the SiPM and the preamplifier, i.e., the interconnections and the input impedance of the current buffer; $G_3(s)$ represents the current-to-voltage transfer function of the preamplifier.

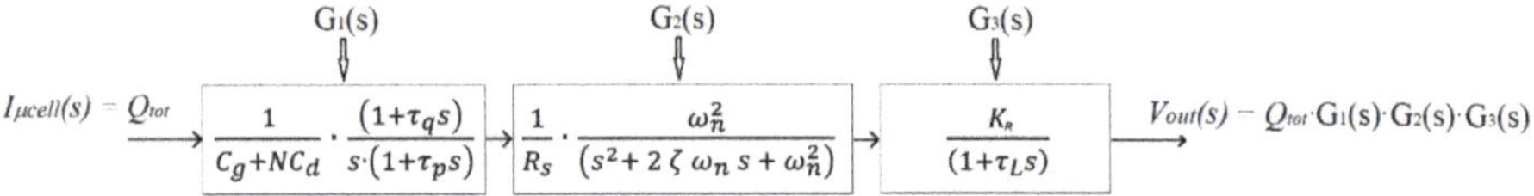

Figure 4. Simplified block diagram of the system.

Thus, expression (11) allows isolating the contribution of each of the main blocks of the detection chain to the shaping of the initial current pulse.

The results provided by expressions (9) and (11) have been compared, using the set of electrical parameters of the SiPM S10931-050P from Hamamatsu reported in Table 1, extracted by applying the procedure described in Licciulli [22].

Table 1. Electrical parameters of the SiPM used in the simulations.

R_q	182.75 kΩ
C_q	17.72 fF
C_d	75.17 fF
C_g	36.85 pF
R_{sub}	22.9 Ω
N	3600

The inverse Laplace transforms of Expressions (9) and (11) are plotted in Figure 5. They represent, respectively, the response of the complete model and the response of its approximation, valid for high frequencies, when just one micro-cell of the SiPM undergoes avalanche breakdown with the following set of parameters: L_{par} = 10 nH, R_{in} = 10 Ω, C_{in} = 1 pF, K_R = 1.2 kΩ, and BW = $1/2\pi\tau_L$ = 1 GHz.

The inset in Figure 5 proves that the approximation of $V_{out}(t)$ and the complete model fit almost perfectly in the region of interest for time pickoff, with only marginal deviations. The two expressions start to diverge only in proximity of the peak of the exact model. It is worth noting that good fittings are obtained regardless of any realistic choice of the parameter values.

As already mentioned, the slope of the response at the chosen threshold V_{TH} is a key parameter for time resolution: once established the noise level, the steeper the waveform at the threshold crossing point, the lower the jitter.

Figure 6 shows the time derivatives of the waveforms in Figure 5 in the portion of their rising edge, confirming the validity of approximation (11).

In general, the slope of the response exhibits a strong dependence on the value of the series inductance. Figure 7 shows that also a small 10 nH inductance remarkably affects the shape and the amplitude of the slope of the response. It is also apparent that the approximate model can predict the slope of the pulse even in presence of such a small inductance.

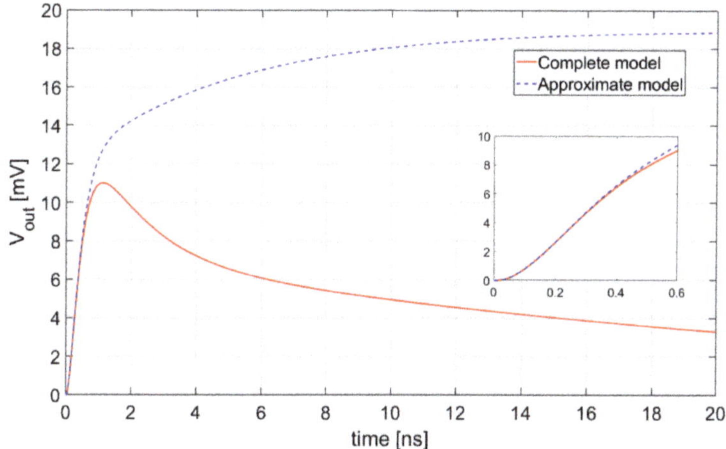

Figure 5. Comparison between the waveforms $V_{out}(t)$ obtained from Equations (9) (complete model) and (11) (approximate model). The inset shows the early part of the transient response.

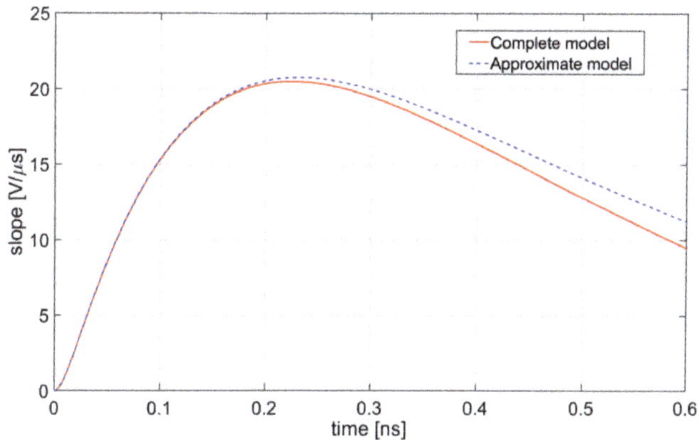

Figure 6. Time derivatives of the response V_{out} (t): comparison between the complete model in Equation (9) and its approximation in Equation (11).

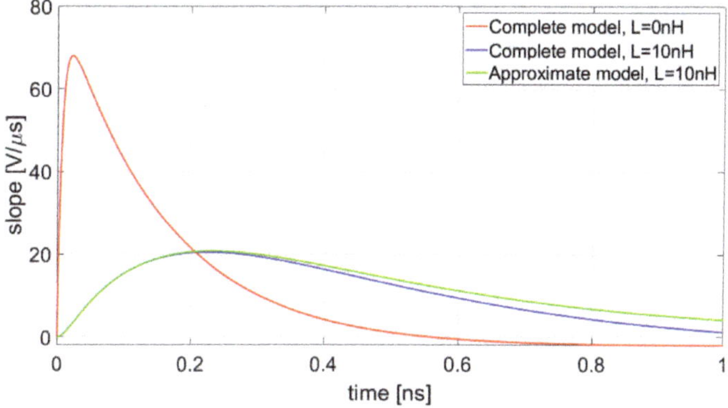

Figure 7. Slope of output pulses obtained with no series inductance and with a 10 nH inductance included in both the complete and approximate models.

4. A Comprehensive Analysis Including the Front-End Bandwidth

To account for the effect of the finite bandwidth of the front-end upon the shape of the response and its slope, in Equation (11), the preamplifier has been described as a single pole transfer function, as reported hereinafter for convenience:

$$V_{I,\,out}(s) \cong V_0 \cdot \frac{\left(1 + \tau_q s\right)}{s\left(1 + \tau_p s\right)} \cdot \frac{\omega_n^2}{R_s\left(s^2 + 2\,\zeta\,\omega_n\,s + \omega_n^2\right)} \cdot \frac{K_R}{\left(1 + \tau_L s\right)}. \tag{12}$$

In Equation (12), the subscript "I" has been added to emphasize that we are referring to a current-mode approach. Here, a transimpedance preamplifier with gain $K_R = A_i \cdot R_L$ and cut-off frequency set by the dominant time constant τ_L is used to convert the current pulse $I_{in}(t)$ into the output voltage $V_{I,out}(t)$, as illustrated in Figure 1. As already pointed out, in case a voltage-mode readout is adopted, (see Figure 3), the gain K_R is replaced with the product $K_R' = A_v \cdot R_{in}$ and the time constant τ_L is replaced with τ_{AMP}, resulting in the following expression of the output voltage pulse:

$$V_{V,\,out}(s) \cong V_0 \cdot \frac{\left(1 + \tau_q s\right)}{s\left(1 + \tau_p s\right)} \cdot \frac{\omega_n^2}{R_s\left(s^2 + 2\,\zeta\,\omega_n\,s + \omega_n^2\right)} \cdot \frac{A_v \cdot R_{in}}{\left(1 + \tau_{AMP} s\right)} \tag{13}$$

By observing that, in practical cases, $\tau_{in} = R_{in} C_{in} \ll \frac{L_{par}}{R_{par}}$, the term $2\,\zeta\,\omega_n$ can be approximated to $\frac{1}{\tau_{in}}$. Thus, the second order contribution that appears in Equations (12) and (13) can be further simplified as follows:

$$G_2(s) = \frac{\omega_n^2}{R_s\left(s^2 + 2\,\zeta\,\omega_n\,s + \omega_n^2\right)} \cong \frac{1}{R_s} \frac{\frac{1}{\tau_{in} \cdot \tau_A}}{\left(s^2 + \frac{s}{\tau_{in}} + \frac{1}{\tau_{in} \cdot \tau_A}\right)}, \tag{14}$$

where $\tau_A = \frac{L_{par}}{R_s}$ and the two poles of the system are respectively equal to $-1/\tau_A$ and $-1/\tau_{in}$.

The time constant τ_{in} is very small compared with τ_A (typically $\tau_A/\tau_{in} > 10$) and, consequently, the faster exponential term associated with τ_{in} decays to zero almost instantaneously, compared with the slower term associated with τ_A.

Therefore, a dominant pole approximation can be considered for $G_2(s)$ and Equation (14) can be rearranged as

$$G_2(s) \cong \frac{1}{R_s} \cdot \frac{\frac{1}{\tau_{in} \cdot \tau_A}}{\left(s + \frac{1}{\tau_{in}}\right)\left(s + \frac{1}{\tau_A}\right)} \cong \frac{1}{R_s} \cdot \frac{1}{\left(1 + \tau_A s\right)}. \tag{15}$$

Furthermore, at high frequencies, which describe well the leading edge of the output pulse we are interested in, the following assumption can be made:

$$\frac{1 + \tau_q s}{1 + \tau_p s} \cong \frac{\tau_q}{\tau_p} = \alpha,$$

leading to further simplification of Equation (12), which becomes

$$V_{I,out}(s) \cong \alpha \cdot V_0 \cdot \frac{1}{R_s} \cdot \frac{1}{s} \cdot \frac{1}{\left(1 + \tau_A s\right)} \cdot \frac{K_R}{\left(1 + \tau_L s\right)} \tag{16}$$

The Laplace transform of the slope of the output pulse, $Slope_{V,I}(s)$, is obtained by multiplying expression (16) by the variable s:

$$Slope_{V,I}(s) = \alpha \cdot V_0 \cdot \frac{K_R}{R_s} \cdot \frac{1}{\tau_A \tau_L} \cdot \left(\frac{a}{s + \frac{1}{\tau_A}} + \frac{b}{s + \frac{1}{\tau_L}}\right), \tag{17}$$

where $a = -\frac{\tau_A \cdot \tau_L}{\tau_L - \tau_A}$ and $b = \frac{\tau_A \cdot \tau_L}{\tau_L - \tau_A}$.

Equation (17) can be conveniently used to derive simple expressions for both the time corresponding to the maximum slope, t_{MAX_S}, and the value of the maximum slope of the output pulse as a function of the most relevant parameters involved.

The inverse Laplace transform of Equation (17) leads to the final expression for the slope of the output voltage signal in the time domain:

$$\text{Slope}_{V,I}(t) = \mathcal{L}^{-1}\{\text{Slope}_{V,I}(s)\} = \alpha \cdot V_0 \cdot \frac{K_R}{R_s} \cdot \frac{1}{\tau_A \cdot \tau_L} \cdot \left(a \cdot e^{-\frac{t}{\tau_A}} + b \cdot e^{-\frac{t}{\tau_L}}\right). \tag{18}$$

Taking the time derivative of $\text{Slope}_{V,I}(t)$, equating it to zero and solving for the time variable t, the time t_{MAX_S}, expressed as a function of τ_A, τ_L, L_{par}, and R_s is

$$t_{MAX_S} = \frac{\tau_A \cdot \tau_L}{\tau_A - \tau_L} \cdot \ln\left(\frac{\tau_A}{\tau_L}\right) = \frac{\tau_A}{\theta - 1} \cdot \ln\theta, \tag{19}$$

where $\theta = \frac{\tau_A}{\tau_L}$ is a normalization variable, depending on $\tau_A = L_{par}/R_s$ and τ_L. The maximum slope, i.e., Equation (18) evaluated for $t = t_{MAX_S}$, is the following:

$$\text{Slope}_{V,I}(t_{MAX_S}) = \alpha \cdot V_0 \cdot \frac{K_R}{R_s} \cdot \frac{1}{\tau_A} \cdot e^{-\frac{t_{MAX_S}}{\tau_A}}, \tag{20}$$

and, replacing t_{MAX_S} with Equation (19) and the transimpedance gain K_R with the product $A_i \cdot R_L$, the following expression results:

$$\text{Slope}_{V,I}(t_{MAX_S}) = \alpha \cdot V_0 \cdot \frac{A_i \cdot R_L}{L_{par}} \cdot \theta^{\frac{1}{1-\theta}}. \tag{21}$$

The corresponding expression for the voltage-mode approach is

$$\text{Slope}_{V,V}(t_{MAX_S}) = \alpha \cdot V_0 \cdot \frac{A_v \cdot R_{in}}{L_{par}} \cdot \theta^{\frac{1}{1-\theta}}, \tag{22}$$

where the dominant time constant τ_{AMP} of the voltage amplifier replaces τ_L in the parameter θ.

Equations (21) and (22) are very simple expression which describe the dependence of the maximum slope of the output pulse on the parameter θ, which, in turns, depends on the time constants τ_A, i.e., the ratio L_{par}/R_s, and the dominant time constant of the preamplifier.

It is apparent that the maximum slopes obtained in case of application of a current-mode or voltage-mode front-end approach share the same dependence on the parameter θ. However, Equation (22) exhibits also an explicit dependence of the maximum slope on the input resistance of the preamplifier R_{in}. Consequently, if a voltage-mode approach is used, the maximum slope of the output pulse tends to increase when the input resistance increases, whereas the opposite happens in case the current-mode approach is adopted.

These conclusions are illustrated by the following Figures 8 and 9, which compare the exact slopes, obtained using the complete models (12) and (13), with their corresponding approximations (21) and (22), respectively. In both Figures 8 and 9 the maximum slope is represented as a function of R_{in} for four values of the inductance L_{par}. The plots prove that the proposed lower order approximation of the complete system gives very accurate results.

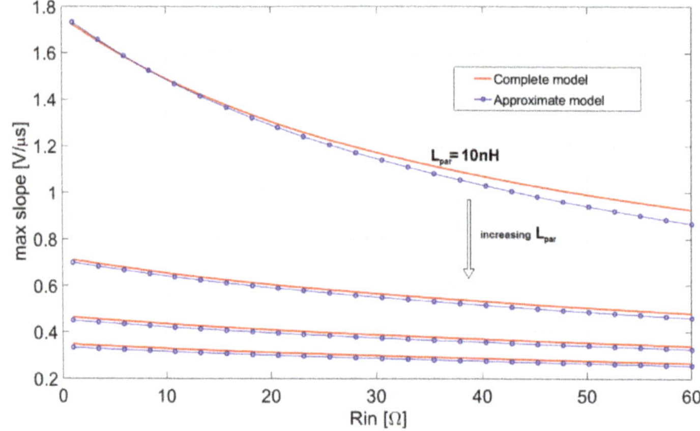

Figure 8. Maximum slope of the leading edge of the output pulse for the current-mode approach (12), compared with its approximation (21), with L_{par} = 10 nH, 40 nH, 70 nH, 100 nH, C_{in} = 0.5 pF and BW = 0.5 GHz.

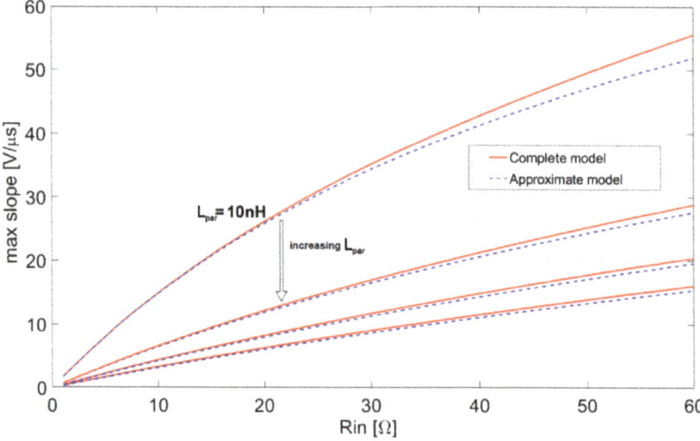

Figure 9. Maximum slope of the leading edge of the output pulse for the voltage-mode approach (13) compared with its approximation (22), with L_{par} = 10 nH, 40 nH, 70 nH, 100 nH, C_{in} = 0.5 pF, and BW = 0.5 GHz.

The same conclusions can be drawn observing the plots in Figure 10. Here, the output waveforms resulting from both a current-mode approach and a voltage-mode approach, obtained with the complete model, are shown for different values of R_{in}. As R_{in} increases, the system becomes faster for both the current-mode and the voltage-mode, thanks to the decreasing value of the time constant $\tau_A = L_{par}/(R_{in} + R_{par})$. This causes the decrease of the peaking time observed in both graphs of Figure 10 for increasing values of R_{in}. However, in the current-mode case, increasing the input resistance causes also a decrease of the amplitude of the current pulse $I_{in}(t)$ and, consequently, of the output voltage pulse. This results in a net reduction of the maximum output slope, despite the decrease of the peaking time. Instead, in the voltage-mode case, the increase of R_{in} also causes an increase of the amplitude of the output pulse and, consequently, of its maximum slope.

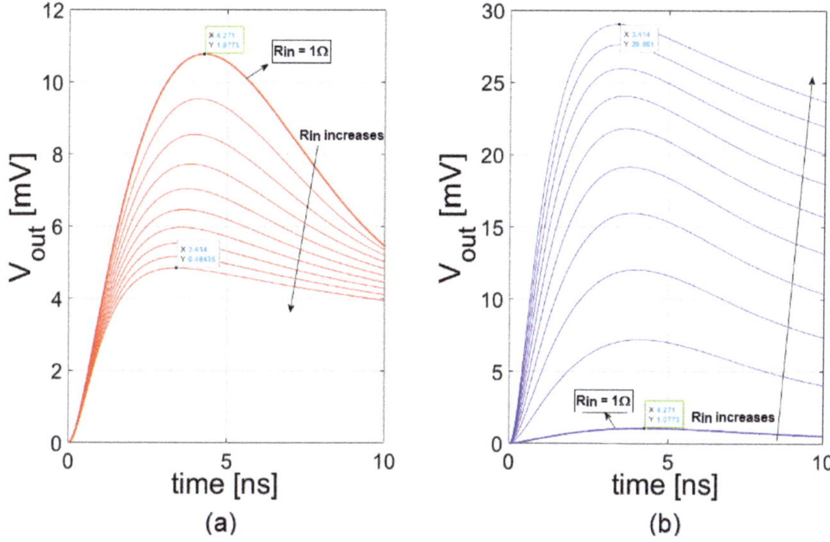

Figure 10. The current-mode (**a**) and voltage-mode (**b**) responses as functions of the input resistance when L_{par} = 70 nH, C_{in} = 0.5 pF, and BW = 0.5 GHz, obtained using the complete model.

A clear advantage of the current-mode approach is that, owing to the absence of internal high impedance nodes, it is preferable compared to a voltage-mode approach in case large bandwidth is needed. Let us consider, for instance, a simple current buffer realized by means of the common base (CB) amplifier depicted in Figure 11a. The transimpedance gain of the circuit is given by the load resistance R_L, whereas the time constant associated to the load capacitance $\tau_L = R_L C_L$ determines its bandwidth, according to the system shown in Figure 1. Moreover, a basic voltage-mode implementation of the front-end using the same active device is the common-emitter (CE) amplifier shown in Figure 11b.

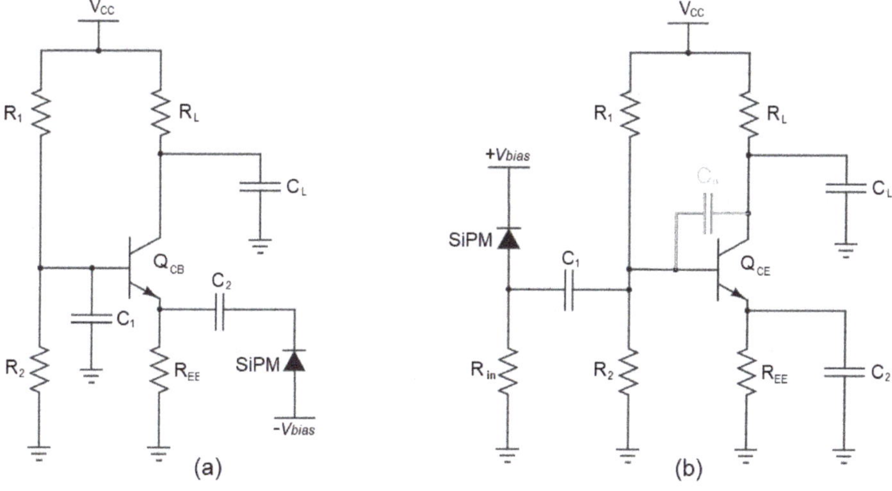

Figure 11. Two readout approaches: (**a**) a BJT in the common base configuration; (**b**) the same BJT in the common emitter configuration.

If the open-circuit time constant method is applied to estimate the dominant time constant τ_{AMP} of the CE amplifier, we obtain

$$\tau_{AMP} \cong R_L C_L + (R_{in}\|r_\pi)C_\pi + R_{in}C_\mu A_V, \tag{23}$$

where $A_V = g_{mCE}R_L$.

To make the bandwidth of the CE equal to the bandwidth of the current buffer, both the second and the third terms in Equation (23) should be negligible. Under the assumption that the input resistance is small enough to make the second term very small, by imposing

$$R_{in}C_\mu g_{mCE}R_L \ll R_L C_L \tag{24}$$

the following condition on the value of R_{in} is obtained:

$$R_{in} \ll \frac{1}{g_{mCE}} \cdot \frac{C_L}{C_\mu}. \tag{25}$$

Condition (25) states that, in practice, in this example of voltage-mode circuit it is impossible to exploit the increase of the input resistance to increase the slope of the output pulse. In fact, this would result in an unavoidable penalty in terms of bandwidth, in comparison to the corresponding current-mode circuit. Increasing the transconductance would have a beneficial effect on the voltage gain of the circuit, thus on the response slope. However, this would also require a further decrease of the input resistance to fulfil the condition (25); otherwise, the bandwidth would be compromised, thus neutralizing the benefits in terms of slope. Instead, in the CB case, increasing the transconductance, and hence the power consumption, would decrease the input resistance of the current-mode circuit, thus increasing the slope of the output pulse, without affecting the bandwidth of the preamplifier.

Another possible example of voltage-mode front-end is reported in Figure 12, based on a simple op-amp non-inverting configuration.

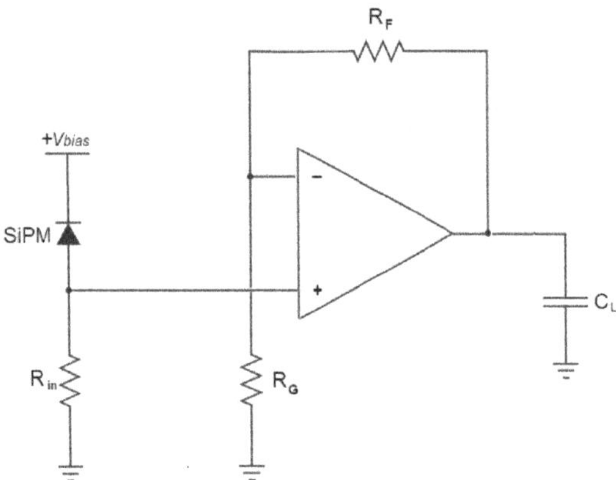

Figure 12. A voltage-mode preamplifier, based on an op-amp.

In this case, the closed-loop bandwidth is expressed as

$$BW = \frac{GBW}{A_V}, \tag{26}$$

where GBW is the gain-bandwidth product of the op-amp, and $A_V = 1 + R_F/R_G$ is the DC closed-loop voltage gain.

Referring to Equations (21) and (22), the maximum slopes obtained with the two front-end approaches are the same if $A_V R_{in} = A_I R_L$ and also if $BW = 1/2\pi\tau_L = 1/2\pi R_L C_L$. Therefore, we must have

$$R_{in} = \frac{A_I R_L}{A_V} = \frac{A_I R_L}{GBW} \cdot BW = \frac{A_I}{2\pi \cdot GBW \cdot C_L}. \tag{27}$$

Equation (27) states that if we want to obtain advantages in terms of maximum slope of the output signal with the voltage-mode circuit in Figure 12, with respect to the current-mode circuit in Figure 11a, its input resistance R_{in} must be very large. In fact, since C_L is of the order of few pF, to keep the value of R_{in} provided by Equation (27) within reasonable limits, i.e., around few hundreds of Ohms, huge values of GBW would be needed. If R_{in} is increased beyond these limits, the time constant τ_{in} cannot be considered negligible any longer, slowing down the rise time of the output pulse, due to the contribution of this further time constant. Moreover, with large values of R_{in}, the duration of the long tail of the output pulse increases, causing pile-up problems and worsening the timing accuracy of the circuit, due to baseline fluctuations [25,26]. In conclusion, also for this further example of voltage-mode preamplifier, exploiting the increase of the input resistance R_{in} to improve the timing performance of the detection system is not practically feasible.

5. Experimental Tests and Results

To confirm the validity of the proposed analysis with experimental results, a printed circuit board (PCB) has been designed to readout the response of the S10931-050P SiPM from Hamamatsu (see Table 1). In this board, a fast RF BJT configured as a common base current buffer, as shown in Figure 11a, is used in the very front-end. Figure 13 shows the PCB with the 3×3 mm^2 SiPM and the experimental setup, enclosed in dark box.

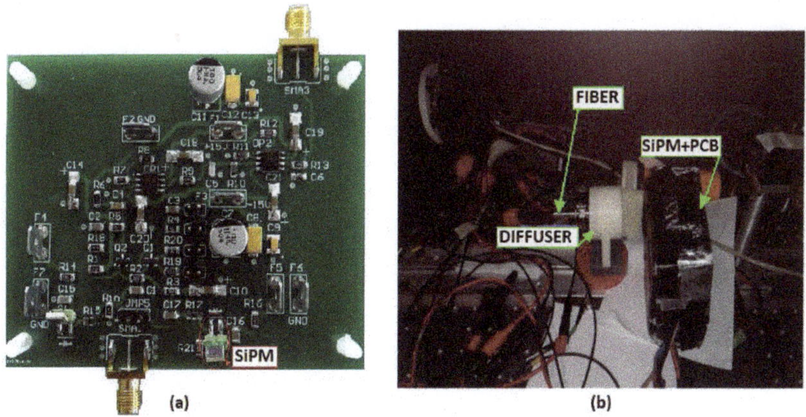

Figure 13. Pictures of (**a**) the printed circuit board used for the experiments and (**b**) the experimental setup in the dark box.

The SiPM can be coupled to the input of the preamplifier directly, using a zero-ohm resistor, or by interposing a small discrete inductor with two possible values, L = 51 nH or L = 100 nH. Moreover, four possible different values of R_{in} can be selected, i.e., 10 Ω, 18 Ω, 33 Ω, and 50 Ω, by changing the emitter resistor R_{EE} that sets the DC current of the transistor, as shown in Figure 11. A large-bandwidth voltage amplifier, cascaded to the front-end, has been used to increase the amplitude of its output pulse. A 380 nm fast pulsed laser source has been used to generate light flashes with FWHM duration of 50 ps. The light is sent towards the SiPM sensitive surface by means of an optical fiber and an optical diffuser has been interposed between the fiber and the detector, to maximize the probability of the detection of single-photon events.

Among all the responses of the circuit to the laser flashes, acquired in coincidence with the laser trigger, only the ones corresponding to single-photon events have been selected. For each possible circuit arrangement in terms of input resistance and series inductance, a noiseless 'golden' waveform has been extracted, by aligning and averaging the acquired pulses. Examples of such waveforms obtained with L = 51 nH and different values of R_{in} are reported in Figure 14.

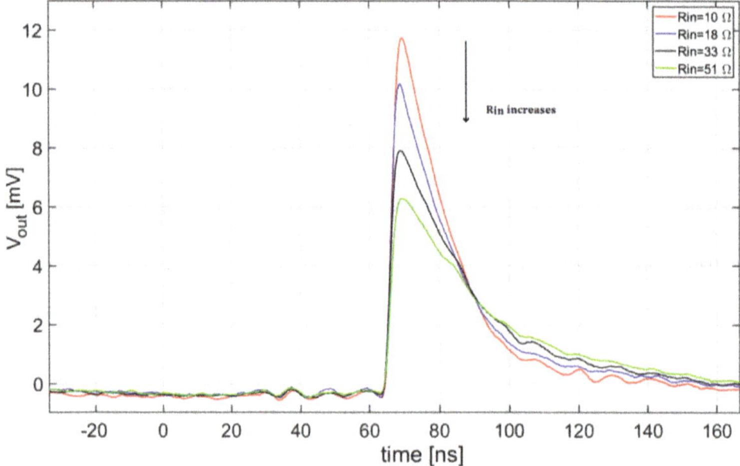

Figure 14. Examples of 'golden' pulse waveforms, obtained with L = 51 nH and different values of R_{in}.

The derivative of each 'golden' pulse and its maximum value have been evaluated. Figure 15 shows an example of such derivative, obtained with L = 51 nH and R_{in} = 33 Ω.

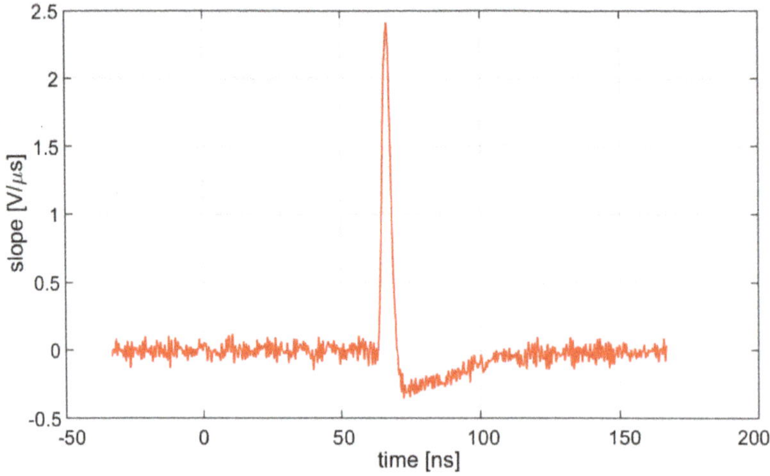

Figure 15. An example of time derivative of a 'golden' pulse, obtained with L = 51 nH and Rin = 33 Ω.

Finally, the peak values of the slopes have been plotted as a function of the input resistance for the three possible values of series inductance, as shown in Figure 16. The fitting curves of the experimental data and the corresponding behavior of the maximum slope obtained respectively by means of the proposed mathematical model and by SPICE simulations of the circuit are also reported in Figure 16.

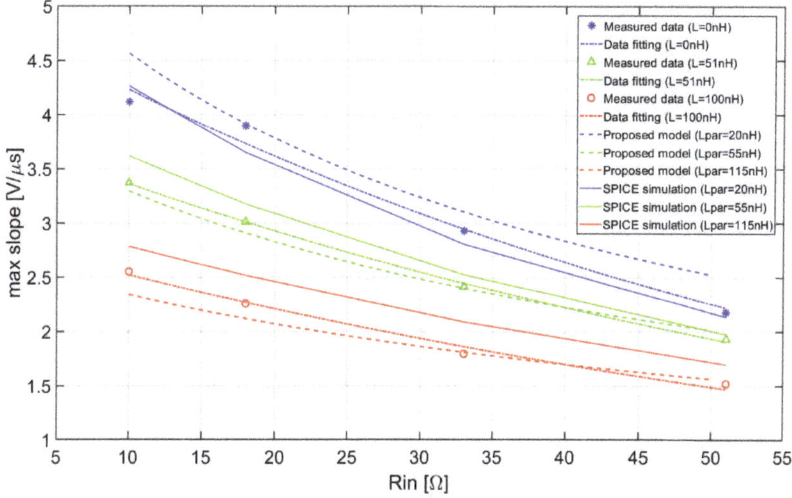

Figure 16. Maximum slopes of the output pulse of the front-end as a function of the input resistance R_{in} for three values of parasitic inductance: comparison among measured data, proposed model predictions, and SPICE simulations.

It is apparent from Figure 16 that the behavior of the measured data basically reproduces the one predicted by the approximate analytical model and by the SPICE simulations. Small corrections of the series inductance have been introduced in the model with respect to the ideal value of the inductance L. These corrections are needed to take account of the unavoidable parasitic contributions associated to the zero-ohm resistor and the interconnections, slightly variable when a different physical inductor is interposed between the SiPM and the front-end. As also shown in Figure 16, these corrections increase the total value of the inductance L_{par} to 20 nH, 55 nH, and 115 nH, respectively in case the zero-ohm resistor, the 51 nH and the 100 nH inductors are inserted. Whatever the value of the series inductance, when the input resistance R_{in} of the CB amplifier is increased, the maximum slope decreases, as predicted by the proposed model.

Figure 17 shows the time jitter as a function of the input resistance for the three different values of series inductance.

The jitter has been evaluated using the time spread distribution of the delay between the trigger pulse of the laser and the time when the output signal resulting from the detection of a single photo-electron event reaches the threshold. The threshold has been fixed at the level corresponding to the maximum slope of the 'golden' pulse. The curves in Figure 17 include the contributions of the measurement setup to the total jitter.

Considering low and moderate values of the series inductance, the jitter is poorly dependent on the input resistance and slightly increases when R_{in} increases. This behavior is consistent with both experimental data reported in Figure 16 and Equation (21), since the slope decreases when R_{in} increases and, in our circuit, the electronic noise is dominated by the contribution associated with the resistor R_L (see Figure 11), thus it does not change significantly. With higher inductance values (green curve), the time resolution rapidly degrades when R_{in} increases. This suggests that, when a current-mode approach is used, choosing low values for the input resistance of the front-end electronics has relevant advantages in terms of low input jitter especially when the parasitic interconnection inductance is relatively large. Conversely, when the SiPM is coupled directly to the front-end with very short interconnections and L_{par} is very small, the influence of R_{in} on the timing performance of the circuit is, to some extent, reduced. In this case, the advantages of the decrease of R_{in} must be compared to possible drawbacks, such as, for instance, an increase of the power consumption of the circuit.

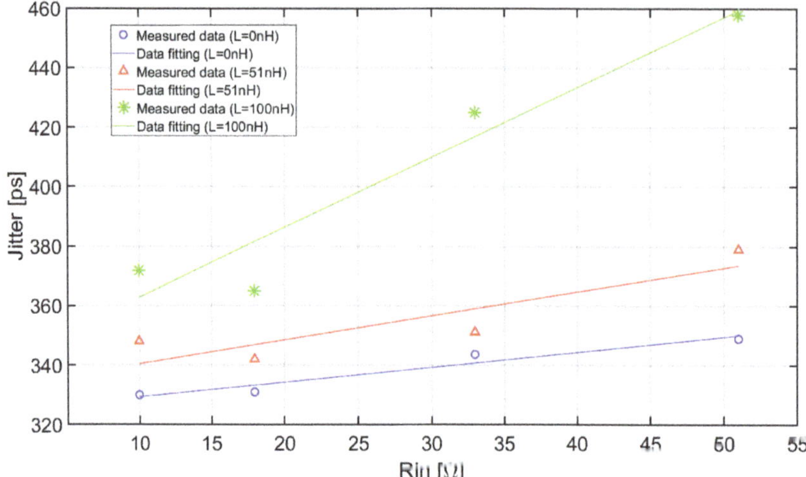

Figure 17. Time jitter as a function of R_{in} and L. The light continuous curves are first order exponential fittings of the measured data points.

6. Conclusions

Analytical expressions for the single-photon response of a SiPM-based detection system are already available in the literature. However, these models either do not account for relevant factors, such as the parasitic inductance interposed between the detector and the front-end electronics, or are based on inaccurate electric models of the SiPM.

Starting from a complete model of the detector and considering reasonable assumptions, we propose a new factorized expression of the transfer function of the system, able to accurately reproduce its behavior in the initial transient of the response. In this expression, the individual contributions of the SiPM, the interconnection parasitic components, and the front-end electronics have been distinguished as independent blocks.

Based on this model, a simple but accurate closed-form equation for the single-photon response as a function of the main system parameters has been derived. As is well known, the slope of the response has a relevant influence on the timing performance of the system, when leading edge discrimination is the chosen time pickoff technique. Our simplified model reproduces with good accuracy the slope of the initial transient of the response provided by the complete model of the system, when its parameters are varied.

The proposed analysis allows to easily perform a direct comparison of the single-photon response obtained with different front-end architectures. The results of this comparison show that a current-mode readout approach is preferable to a voltage-mode one in terms of timing accuracy, with typical values of the parameters involved.

Moreover, practical design guidelines for the front-end electronics can be devised by means of the proposed analytical model of the response. For instance, when a current-mode approach is chosen, we found that extremely low values of the input resistance of the preamplifier are particularly convenient only in presence of increasing values of the parasitic interconnection inductance.

Last, experimental tests, carried out on an example of current-mode front-end circuit coupled to a 3×3 mm^2 SiPM, have been used to validate the results of our analysis. The behavior of the real readout system has been reproduced with good accuracy by our approximate model, in terms of maximum slope of the single-photon response and resulting time jitter.

Author Contributions: Conceptualization, C.M., P.A.P.C., and S.P.; formal analysis and methodology, P.A.P.C. and C.M.; software validation, P.A.P.C. and S.P; experimental validation and data curation, M.D.G. and P.A.P.C.;

writing—original draft preparation, P.A.P.C.; writing—review and editing, all authors; supervision, project administration, C.M. and P.A.P.C. All authors have read and agreed to the published version of the manuscript.

Funding: This research was funded by Politecnico di Bari, grant "Fondo di Ricerca dell'Ateneo". The APC was funded by Politecnico di Bari.

Acknowledgments: The authors would like to thank prof. Nicola Giglietto, Politecnico di Bari, prof. Francesco Giordano, Università di Bari and dott. Francesco Licciulli, INFN Sezione di Bari, for their valuable support.

Conflicts of Interest: The authors declare no conflict of interest.

References

1. Lamprou, E.; Aguilar, A.; Gonzalez-Montoro, A.; Monzo, J.; Canizares, G.; Vidal, L.F.; Hernandez, L.; Iranzo, S.; Marti, R.; Sanchez, S.; et al. Progress Report for an Accurate PET Detector Based on SiPMs and the TOFPET ASIC. In Proceedings of the 2017 IEEE Nuclear Science Symposium and Medical Imaging Conference (NSS/MIC), Atlanta, GA, USA, 21–28 October 2017; pp. 1–3.

2. Cozzi, G.; Busca, P.; Carminati, M.; Fiorini, C.; Montagnani, G.L.; Acerbi, F.; Gola, A.; Paternoster, G.; Piemonte, C.; Regazzoni, V.; et al. High-Resolution Gamma-Ray Spectroscopy With a SiPM-Based Detection Module for 1" and 2" LaBr3:Ce Readout. *IEEE Trans. Nucl. Sci.* **2018**, *65*, 645–655. [CrossRef]

3. Becker, W. Overview of Photon Counting Techniques. In *Advanced Time-Correlated Single Photon Counting Techniques*, 1st ed.; Springer: Berlin, Germany, 2005.

4. Wei, L.; Ho, D.; Yan, W. Recent Advances in Fluorescence Lifetime Analytical Microsystems: Contact Optics and CMOS Time-Resolved Electronics. *Sensors* **2017**, *17*, 2800. [CrossRef] [PubMed]

5. Adamo, G.; Busacca, A. Time of Flight measurements via two LiDAR systems with SiPM and APD. In Proceedings of the 2016 AEIT International Annual Conference (AEIT), Capri, Italy, 5–7 October 2016; pp. 1–5.

6. Pantoli, L.; Barile, G.; Leoni, A.; Muttillo, M.; Stornelli, V. Electronic Interface for Lidar System and Smart Cities Applications. *J. Commun. Softw. Syst.* **2019**, *15*, 118–125. [CrossRef]

7. Cates, J.W.; Gundacker, S.; Auffray, E.; Lecoq, P.; Levin, C.S. Improved single photon time resolution for analog SiPMs with front end readout that reduces influence of electronic noise. *Phys. Med. Biol.* **2018**, *63*, 185022. [CrossRef] [PubMed]

8. Acerbi, F.; Ferri, A.; Gola, A.; Cazzanelli, M.; Pavesi, L.; Zorzi, N.; Piemonte, C. Characterization of Single-Photon Time Resolution: From Single SPAD to Silicon Photomultiplier. *IEEE Trans. Nucl. Sci.* **2014**, *61*, 2678–2686. [CrossRef]

9. Acerbi, F.; Paternoster, G.; Gola, A.; Regazzoni, V.; Zorzi, N.; Piemonte, C. High-Density Silicon Photomultipliers: Performance and Linearity Evaluation for High Efficiency and Dynamic-Range Applications. *IEEE J. Quantum Electron.* **2018**, *54*, 1–7. [CrossRef]

10. Spanoudaki, V.C.; Levin, C.S. Photo- Detectors for Time of Flight Positron Emission Tomography (ToF-PET). *Sensors* **2010**, *10*, 10484–10505. [CrossRef] [PubMed]

11. Marano, D.; Grasso, A.D. Novel straightforward and effective extraction methodology for SiPM model parameters. In Proceedings of the 2017 European Conference on Circuit Theory and Design (ECCTD), Catania, Italy, 4–6 September 2017; pp. 1–4.

12. Marano, D.; Bonanno, G.; Garozzo, S.; Grillo, A.; Romeo, G. New Improved Model and Accurate Analytical Response of SiPMs Coupled to Read-Out Electronics. *IEEE Sens. J.* **2015**, *16*, 19–21. [CrossRef]

13. Ciciriello, F.; Corsi, F.; Licciulli, F.; Marzocca, C.; Matarrese, G. Assessing the time resolution of an integrated front-end for solid state radiation detectors. In Proceedings of the 2011 IEEE Nuclear Science Symposium Conference Record, Valencia, Spain, 23–29 October 2011; pp. 1678–1682.

14. Huizenga, J.; Seifert, S.; Schreuder, F.; Van Dam, H.; Dendooven, P.; Löhner, H.; Vinke, R.; Schaart, D.R. A fast preamplifier concept for SiPM-based time-of-flight PET detectors. *Nucl. Instrum.Methods Phys. Res. Sect. A Accel. Spectrom. Detect. Assoc. Equip.* **2012**, *695*, 379–384. [CrossRef]

15. Fernandez-Tenllado, J.; Ballabriga, R.; Campbell, M.; Gascon, D.; Gomez, S.; Mauricio, J. Optimal design of single-photon sensor front-end electronics for fast-timing applications. In Proceedings of the 2019 IEEE Nuclear Science Symposium and Medical Imaging Conference (NSS/MIC), Manchester, UK, 26 October–2 November 2019; pp. 1–5.

16. Ciciriello, F.; Corsi, F.; Licciulli, F.; Marzocca, C.; Matarrese, G. Comparing front-end alternatives for SiPM's in single-photon time resolution applications. In Proceedings of the 2015 IEEE Nuclear Science Symposium and Medical Imaging Conference (NSS/MIC), San Diego, CA, USA, 31 October–7 November 2015; pp. 1–8.

17. Calò, P.P.; Ciciriello, F.; Petrignani, S.; Marzocca, C. SiPM readout electronics. *Nucl. Instruments Methods Phys. Res. Sect. A Accel. Spectrom. Detect. Assoc. Equip.* **2019**, *926*, 57–68. [CrossRef]

18. Corsi, F.; Dragone, A.; Marzocca, C.; Del Guerra, A.; Delizia, P.; Dinu, N.; Piemonte, C.; Boscardin, M.; Betta, G.-F.D. Modelling a silicon photomultiplier (SiPM) as a signal source for optimum front-end design. *Nucl. Instrum. Methods Phys. Res. Sect. A Accel. Spectrom. Detect. Assoc. Equip.* **2007**, *572*, 416–418. [CrossRef]

19. Ciciriello, F.; Corsi, F.; Licciulli, F.; Marzocca, C.; Matarrese, G.; Bisogni, M.G.; Del Guerra, A. Designing the front-end electronics of a SiPM based γ-ray detection system for optimal time resolution. In Proceedings of the 2012 IEEE Nuclear Science Symposium and Medical Imaging Conference Record (NSS/MIC), Anaheim, CA, USA, 29 October–3 November 2012; pp. 827–833.

20. Ciciriello, F.; Corsi, F.; Licciulli, F.; Marzocca, C.; Matarrese, G.; Del Guerra, A.; Bisogni, M.G. Accurate modeling of SiPM detectors coupled to FE electronics for timing performance analysis. *Nucl. Instrum. Methods Phys. Res. Sect. A Accel. Spectrom. Detect. Assoc. Equip.* **2013**, *718*, 331–333. [CrossRef]

21. Marano, D.; Bonanno, G.; Garozzo, S.; Grillo, A.; Romeo, G. A New Simple and Effective Procedure for SiPM Electrical Parameter Extraction. *IEEE Sens. J.* **2016**, *16*, 3620–3626. [CrossRef]

22. Licciulli, F.; Marzocca, C. Parameter Extraction Method for the Electrical Model of a Silicon Photomultiplier. *IEEE Trans. Nucl. Sci.* **2016**, *63*, 2517–2526. [CrossRef]

23. Di Lorenzo, S.C.; Callier, S.; Fleury, J.; Dulucq, F.; De La Taille, C.; Chassard, G.M.; Raux, L.; Seguin-Moreau, N. SPIROC: Design and performances of a dedicated very front-end electronics for an ILC Analog Hadronic CALorimeter (AHCAL) prototype with SiPM read-out. *J. Instrum.* **2013**, *8*, C01027. [CrossRef]

24. Fischer, P.; Peric, I.; Ritzert, M.; Koniczek, M. Fast Self Triggered Multi Channel Readout ASIC for Time- and Energy Measurement. *IEEE Trans. Nucl. Sci.* **2009**, *56*, 1153–1158. [CrossRef]

25. Gundacker, S.; Turtos, R.M.; Auffray, E.; Paganoni, M.; Lecoq, P. High-frequency SiPM readout advances measured coincidence time resolution limits in TOF-PET. *Phys. Med. Biol.* **2019**, *64*, 055012. [CrossRef] [PubMed]

26. Gola, A.; Piemonte, C.; Tarolli, A. Analog Circuit for Timing Measurements with Large Area SiPMs Coupled to LYSO Crystals. *IEEE Trans. Nucl. Sci.* **2013**, *60*, 1296–1302. [CrossRef]

© 2020 by the authors. Licensee MDPI, Basel, Switzerland. This article is an open access article distributed under the terms and conditions of the Creative Commons Attribution (CC BY) license (http://creativecommons.org/licenses/by/4.0/).

Article

Silicon Photomultiplier Sensor Interface Based on a Discrete Second Generation Voltage Conveyor

Vincenzo Stornelli *, Leonardo Pantoli, Gianluca Barile, Alfiero Leoni and Emanuele D'Amico

Department of Industrial and Information Engineering and Economics (DIIEE), University of L'Aquila, Piazzale Pontieri 1, Monteluco di Roio, I 67100, 67100 L'Aquila, Italy; leonardo.pantoli@univaq.it (L.P.); gianluca.barile@univaq.it (G.B.); alfiero.leoni@univaq.it (A.L.); emanuele.damico@student.univaq.it (E.D.)
* Correspondence: vincenzo.stornelli@univaq.it

Received: 5 February 2020; Accepted: 2 April 2020; Published: 5 April 2020

Abstract: This work presents the design of a discrete second-generation voltage conveyor (VCII) and its capability to be used as electronic interface for silicon photomultipliers. The design addressed here exploits directly at the transistor level, with commercial components, the proposed interface; the obtained performance is valuable considering both the discrete elements and the application. The architecture adopted here realizes a transimpedance amplifier that is also able to drive very high input impedance, as usually requested by photons detection. Schematic and circuital design of the discrete second-generation voltage conveyor is presented and discussed. The complete circuit interface requires a bias current of 20 mA with a dual 5V supply voltage; it has a useful bandwidth of about 106 MHz, and considering also the reduced dimensions, it is a good candidate to be used in portable applications without the need of high-cost dedicated integrated circuits.

Keywords: current mode; sensor interface; silicon photomultiplier; transimpedance amplifier; voltage current conveyor

---

1. Introduction

Discrete components are still the primary solution for many electronic applications [1]. When a low integration factor can be adopted or system dimensions are not critical, hybrid circuits can still obtain good performance and synthesize the desired electrical behavior, because, from a technology point of view, commercially available electronic components are able to provide almost state-of-the-art performances. Additionally, in integrated systems, the capability to design and test preliminary hybrid solutions is usually beneficial for the designer in order to validate the circuit functionality [2,3].

From different points of view, silicon photomultipliers (SiPMs) are becoming an enabling technology in several fields (physical, medical, automotive applications, etc.) replacing the traditional photomultiplier tubes, especially in portable sensor applications. The technology advancements in recent times have achieved a high sensitivity and detection capability in compact dimensions and also require a reduced power consumption. Even though SiPMs represent a new powerful solution in the realization of photon detectors, on the one hand, on the other hand, they stress the electronics, requiring advanced interfaces with strict performances. This is due to the large parasitics provided by the SiPM, in particular, the output capacitance that increases even more when considering SiPM arrays, a typical solution adopted when a larger sensitive area is required.

In the literature, some SiPM sensor interfaces have been presented in the last few years [4–7], and this topic is currently a sensible task due to a large number of possible applications, as previously said. These works usually present electronic solutions at an integrated level, and they are mainly based on a voltage-design approach [8–11]. To the best of our knowledge, no hybrid solutions have been presented for this purpose for practical applications. In addition, previously presented works usually present voltage-mode design, because it is preferable for noise performance, even if it lacks velocity with respect to current-mode circuits [12]. Only a few solutions in mixed voltage/current mode have already been presented in the literature [13–16], the latter proposed by the same authors [15,16].

The discrete, full commercial component-based sensor interface proposed here achieves many benefits. First of all, it is a mixed voltage-mode current-mode design, and thanks to this choice, it is able to merge the two different design conceptions, taking advantage of both of them. In addition, it is the first discrete hybrid interface designed at the transistor level, making use of the so-called second-generation voltage conveyor (VCII) [17–20]. To the best of our knowledge, no other discrete VCIIs have been presented in the literature up to now. The proposed circuit was fabricated, and a smart prototype board was tested in our laboratory with promising results that are reported here, providing a feasible demonstration of its usability in practical applications. In order to test the interface, the sensor was emulated by means of an equivalent circuit as discussed in this paper, allowing us to easily mimic different working conditions.

In the following, Section 2 summarizes and makes some remarks on the VCII characteristics, Section 3 illustrates the proposed solution and describes the achievable performances, while Section 4 presents measurement results and the overall characteristics of the sensor interface. Finally, in Section 5, conclusions are drawn.

2. VCII Characteristics

The VCII is a three-port device, presented for the first time in [17] and conceived for duality from the current conveyor [21–23]. It has two input ports, X and Y, and an output port Z, as shown in Figure 1. It is logically representable as a current buffer between Y and X terminals and a subsequent voltage buffer between the X and Z ports. The peculiarity of the VCII that makes it suitable for the proposed application is that it inherently acts as a transimpedance amplifier between the Y and Z terminals, obtaining at the same time low input and output impedances. The relationship between external terminals can be summarized as in Equation (1), where α is the voltage gain between the X and Z ports and β is the current gain between Y and X terminals, while $r_{x,y,z}$ are the parasitic impedances relative to the corresponding X, Y, and Z terminals. Ideally, α and β are equal to one (β in absolute value) and r_x is infinite, while r_y and r_z are equal to zero.

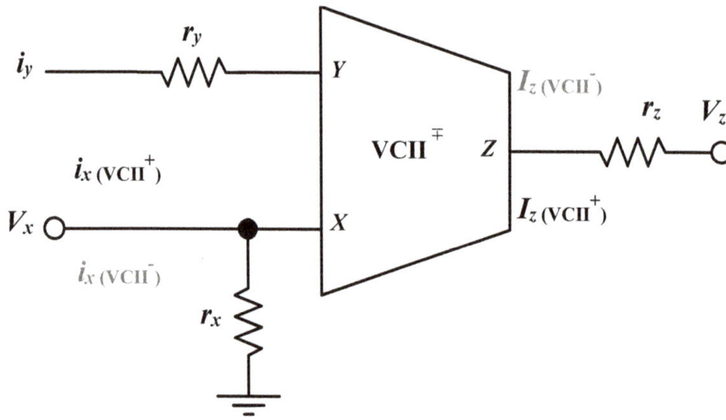

Figure 1. Block representation of the second-generation voltage conveyor (VCII) with its parasitic port impedances.

$$\begin{bmatrix} i_x \\ v_y \\ v_z \end{bmatrix} = \begin{bmatrix} \frac{1}{r_x} & \pm\beta & 0 \\ 0 & r_y & 0 \\ \alpha & 0 & r_z \end{bmatrix} \begin{bmatrix} V_X \\ I_Y \\ I_Z \end{bmatrix} \tag{1}$$

Further advantages of this building block are the capability to ensure faster operation with respect to traditional purely voltage-mode circuits, to maintain good performance up to a relatively high frequency, and to provide a constant transimpedance gain regardless of the operational bandwidth. The latter assumption can be justified by considering and discussing the configuration proposed in Figure 2. Assuming that $r_x \gg R$, and by considering Equation (1), we have the following:

$$V_x = I_x R = \pm \beta I_{in}(R r_x) \cong \pm \beta I_{in} R \tag{2}$$

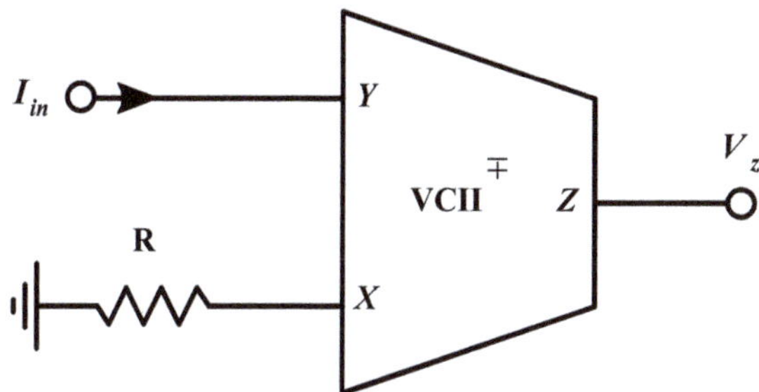

Figure 2. VCII configuration as current to voltage converter.

Then, combining Equation (2) with Equation (1), the output voltage at the Z terminal can be obtained as in Equation (3):

$$V_{out} = V_z = \alpha V_x = \pm \beta \alpha R I_{in} \tag{3}$$

and so, the final version of the transimpedance transfer function, F_{TI} (which is the transimpedance gain) is synthesized as follows:

$$F_{TI} = \frac{V_{out}}{I_{in}} = \pm \alpha \beta R \tag{4}$$

As evident from Equation (4), the gain linearly depends only on the value of external resistor R, disregarding the parameters α and β that are almost equal to one, if the circuit is properly designed.

3. Transistor-Level Hybrid Interface

The sensor interface proposed here is realized with the described VCII acting as a transimpedance amplifier and implemented completely at the transistor level with discrete components. The simplified schematic of the proposed VCII is reported in Figure 3.

The current sources are physically implemented with current mirrors whose reference current is obtained by the provided bias voltage. The input stage is the so-called regulated common-gate [8,24], in which the traditional common source amplifier is implemented by a differential amplifier, whose differential pair consists of the transistors M_1 and M_2. The Y terminal of the VCII is considered to be the inverting input of the differential pair, while the non-inverting input is grounded; in this way, better rejection of the common-mode DC input noise is achieved. The input current incoming from the Y terminal is then mirrored on the X terminal with the current mirror implemented by M_6 and M_7 and the current sources I_{bias2} and I_{bias3}. In this manner, the input current buffer is realized between the Y and X terminals.

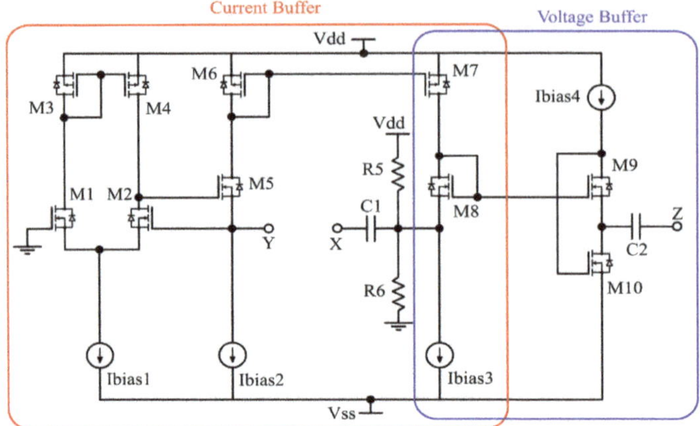

Figure 3. Simplified schematic of the proposed discrete VCII.

Conversely, the output section of the schematic implements the voltage mirroring between the X and Z ports. The voltage buffer is obtained by the so-defined flipped voltage follower (FVF), which, compared to a traditional common-drain circuit, realizes a feedback on the bias line of the output transistor M_9. In this way, it is possible to obtain a constant bias current and thus a fixed drain-source voltage for this transistor, achieving the desired voltage buffering action. Regarding the stability design criterion of the FVF block, we have taken into account the already developed theory reported in [25]. The current buffer block design also follows general stability consideration, because the two blocks are cascaded and there is no feedback loop between them. In addition, in assessing the noise impact, the designer should consider both current and voltage noise contributions. In the proposed front-end circuitry, the main noise contributions are due to the two bias currents Ibias2 and Ibias3, as well as the flipped voltage follower-biasing architecture.

For completeness, in Figure 4 the complete schematic of the proposed VCII is depicted, where the resistors R_1 and R_2 are used to properly bias the transistor M_8, while C_1 and C_2 are AC coupling capacitors.

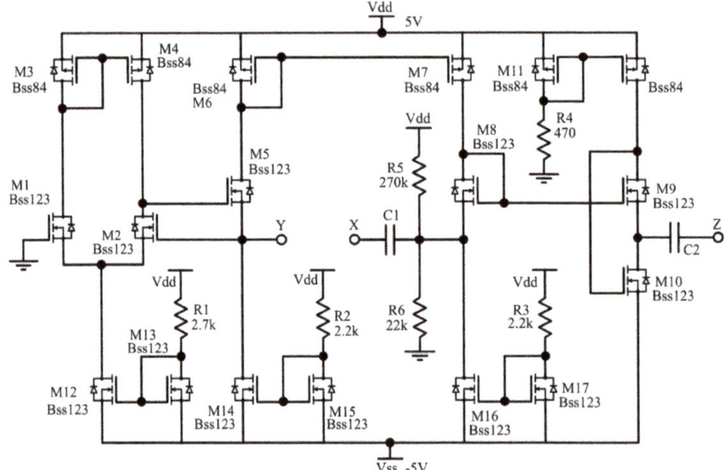

Figure 4. Complete schematic of the defined VCII in SPICE Environment.

By considering the proposed VCII as a transimpedance amplifier, it is possible to evaluate also input and output impedances that can be calculated with Equations (5) and (6), respectively.

$$R_{in} = \frac{1}{g_m(M_5)g_m(M_2)(r_o(M_2)//r_o(M_4))} \tag{5}$$

$$R_{out} = \frac{1}{g_m(M_9)g_m(M_{10})r_o(M_{10})} \tag{6}$$

The circuit was simulated in LTSPICE Environment and completely realized by using discrete metal-oxide semiconductor (MOS) transistors. The selected devices are the N-channel BSS123 and the P-channel BSS84, both from ON Semiconductor. They have fast switching performances and are particularly suited for low-voltage and low-current applications. The complete schematic of the circuit is reported in Figure 4. Performances of the VCII have been optimized by varying current and bias voltage and by analyzing α and β parameters with the goal to make them close to unity in the largest possible bandwidth. Simulated results are shown in Figures 5 and 6, respectively; they demonstrate the capability of the proposed circuit to effectively work even at very high frequencies thanks to the large bandwidth. In more detail, Figure 5a,b shows the magnitude and phase of the transfer function of the voltage buffer section of the interface, demonstrating for α a -3dB bandwidth of about 55 MHz; while Figure 6a,b illustrates the characteristics of the circuit section operating as a current buffer: the −3dB bandwidth for the β parameter is about 33 MHz.

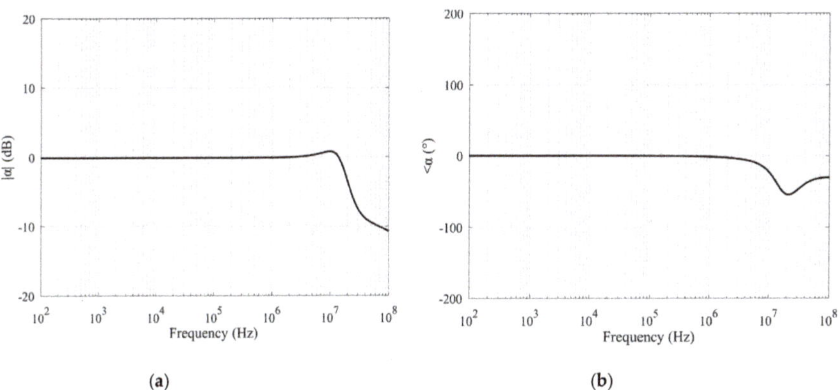

(a)   (b)

Figure 5. Voltage buffer characteristics: magnitude (**a**) and phase (**b**) of α.

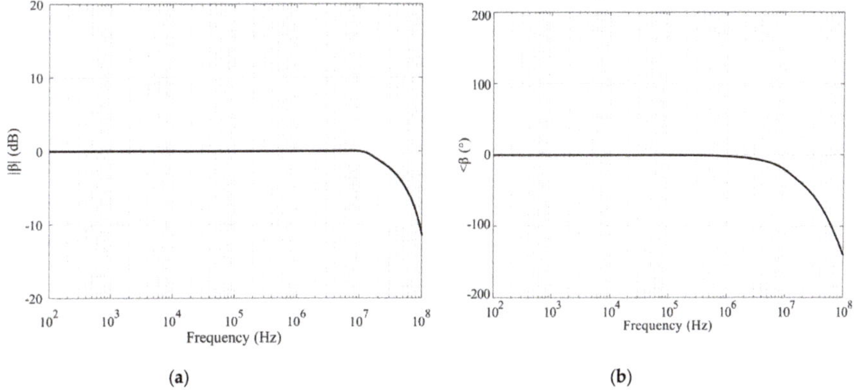

(a)   (b)

Figure 6. Voltage buffer characteristics: magnitude (**a**) and phase (**b**) of β.

By connecting an external resistor on the X terminal of the described VCII (as shown in Figure 2) and considering the overall transfer function between Y and Z terminals, the transimpedance characteristic can be evaluated and optimized, it being the main design parameter in order to maximize the performances for the proposed interface in term of bandwidth and response time. Figure 7a,b shows the complete transfer function V_{out}/I_{in} as magnitude and phase charts, respectively. They have been obtained considering a 100 Ω resistive load on the X terminal, even if, as will be better discussed later, this component will determine mainly the gain value of the interface. Even so, in these operating conditions, the results clearly show an almost constant gain of 42 dB and a useful bandwidth of about 106 MHz.

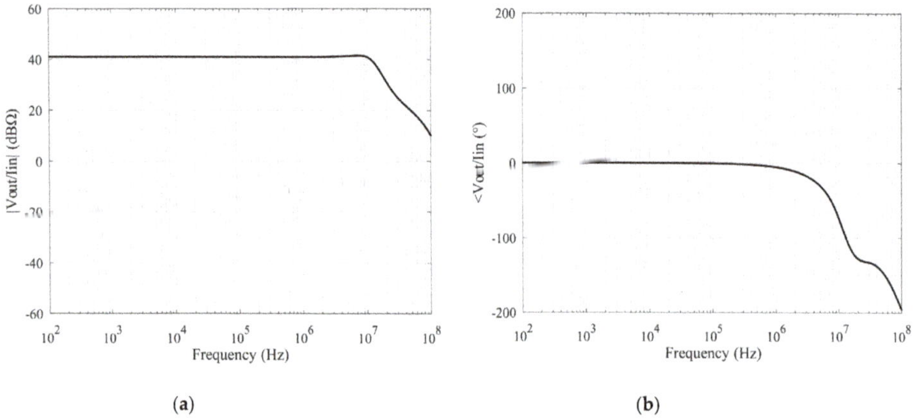

(a)   (b)

Figure 7. Transimpedance characteristics: magnitude (**a**) and phase (**b**).

4. SiPM Description

In Figure 8 (left side), a simplified circuital representation of a single SiPM elementary cell is depicted. It is mainly composed of a single-photon avalanche diode (SPAD), in series with a damping resistor $R_{DAMPING}$. The single cell, also called a pixel, is connected in parallel with other identical structures, with a common electrode structure, forming a multi-pixel photon counter matrix. The SPAD here is intended to operate in reverse bias, setting the supply voltage V_{BIAS} above the breakdown threshold of the photodiode. Consequently, the SPAD enters into a very unstable state of operation, called Geiger mode, which represents the core of the sensor functionality. In this state, when a single photon hits the SPAD sensitive area, its energy is transferred to an electron–hole couple generation, which starts a chain reaction for which, other electron–hole couples are created because of the high electric field imposed by the external reverse bias voltage. This leads to the formation of a high current flow, which, however, is limited by the damping resistor, which quenches the self-sustained phenomenon, thus restoring the Geiger state. From a circuital point of view (Figure 8, right side), the SPAD can be described as a small resistance R_S in series with a voltage generator, which represents the diode breakdown voltage, in parallel with the junction Capacitor C_J. The switch S is introduced to simulate the occurrence of a photon in the sensitive area.

At regime, the capacitor C_J is charged at V_{BIAS}, and the Geiger mode is active. When a single photon is absorbed by the pixel, the switch S closes, and the junction capacitance starts discharging quickly through the resistor R_S. As the capacitor discharges, the voltage decreases towards zero and is restored by the external bias voltage through the damping resistor, thus quenching the avalanche effect. As a final result, this process produces a current pulse, where the rise time is led by the diode equivalent series resistance R_S, having a time constant $\tau = R_S C_J$, while for the fall time, the main contribution is given by the damping resistor, having a time constant of $\tau = R_{DAMPING} \cdot C_J$. Therefore, the peak value of the current pulse I_{peak} can be computed as follows:

$$I_{peak} = \frac{V_{BIAS} - V_{BD}}{R_{DAMPING} + R_s} \tag{7}$$

and it can be modified only by changing the bias voltage value V_{BIAS}, with the other contributions fixed by the SiPM technology. If this single pixel is connected in parallel with other identical cells, it is clear that the corresponding total parasitic capacitance C_{TOT} of the SiPM is equal to the following:

$$C_{TOT} = NC_J \tag{8}$$

where N is the number of pixels connected in parallel. Therefore, the total capacitance could be large, up to thousands of pF, depending on the SiPM model and the number of cells constituting the sensor matrix, and this represents the most critical aspect for a front-end circuitry, because it can considerably degrade the interface bandwidth, if not properly designed. In the proposed VCII-based solution, the input impedance R_{in} at the Y node, as shown in Equation (5), is considerably less than 1. Therefore, the parasitic capacitance of the SiPM, expressed as $1/(s \cdot C_{TOT})$, will be greater than the transimpedance amplifier input impedance even at higher frequencies, ensuring a large bandwidth.

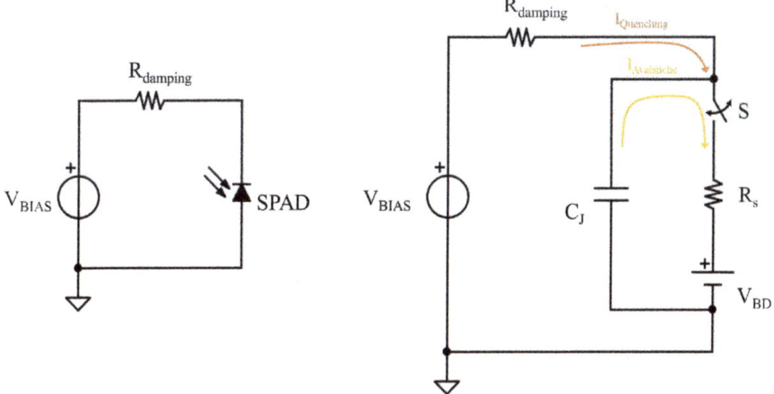

Figure 8. Silicon photomultiplier (SiPM) equivalent circuit.

5. Results and Measurements

The sensor interface that has been designed and discussed in Section 3 was fabricated as a discrete prototype. The complete schematic, designed in LTSPICE Environment, was optimized at the layout level through Autodesk Eagle software. The circuit was organized in a compact double-sided printed circuit board (PCB); the transistors NMOS BSS123 and PMOS BSS84 have a SOT23 package, while for passive components, a 0402-inch socket was chosen. The substrate was FR4 with a 0.7 mm thickness, and the board size was 40 mm × 40 mm. Figure 9a,b shows the top and bottom sides of the realized prototype board.

The interface was tested both in time and frequency domains for complete characterization. The stimulus for time-domain measurements was generated by using the Keysight 33600A Signal Generator, which can define a single pulse or a pulse train with a shape factor and characteristics similar to that provided by a typical SiPM sensor [26]. Both the time domain and frequency domain measurements were performed utilizing the digital oscilloscope and signal analyzer InfiniiVision MSOX3054T provided again by Keysight Technologies. DC power supply and current probes (Keysight E36313A) were also used. In order to emulate real SiPM electric behavior, a simple conditioning circuitry was added between the signal generator and the proposed front-end system. In particular, the generated voltage pulses were converted into current pulses by means of a resistor, while the produced current signal was then buffered by means of the commercial current buffer AD844 from Analog Devices. Finally, a signal conditioning circuit with a 320 pF shunt capacitor was added at the output of the current buffer to reproduce the sensor parasitic capacitance effect. The characteristics of the commercial SiPM S13360-3050 from Hamamatsu were considered as a reference for the signal conditioning circuit definition. In Figure 10, a block diagram of the complete test bench used for measurements is reported. The device under test (D.U.T.) block refers to the proposed sensor interface, whose schematic is shown in Figure 4.

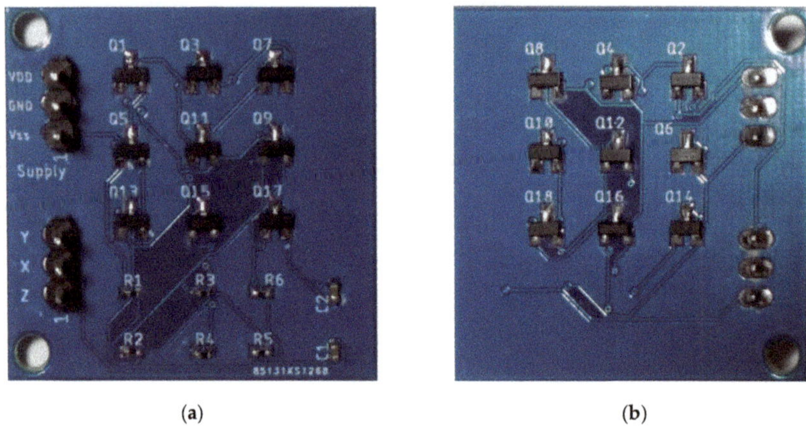

(a)　　　　　　　　　　　　　　　　(b)

Figure 9. Discrete prototype board: top (**a**) and bottom (**b**) views.

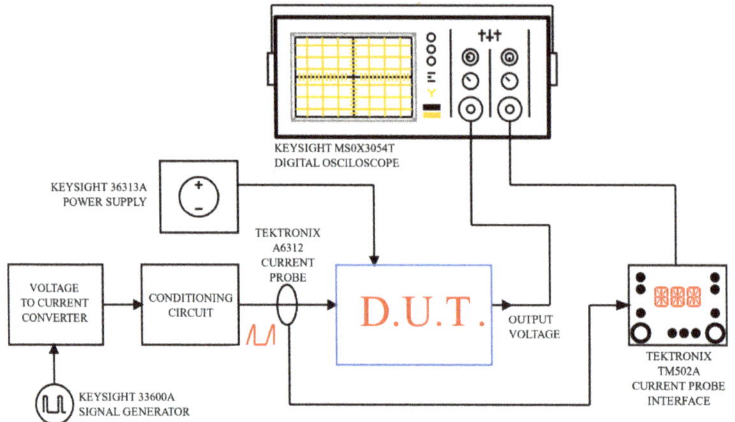

Figure 10. Test bench scheme of the prototype board.

At first, the functionalities of the base building block, the VCII shown in Figure 3, were tested. Both α and β characteristics were evaluated by considering a voltage/current input signal, respectively. The analysis was done by setting a relatively small amplitude, thereby preserving the circuit operation in the linear regime, and both were kept at a fixed and variable frequency. In this manner, both the values and bandwidths of these parameters were evaluated directly in the time domain, and the data confirmed the simulation results with good accuracy. The operational bandwidth of α and β are reported in Figure 11. They were obtained by sweeping the frequency of the input source and iterating the measurements each time. The transimpedance characteristics were also verified with a circuital setup organized as in Figure 2.

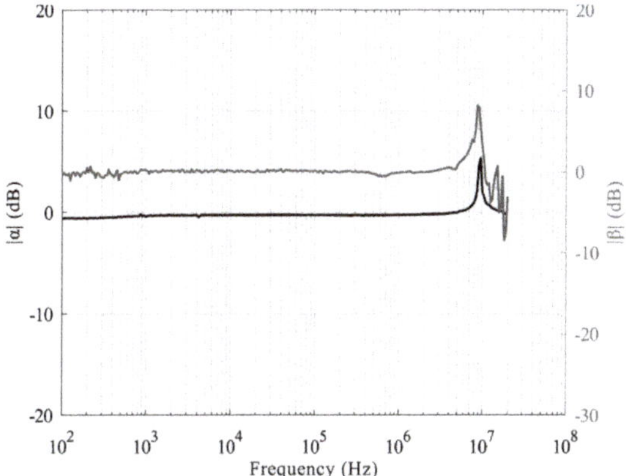

Figure 11. Measured transfer function in magnitude of both α (left axis) and β (right axis).

The transimpedance gain remains constant at a value of value 39.8 dB, while the operation bandwidth is 150 MHz when considering an external 100 Ω resistor on the X terminal. The achieved measurements are shown in Figure 12. Comparing them with simulations (Figure 7), there is good agreement, also confirming the accuracy of the spice models of the active devices beyond the feasibility of the design. In addition, by changing the cited resistor, it is possible to modify the transimpedance gain, as shown in Figure 13. This is a further advantage that justifies the use of the proposed architecture.

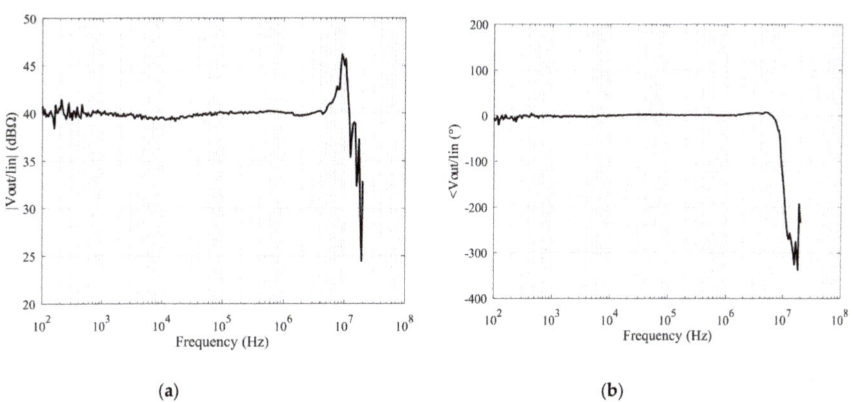

(a)

(b)

Figure 12. Measured transimpedance characteristics: magnitude (**a**) and phase (**b**).

Finally, the interface was also tested with respect to short pulses or pulse trains. So, time-domain signals reproducing the real signals coming from SiPMs were defined by using the aforementioned signal generator. Pulses of different duration were considered, and the results demonstrate the capability of the circuit to properly detect and follow short pulses up to 80 ns, as shown in Figure 14, while in Figure 15, an incoming pulse train at different amplitudes is shown together with the output signal. As reported, the proposed circuit shows good sensitivity and a good settling time at different amplitude levels, making it possible to use the interface in real-world applications when a fast detection of several pulses in a short time is a typical circumstance. In Table 1, a comparison with different literary solutions is reported. It is worth noting that the solution we have proposed, to the best of our knowledge, is the first transistor-only VCII discrete interface that can be used in practical applications with silicon photomultipliers. As a result, we have been able to include in Table 1 only equivalent integrated solutions. Some transimpedance amplifiers in both CMOS and SiGe technologies are shown.

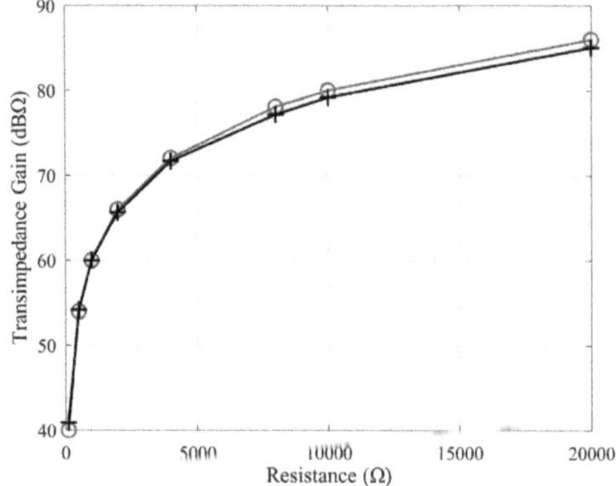

Figure 13. Ideal (grey trace) and measured (black trace) transimpedance gain at different values of the external X terminal gain resistor.

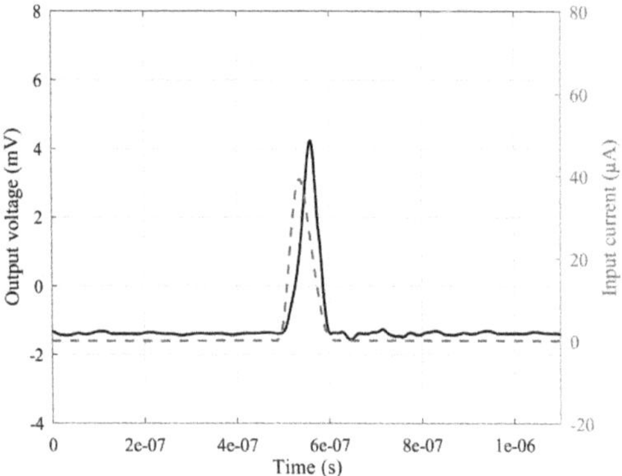

Figure 14. Input current pulse with a duration of 80 ns (grey trace, right axis) and output voltage signal of the defined SiPM interface (black trace, left axis).

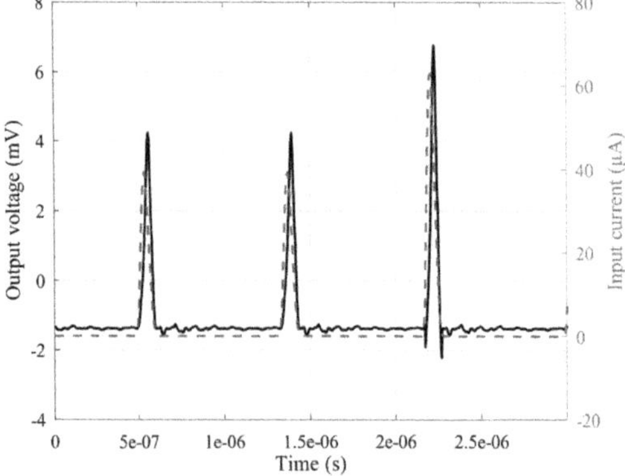

Figure 15. Output voltage signal of the circuit (black trace, left axis) when considering a current pulse train with different amplitudes and repetition time (grey trace, right axis).

Table 1. Comparison with different literary solutions.

Refs.	Technology	Supply	Power	Transimpedance Gain	Bandwidth	Noise
[8]	CMOS 130 nm	1.2 V	0.34 μW	100 dB	10 MHz	2.7 mVrms (output)
[27]	CMOS 350 nm	3.3 V	0.68 μW	100 dB	50 MHz	1300 e- (ENC)
[28]	CMOS 350 nm	3.3 V	0.68 μW	500	150 Hz	2 uVrms (input)
[29]	CMOS 350 nm	3.3 V	0.68 μW	/	/	6.9 mVrms (output)
[30]	SiGe 130 nm	−3.2 V	82 μW	56 dB	45 GHz	$30.6\,\text{pA}/\sqrt{Hz}$
This work	Discrete	±5 V	200 mW	42 dB	106 MHz	$9\,\text{mV}_{\text{RMS}}$ (output)

Obviously, our work has a large power consumption because it is realized with hybrid components and the power consumption is strictly dependent on the technology and the required biasing conditions. It is worth noting that in many applications, the power consumption is not critical, and the sensor interface minimally affects the total power consumption of the complete system. It is also important to remark that the main novelties are the realization of a functional interface with discrete elements, the ease of realization, and the very low cost. To conclude, even though we present a discrete solution, our results are comparable with integrated interfaces, so the overall performance of the proposed solution is credible.

6. Conclusions

In this paper, a fully discrete MOS transistor-only VCII was presented for the first time. It was designed, realized with commercial components, and tested with successful results. The capability to use the proposed circuit as a sensor interface for silicon photomultipliers was also evaluated and tested. The proposed solution has compact dimensions and low power consumption and is able to provide an agile response for quick incoming signals. The system was tested with both short current input pulses and pulse trains, reproducing the operative conditions of SiPM systems. As shown, the proposed interface provides stable characteristics, demonstrating its feasibility to be used in practical applications.

Author Contributions: Conceptualization: V.S., L.P., G.B., A.L., and E.D.; data curation, G.B., A.L., and E.D.; validation, V.S. and L.P.; writing—original draft, L.P., V.S., A.L., and G.B.; Writing—review and editing, V.S., G.B., A.L., and L.P. All authors have read and agreed to the published version of the manuscript.

Funding: This research received no external funding.

Acknowledgments: The authors would like to thank Stefano Ricci and Andrea Pelliccione of the Electronics Laboratory of the University of L'Aquila for fabricating the prototype and for their support.

Conflicts of Interest: The authors declare no conflicts of interest.

References

1. Mandache, L.; Ivanov, S.; Prejbeanu, R.; Roman, M.; Sirbu, I.; Ivanov, V.; Radu, A. Reliable Time-Domain Analysis of Nonlinear Lumped Circuits with Topology Issues, Suitable for Transients of AC Drives. In Proceedings of the 2019 International Conference on Electromechanical and Energy Systems (SIELMEN), Craiova, Romania, 9–11 October 2019.

2. Whittaker, T.; Whittow, W.; Vardaxoglou, J. Artificially Engineered Capacitors for Discrete High-Frequency Electronic Circuitry. *IEEE Trans. Microw. Theory Tech.* **2020**, *68*, 74–86. [CrossRef]

3. Stomelli, V.; Leoni, A.; Ferri, G.; Errico, V.; Ricci, M.; Pallotti, A.; Saggio, G. A Multi-Source Energy Harvesting Sensory Glove Electronic Architecture. In Proceedings of the 2018 3rd International Conference on Smart and Sustainable Technologies (SpliTech), Split, Croatia, 26–29 June 2018; pp. 1–4.

4. Cozzi, G.; Busca, P.; Carminati, M.; Fiorini, C.; Gola, A.; Piemonte, C.; Regazzoni, V. Development of a SiPM-based detection module for prompt gamma imaging in proton therapy. In Proceedings of the 2016 IEEE Nuclear Science Symposium, Medical Imaging Conference and Room-Temperature Semiconductor Detector Workshop (NSS/MIC/RTSD), Strasbourg, France, 29 October–6 November 2016.

5. Busca, P.; Fiorini, C.; Occhipinti, M.; Trigilio, P.; Nagy, K.; Bukki, T.; Czeller, M.; Nyitrai, Z.; Piemonte, C.; Ferri, A.; et al. A SiPM-based detection module for SPECT/MRI systems. In Proceedings of the 2015 IEEE Nuclear Science Symposium and Medical Imaging Conference (NSS/MIC), San Diego, CA, USA, 31 October–7 November 2015.

6. Goertzen, A.; Zhang, X.; McClarty, M.; Berg, E.; Liu, C.-Y.; Kozlowski, P.; Retiere, F.; Ryner, L.; Sossi, V.; Stortz, G.; et al. Design and Performance of a Resistor Multiplexing Readout Circuit for a SiPM Detector. *IEEE Trans. Nucl. Sci.* **2013**, *60*, 1541–1549. [CrossRef]

7. Sacco, I.; Fischer, P.; Ritzert, M.; Peric, I. A low power front-end architecture for SiPM readout with integrated ADC and multiplexed readout. *J. Instrum.* **2013**, *8*, C01023. [CrossRef]

8. Silva, M.; Oliveira, L. Regulated Common-Gate Transimpedance Amplifier Designed to Operate with a Silicon Photo-Multiplier at the Input. *IEEE Trans. Circuits Syst. I Regul. Pap.* **2014**, *61*, 725–735. [CrossRef]

9. Huizenga, J.; Seifert, S.; Schreuder, F.; van Dam, H.; Dendooven, P.; Löhner, H.; Vinke, R.; Schaart, D. A fast preamplifier concept for SiPM-based time-of-flight PET detectors. *Nucl. Instrum. Methods Phys. Res. Sect. A Accel. Spectrom. Detect. Assoc. Equip.* **2012**, *695*, 379–384. [CrossRef]

10. Katz, A.; Eshkoli, A.; Feningstein, A.; Jakobson, C.; Nemirovsky, Y. Design considerations of CMOS Si photomultiplier for long range LIDAR. In Proceedings of the 2017 IEEE International Conference on Microwaves, Antennas, Communications and Electronic Systems (COMCAS), Tel-Aviv, Israel, 13–15 November 2017.

11. Dorosz, P.; Baszczyk, M.; Kucewicz, W.; Mik, L. Low-Power Front-End ASIC for Silicon Photomultiplier. *IEEE Trans. Nuclear Sci.* **2018**, *65*, 1070–1078. [CrossRef]

12. Ciciriello, F.; Corsi, F.; Licciulli, F.; Marzocca, C.; Matarrese, G. Time performance of voltage-mode vs. current-mode readouts for SiPM's. In Proceedings of the 2015 6th International Workshop on Advances in Sensors and Interfaces (IWASI), Gallipoli, Italy, 18–19 June 2015.

13. Orita, T.; Koyama, A.; Yoshino, M.; Kamada, K.; Yoshikawa, A.; Shimazoe, K.; Sugawara, H. The current mode Time-over-Threshold ASIC for a MPPC module in a TOF-PET system. *Nucl. Instrum. Methods Phys. Res. Sect. A Accel. Spectrom. Detect. Assoc. Equip.* **2018**, *912*, 303–308. [CrossRef]

14. Corsi, F.; Marzocca, C.; Foresta, M.; Matarrese, G.; Del Guerra, A.; Marcatili, S.; Llosa, G.; Collazuol, G.; Dalla Betta, G.; Piemonte, C. Preliminary results from a current mode CMOS front-end circuit for silicon photomultiplier detectors. In Proceedings of the 2007 IEEE Nuclear Science Symposium Conference Record, Honolulu, HI, USA, 28 October–3 November 2007.

15. Barile, G.; Leoni, A.; Pantoli, L.; Safari, L.; Stornelli, V. A New VCII Based Low-Power Low-Voltage Front-end for Silicon Photomultipliers. In Proceedings of the 2018 3rd International Conference on Smart and Sustainable Technologies (SpliTech), Split, Croatia, 26–29 June 2018.

16. Pantoli, L.; Barile, G.; Leoni, A.; Muttillo, M.; Stornelli, V. A Novel Electronic Interface for Micromachined Si-Based Photomultipliers. *Micromachines* **2018**, *9*, 507. [CrossRef] [PubMed]

17. Čajka, J.; Vrba, K. The Voltage Conveyor May Have in Fact Found its Way into Circuit Theory. *AEU Int. J. Electron. Commun.* **2004**, *58*, 244–248. [CrossRef]

18. Safari, L.; Barile, G.; Stornelli, V.; Ferri, G. An Overview on the Second Generation Voltage Conveyor: Features, Design and Applications. *IEEE Trans. Circuits Syst. II Express Briefs* **2019**, *66*, 547–551. [CrossRef]

19. Barile, G.; Ferri, G.; Safari, L.; Stornelli, V. A New High Drive Class-AB FVF Based Second Generation Voltage Conveyor. *IEEE Trans. Circuits Syst. II Express Briefs* **2019**. [CrossRef]

20. Safari, L.; Barile, G.; Ferri, G.; Stornelli, V. High performance voltage output filter realizations using second generation voltage conveyor. *Int. J. RF Microw. Comput. Aided Eng.* **2018**, *28*, e21534. [CrossRef]

21. Sedra, A.; Smith, K. A second-generation current conveyor and its applications. *IEEE Trans. Circuit Theory* **1970**, *17*, 132–134. [CrossRef]

22. Ferri, G.; Guerrini, N. *Low Voltage, Low Power CMOS Current Conveyors*; Springer: New York, NY, USA, 2011.

23. Stornelli, V.; Ferri, G.; Pantoli, L.; Barile, G.; Pennisi, S. A rail-to-rail constant-g m CCII for Instrumentation Amplifier applications. *AEU Int. J. Electron. Commun.* **2018**, *91*, 103–109. [CrossRef]

24. Dias, D.; de Melo, F.; Oliveira, L.; Oliveira, J. Regulated common-gate transimpedance amplifier for radiation detectors and receivers. In Proceedings of the 21st International Conference Mixed Design of Integrated Circuits and Systems (MIXDES), Lublin, Poland, 19–21 June 2014.

25. Carvajal, R.; Ramirez-Angulo, J.; Lopez-Martin, A.; Torralba, A.; Galan, J.; Carlosena, A.; Chavero, F. The flipped voltage follower: A useful cell for low-voltage low-power circuit design. *IEEE Trans. Circuits Syst. I Regul. Pap.* **2005**, *52*, 1276–1291. [CrossRef]

26. Martinenghi, E.; Dalla Mora, A.; Contini, D.; Farina, A.; Villa, F.; Torricelli, A.; Pifferi, A. Spectrally Resolved Single-Photon Timing of Silicon Photomultipliers for Time-Domain Diffuse Spectroscopy. *IEEE Photonics J.* **2015**, *7*, 1–12. [CrossRef]

27. Albuquerque, E.; Bexiga, V.; Bugalho, R.; Carriço, B.; Ferreira, C.S.; Ferreira, M.; Godinho, J.; Gonçalves, F.; Leong, C.; Lousã, P.; et al. Experimental characterization of the 192 channel CLEAR-PEM front-end ASIC coupled to a multi-pixel APD readout of LYSO:Ce crystals. *Nucl. Instrum. Methods Phys. Res. Sect. A* **2009**, *598*, 802–814. [CrossRef]

28. Albuquerque, E.; Silva, M. *Project PET, 4th Progress Report: Revised Design of the Front-End ASIC—Analog Processing*; INESC-ID INESC-ID Technical Report 36/2006; INESC-ID: Lisboa, Portugal, 2006.

29. Oliveira, L.B.; Leitão, C.M.; Silva, M.M. Noise performance of regulated cascode transimpedance amplifiers for radiation detectors. *IEEE Trans. Circuits Syst. I Regul. Pap.* **2012**, *59*, 1841–1848. [CrossRef]

30. Giannakopoulos, S.; He, Z.S.; Darwazeh, I.; Zirath, H. Differential common base TIA with 56 dB Ohm gain and 45 GHz bandwidth in 130 nm SiGe. In Proceedings of the 2017 IEEE Asia Pacific Microwave Conference (APMC), Kuala Lumpar, Malaysia, 13–16 November 2017; pp. 1107–1110.

 © 2020 by the authors. Licensee MDPI, Basel, Switzerland. This article is an open access article distributed under the terms and conditions of the Creative Commons Attribution (CC BY) license (http://creativecommons.org/licenses/by/4.0/).

Communication

A Power-Efficient Radiation Sensor Interface with a Peak-Triggered Sampling Scheme for Mobile Dosimeters

Hyuntak Jeon, Injun Choi, Soon-Jae Kweon * and Minkyu Je

School of Electrical Engineering, Korea Advanced Institute of Science and Technology (KAIST), Daejeon 34141, Korea; ht.jeon@kaist.ac.kr (H.J.); injunchoi@kaist.ac.kr (I.C.); mkje@kaist.ac.kr (M.J.)
* Correspondence: hksn0226@kaist.ac.kr

Received: 18 May 2020; Accepted: 5 June 2020; Published: 7 June 2020

Abstract: Radiation sensor interfaces for battery-powered mobile dosimeters must consume low power to monitor the amount of radiation exposure over a long period. This paper proposes a power-efficient radiation sensor interface using a peak-triggered sampling scheme. Since the peak of the analog-to-digital converter's (ADC's) input represents radiation energy, our ADC only operates around the peak value thanks to the proposed sampling scheme. Although our ADC operates with a high sampling frequency, this proposed sampling scheme reduces the power consumption of the sensor interface because of the reduced operation time of the ADC. Our sensor interface does not have signal distortion caused by a conventional shaper because the interface quantizes the peak value using the high sampling frequency instead of the shaper. When the radiation input occurs once every 10 μs, the power consumption of the ADC with the proposed sampling scheme is only about 21.5% of the ADC's power consumption when the ADC continuously operates. In this worst case, the fabricated radiation sensor interface in a 0.18-μm complementary metal-oxide-semiconductor (CMOS) process consumes only 1.11 mW.

Keywords: radiation sensor interface; silicon photomultiplier (SiPM); mobile dosimeter; analog-to-digital converter (ADC)

---

1. Introduction

Nuclear power plants have been the basis of modern industrial development because these plants produce large amounts of highly efficient electricity. However, since the recent Fukushima nuclear accidents, concerns about the safety of nuclear power plants have been raised more than before. The radiation exposure over a certain level may cause biologically harmful effects, such as carcinogenesis [1–3]. As a result, not only the nuclear power plant workers managed by competent agents, but also the ordinary people in daily life are at increased risk of radiation exposure [4].

If people in daily life can easily use mobile dosimeters, they will be able to promptly address harmful situations. Figure 1 shows a block diagram of a mobile dosimeter, which consists of a radiation detector, a sensor interface, a data extraction module, and a power management circuit. Because the required bias voltage of silicon photo-multipliers (SiPMs) is lower than that of other radiation detectors [5], SiPMs are currently being used as radiation detectors. The sensor interface converts the output of the radiation detector to digital outputs, and the data extraction module processes these digital outputs appropriately for transmitting data to data servers or other mobile devices. The power management circuit generates supply voltages of the mobile dosimeter from a small battery for users to make these dosimeters portable in daily life. In such a battery-powered system, it is important to implement each subsystem in a power-efficient manner for a long time of operation. We propose a power-efficient radiation sensor interface because prior ones consume high power in handling high-speed radiation signals.

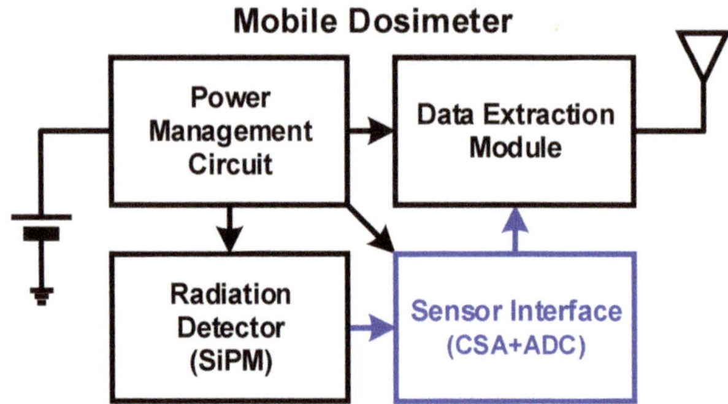

Figure 1. Block diagram of a mobile dosimeter.

2. Prior Radiation Sensor Interface

Figure 2 shows the architecture and operation of prior radiation sensor interfaces [6–11]. It consists of a charge-sensitive amplifier (CSA), a shaper, and an analog-to-digital converter (ADC). The radiation detector converts radiation energy (hv) into charge signals. When radiation particles are injected into the radiation detector, charges are generated in the form of spikes, which have a short period. The accumulated charge amount is proportional to the radiation energy.

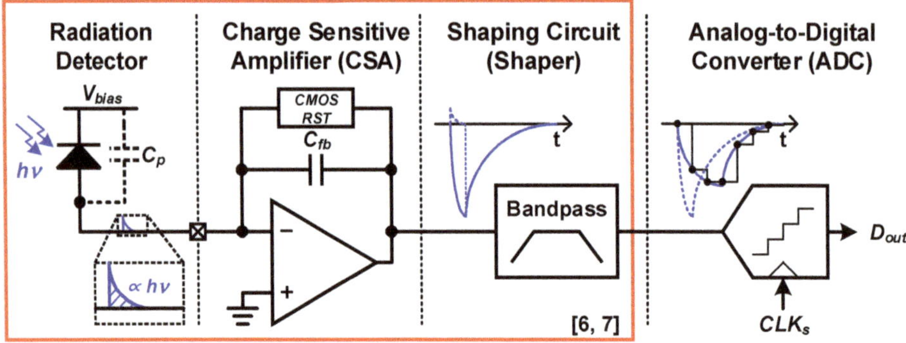

Figure 2. Architecture and operation of prior radiation sensor interfaces.

The CSA integrates the charge spike into the feedback capacitor to generate the voltage signal, which has enough amplitude for the ADC to handle [6,7,11]. Since the parasitic capacitance (C_p) of the radiation detector is much larger than the feedback capacitance (C_{fb}), the voltage loop gain of the CSA is close to zero, and the integration is performed through the open-loop bandwidth of the operational amplifier (OP-AMP). As a result, the conversion from charge to voltage through this type of CSA is much more power-efficient than the conversion through closed-loop amplifiers. In this CSA, the feedback capacitor must be periodically reset through a complementary metal-oxide-semiconductor (CMOS) switch and other digital logic elements to recover the original DC bias point of OP-AMP before integrating the newly generated charge spike.

As shown in Figure 2, since the peak of the CSA's output represents radiation energy, the ADC must quantize the peak value of the CSA's output. When the CSA's output is directly quantized by the ADC, the ADC needs a very high sampling frequency (f_s) due to the short period of the charge spike.

In order to reduce f_s, the shaper is introduced to stretch the period of the CSA's output. When the ADC quantizes the shaper's output rather than the CSA's output, the ADC can quantize the peak value of the CSA's output with a lower f_s. This results in reducing the power consumption of the ADC. However, the shaper, which has a bandpass characteristic, causes signal distortion because the charge spike itself has a wideband characteristic inherently.

3. Proposed Radiation Sensor Interface

We propose a peak-triggered sampling scheme that minimizes the power consumption of the ADC and quantizes the peak without the shaper. Figure 3 shows a block diagram of the proposed radiation sensor interface for the mobile dosimeter. The sensor interface consists of a CSA, a proposed peak-triggered signal generator (PTSG), and a successive approximation register (SAR) ADC with a voltage buffer (BUF). The CSA converts charge spikes to voltage signals, and the SAR ADC quantizes the CSA's output using a sampling clock signal (CLK_{PTS}), which is generated by the PTSG. Since this CLK_{PTS} has logic transitions only around the peak of the CSA's output (V_{out}), the PTSG reduces the operation time of the ADC. The BUF is adopted to precisely sample V_{out} on a large sampling capacitor of the SAR ADC. A more detailed description of operation is as follows.

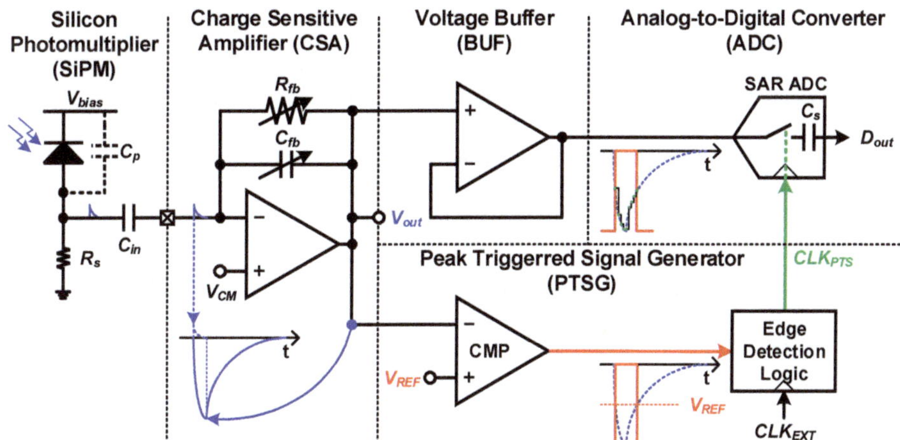

Figure 3. Architecture and operation of the proposed power efficient radiation sensor interface.

Our sensor interface supports a radiation detector based on SiPM with a series resistor and a DC blocking capacitor. The DC blocking capacitor with a capacitance of 1 µF is used to decouple the DC bias points of the SiPM output and the CSA input. We also adopt the open-loop CSA used in [6,7] to convert the charge spike to voltage signals with low power consumption. Instead of the reset switch in Figure 2, our CSA maintains DC operating point through a feedback resistor (R_{fb}) not to miss charge spikes without any complicated digital logic elements [12].

The proposed PTSG, which consists of a comparator (CMP) and an edge detection logic, generates CLK_{PTS} from an external clock signal (CLK_{EXT}). The CMP compares V_{out} and a pre-defined DC reference voltage (V_{REF}). When V_{out} is lower than V_{REF}, the CMP's output equals the supply voltage. In contrast, the CMP's output equals ground when V_{out} is higher than V_{REF}. The edge detection logic passes CLK_{EXT} only when the CMP's output equals the supply voltage. As a result, since the SAR ADC only performs sampling and conversion when radiation particles are injected and V_{out} is near the peak value, the PTSG drastically reduces the averaging power consumption of the SAR ADC by reducing the operation time of the ADC. In addition, the PTSG reduces the power consumption of the data extraction module because our sensor interface does not generate unnecessary data to process and transmit.

When the maximum power consumption of the sensor interface is limited in a mobile dosimeter by battery size, our sensor interface can use a higher f_s than previous sensor interfaces because the operation time of the ADC is reduced. As a result, our sensor interface quantizes the peak value of V_{out} through high f_s instead of the shaper, which causes distortion.

Commercial SiPM has various problems to be considered when actually using it [13]. Therefore, it is important to set V_{REF} in our sensor interface for optimizing the power consumption and accuracy. V_{REF} can be determined after monitoring V_{out} or the ADC's outputs (D_{out}) with radiation check sources, which are harmless to the human body. In addition, the dark current (or count in [13]) should be considered to ensure more accurate operation. The dark current is the main noise source of a SiPM, which is caused by thermal electrons generated in the active volume [13]. Since this dark current is smaller than the spike generated by radiation particles, to prevent waste of the power by the dark current, V_{REF} can be determined by monitoring D_{out} without the check source. That is, when the radiation energy is not changed, V_{REF} should be decreased until there is no sudden change in the D_{out}. Since this dark current is a function of the active area and varies from device to device [13], initial calibration is required for each mobile dosimeter.

Figure 4 shows the simulated power consumption of the SAR ADC with the peak-triggered sampling scheme. The power consumption of the SAR ADC is the product of the average current that flows to the circuit and the driving voltage. It can be seen that as the interval of radiation spike increases, it decreases drastically. The DC bias and the peak value of the CSA's output are about 1 V and 200 mV, respectively. V_{REF} used in this simulation is 0.6 V, and f_s is 4 MHz. When the interval between radiation spikes is 10 µs, the power consumption of the ADC with the proposed sampling scheme is about 0.11 mW, which is about 21.5% of the ADC's power consumption when the ADC continuously operates. The power consumption of the ADC decreases as the interval of the radiation spikes increases. When the interval is 100 µs, the power consumption of the ADC with the proposed sampling scheme is reduced to about 0.01 mW.

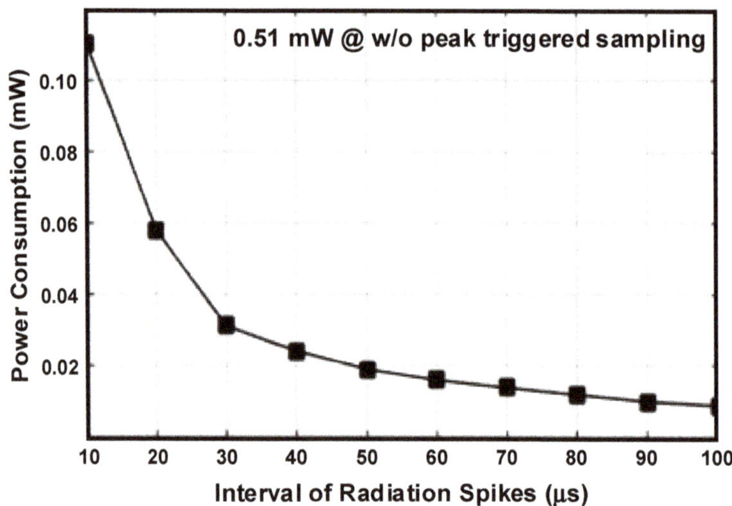

Figure 4. Simulated power consumption of the analog-to-digital converter (ADC) with the peak-triggered sampling scheme according to the interval of radiation spikes.

The OP-AMPs of the CSA and BUF are implemented based on the conventional two-stage OP-AMP structure [14], as shown in Figure 5. Each OP-AMP of the CSA and BUF achieves a simulated open-loop bandwidth of 50 MHz with a 5-pF load and a DC gain of 60 dB while consuming only 0.5 mW. These simulated results confirm good enough performances to satisfy the specifications required for the CSA to interface with the SiPM thanks to the open-loop structure of the CSA as well as for the BUF to drive the large sampling capacitor of the following SAR ADC.

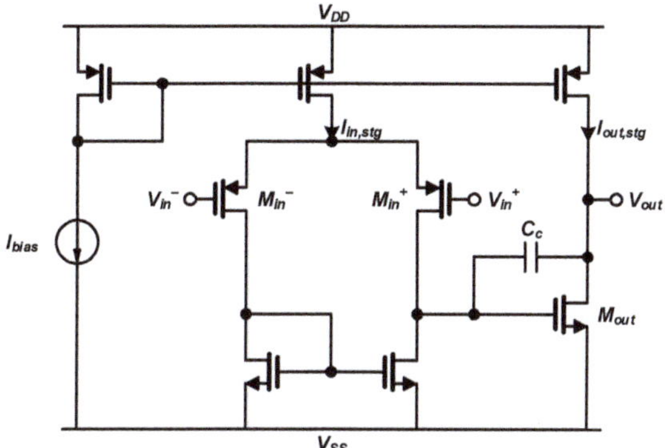

Figure 5. Circuit diagram of the 2-stage operational amplifier (OP-AMP) used in the charge-sensitive amplifier (CSA) and voltage buffer (BUF).

The CMP of the proposed PTSG has to operate to detect the time moment when the V_{out} reaches the same value as the V_{REF}. As shown in Figure 6, the CMP is designed using the cross-coupled hysteresis comparator structure, which is suitable for high-speed operation [15]. Our designed CMP achieves a simulated delay of 10 ns with only 10-μW power consumption. This delay is sufficiently short compared to the time for the CSA's output to reach the peak value. Furthermore, since the power consumption of the CMP is much smaller than other analog circuits, the power consumption of the overall sensor interface can be minimized.

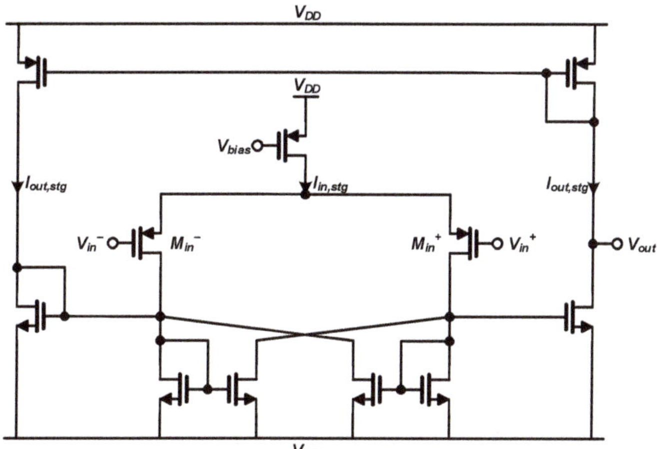

Figure 6. Circuit diagram of the comparator (CMP) used in the proposed peak-triggered signal generator (PTSG).

4. Measurement Results

Figure 7 shows the measured output waveforms of the proposed radiation sensor interface. We generate the input charge spike mimicking the condition that SiPM injects 15 fC of charge into the CSA during a sufficiently short time of about 1 μs. When this charge spike is injected, V_{out} reaches the peak value after 1 μs and recovers to the original DC bias point by R_{fb}. When V_{out} is lower than a

V_{REF} of 0.6 V, CLK_{PTS} exhibits logic transitions with f_s = 4.096 MHz. In contrast, CLK_{PTS} does not have transitions, and the ADC does not perform sampling and conversion when V_{out} is higher than a V_{REF} of 0.6 V. As a result, as shown in Figure 7, the ADC performs sampling and conversion only when a radiation spike is injected and V_{out} is lower than V_{REF}. The peak value of D_{out} shows good agreement with a peak value of V_{out}.

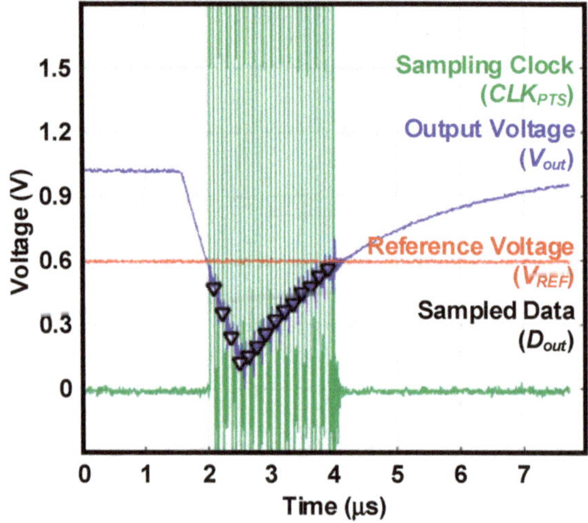

Figure 7. Measured output waveforms with the proposed sampling scheme.

Figure 8 presents the measured performances of the 10 bit SAR ADC in order to show how accurately V_{out} is converted to D_{out} when f_s = 4.096 MHz. Figure 8a shows the measured output spectrum when a 165.625-kHz sine wave is applied to the SAR ADC. The SAR ADC achieves a signal-to-noise and distortion ratio (SNDR) of 53.9 dB and an effective number of bits (ENOB) of 8.65, which means that the ADC has a resolution of 1.8 V/28.65 = 4.44 mV. In addition, as shown in Figure 8b, the differential nonlinearity (DNL) and integral nonlinearity (INL) are measured below 0.5 least significant bit (LSB). That is, this SAR ADC has sufficiently good linearity. As a result, this designed ADC achieves high resolution and obtains reliable data even at high f_s.

Figure 9 shows a die photograph of the proposed radiation sensor interface integrated circuit (IC). The proposed sensor interface is implemented in a 180-nm standard CMOS, and the fabricated IC occupies a small area of 0.715 mm². In particular, the proposed PTSG occupies a much smaller area than the SAR ADC, which occupies a dominant portion of the total area.

A comparison of the performance of the proposed sensor interface with prior interfaces is shown in Table 1. The proposed interface for mobile dosimeters includes an ADC, unlike prior interfaces, and its performance is also described. When the ADC continuously operates, the ADC consumes 0.51 mW. However, when the peak-triggered sampling scheme is used, the power consumption of the ADC is 0.11 mW and 0.01 mW for intervals of 10 µs and 100 µs, respectively. The CSA and BUF are implemented with relatively low power consumption.

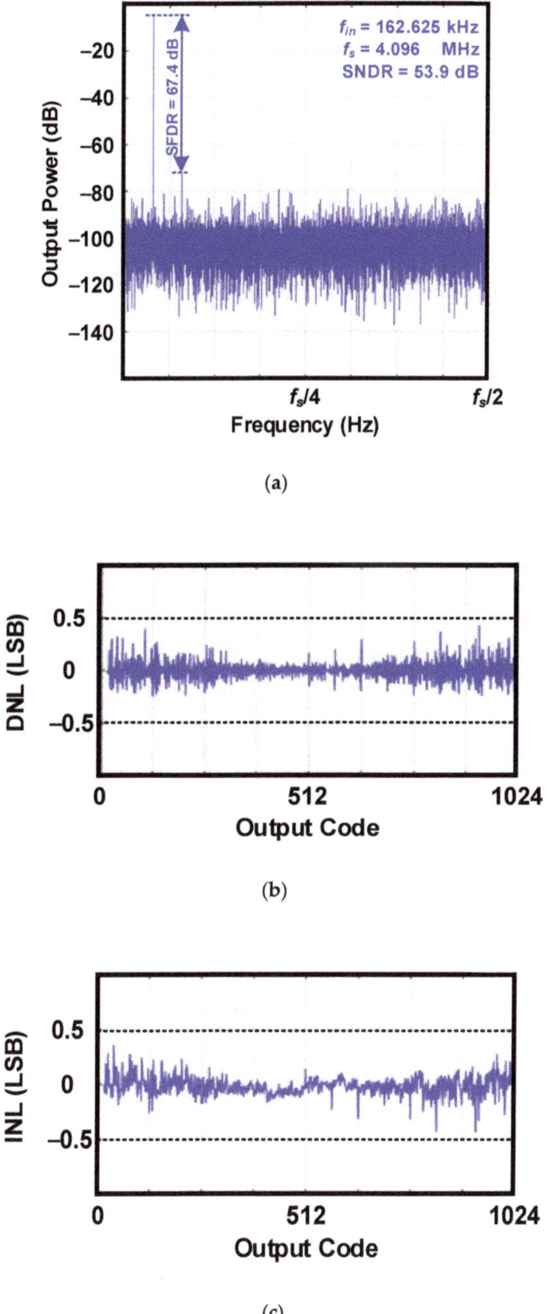

Figure 8. Measured performances of the 10 bit successive approximation register (SAR) ADC: (**a**) Measured output spectrum of digital outputs; (**b**) Measured differential nonlinearity (DNL); (**c**) Measured integral nonlinearity (INL).

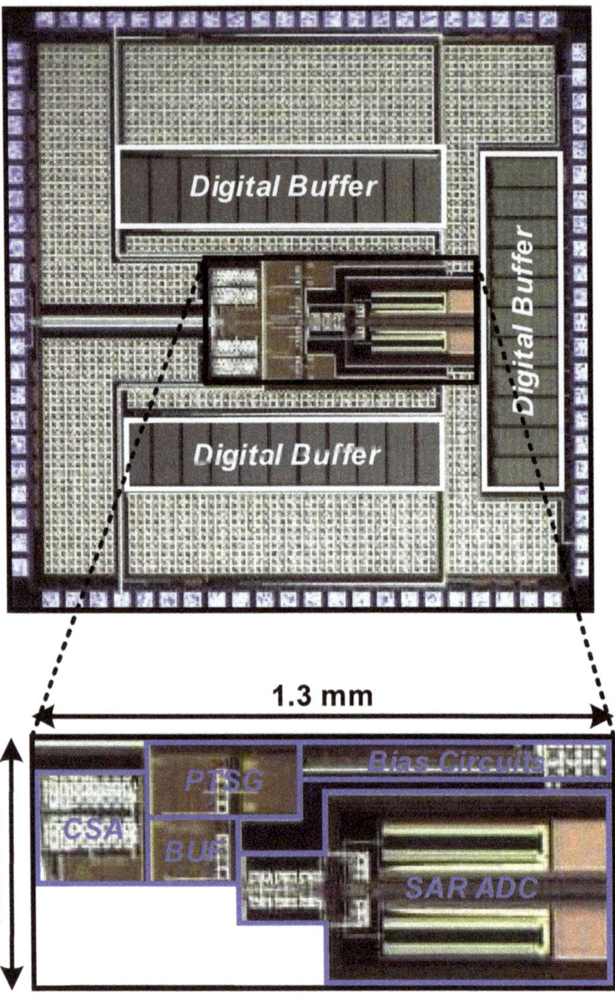

Figure 9. Die photograph of the proposed radiation sensor interface IC.

Table 1. Performance summary and comparison.

Parameters	[6]	[7]	This Work
Process (nm)	500	350	180
Structure	CSA + Shaper	CSA + Shaper	CSA + BUF + ADC
Area/Ch (mm^2)	1	N/A	0.715
Power/Ch (mW)	6	1	1 (CSA+BUF) < 0.11 (ADC)
f_s (MHz)	N/A	N/A	4.096
SNDR (dB)	N/A	N/A	53.9
SFDR (dB)	N/A	N/A	67.4
ENOB (bits)	N/A	N/A	8.65

5. Conclusions

We presented a radiation sensor interface using a peak-triggered sampling scheme. The proposed sampling scheme drastically reduces the power consumption of the ADC because the operation time of the ADC is reduced significantly. In addition, when the maximum allowable power consumption is limited, our sensor interface can use a higher f_s than conventional ones. Therefore, our sensor interface accurately quantizes the peak value of the CSA's output without a shaper, which causes distortion. As a result, our proposed radiation sensor interface that achieves low power consumption and high accuracy is suitable for mobile dosimeters.

Author Contributions: H.J. and I.C. implemented and measured the I.C. H.J. and S.-J.K. analyzed measurement results. H.J. wrote the draft of the manuscript with I.C.'s help. S.-J.K. is the corresponding author, reviewed and edited this manuscript through discussion with M.J. All authors discussed the results and commented on the manuscript. All authors have read and agreed to the published version of the manuscript.

Funding: This research was supported in part by the Convergence Technology Development Program for Bionic Arm (2017M3C1B2085296) through NRF of Korea funded by the Ministry of Science & ICT. In addition, this research was supported by BK21 plus program through the National Research Foundation (NRF) of Korea funded by the Ministry of Education.

Acknowledgments: The EDA tool was supported by the IC Design Education Center (IDEC), Korea.

Conflicts of Interest: The authors declare no conflict of interest.

References

1. Baumann, R.C. Determining the impact of alpha-particle-emitting contamination from the Fukushima-Daiichi disaster on Japanese semiconductor manufacturing sites. *IEEE Trans. Nucl. Sci.* **2012**, *59*, 1186–1196. [CrossRef]

2. Altomonte, H. Japan's nuclear disaster: Its impact on electric power generation worldwide. *IEEE Power Energy Mag.* **2012**, *10*, 94–96. [CrossRef]

3. Cho, G.; Yoo, H.; Lee, D.; Park, J.; Kim, H. Scintillator-based electronic personal dosimeter for mobile application. In *Smart Sensors at the IoT Frontier*; Yasuura, H., Kyung, C.-M., Liu, Y., Lin, Y.-L., Eds.; Springer International Publishing: Cham, Switzerland, 2017; pp. 191–218.

4. Chil, R.; Konstantinous, G.; Fraile, L.M.; Vaquero, J.; Rodriquez, C.; Borromeo, S.; Desco, M.; Udias, J.M.; Vaquero, J.J. Personal dosimetry geolocalized system for radiation monitoring. In Proceedings of the IEEE Nuclear Science Symposium, Medical Imaging Conference and Room-Temperature Semiconductor Detector Workshop (NSS/MIC/RTSD), Strasbourg, France, 29 October–6 November 2016.

5. ON Semiconductor: Datasheet: Arrayj Series Silicon Photomultiplier (SiPM) High Fill-Factor Arrays, Rev.6. Available online: https://www.onsemi.com/pub/Collateral/ARRAYJ-SERIES-D.PDF (accessed on 7 June 2020).

6. Krieger, B.; Ewell, K.; Ludewigt, B.A.; Maier, M.R.; Markovic, D.; Milgrome, O.; Wang, Y.J. An 8 × 8 Pixel IC for X-Ray Spectroscopy. *IEEE Trans. Nucl. Sci.* **2001**, *48*, 493–498. [CrossRef]

7. Noulis, T.; Siskos, S.; Sarrabayrouse, G.; Bary, L. Advanced low-noise X-ray readout ASIC for radiation sensor interfaces. *IEEE Trans. Circuits Syst.* **2008**, *55*, 1854–1862. [CrossRef]

8. FTLAB: Datasheet: GDK101 Gamma Radiation Sensor Module, ver1.5. Available online: http://allsmartlab.com/eng/294-2/ (accessed on 7 June 2020).

9. Vanhavere, F.; Marlein, B.; Nascimento, L.F.; and Lövestam, G. Testing three types of active personal neutron dosemeters for application in a nuclear research centre. *Radiat. Meas.* **2010**, *45*, 1593–1596. [CrossRef]

10. Tsujimura, N.; Yoshida, T.; Takada, C. Responses of commercially available neutron electronic personal dosemeters in neutron fields simulating workplaces at MOX fuel fabrication facilities. *Radiat. Meas.* **2011**, *46*, 1720–1723. [CrossRef]

11. Lee, C.; Cho, G.; Unruh, T.; Hur, S.; Kwon, I. Integrated circuit design for radiation-hardened charge-sensitive amplifier survived up to 2 Mrad. *Sensors* **2020**, *20*, 2654. [CrossRef] [PubMed]

12. Knoll, G.F. *Radiation Detection and Measurement*, 3rd ed.; John Wiley & Sons, Inc.: New York, NY, USA, 2010.

13. sensL: Introduction to SiPM TECHNICAL NOTE. Available online: https://www.sensl.com/downloads/ds/TN%20-%20Intro%20to%20SPM%20Tech.pdf (accessed on 7 June 2020).

14. Gray, P.R.; Meyer, R.G. MOS operational amplifier design–A tutorial overview. *IEEE J. Solid State Circuits* **1982**, *17*, 969–982. [CrossRef]
15. Bult, K.; Buchwald, A. An embedded 240-mW 10-b 50-MS/s CMOS ADC in 1-mm^2. *IEEE J. Solid State Circuits* **1997**, *32*, 1887–1895. [CrossRef]

© 2020 by the authors. Licensee MDPI, Basel, Switzerland. This article is an open access article distributed under the terms and conditions of the Creative Commons Attribution (CC BY) license (http://creativecommons.org/licenses/by/4.0/).

 *sensors*

Article

Development of Microcontroller-Based System for Background Radiation Monitoring [†]

Andriy Holovatyy [1], Vasyl Teslyuk [2], Natalia Kryvinska [3,*] and Artem Kazarian [2]

[1] Department of Computer Aided Design Systems, Lviv Polytechnic National University, 79013 Lviv, Ukraine; andrii.i.holovatyi@lpnu.ua

[2] Department of Automated Control Systems, Lviv Polytechnic National University, 79013 Lviv, Ukraine; vasyl.m.teslyuk@lpnu.ua (V.T.); artem.h.kazarian@lpnu.ua (A.K.)

[3] Department of Information Systems, Faculty of Management, Comenius University in Bratislava, 25 82005 Bratislava, Slovakia

* Correspondence: natalia.kryvinska@uniba.sk

† This manuscript is extension version of the conference paper: Holovatyy, A.; Teslyuk, V.; Lobur, M.; Sokolovskyy, Y.; Pobereyko, S. Development of Background Radiation Monitoring System Based on Arduino Platform". In Proceedings of the 13th International Scientific and Technical Conference on Computer Sciences and Information Technologies (CSIT 2018), Lviv, Ukraine, 11–14 September 2018.

Received: 23 November 2020; Accepted: 17 December 2020; Published: 20 December 2020

Abstract: An appearance of radiometers and dosimeters on free sale made it possible to provide better radiation safety for citizens. The effects of radiation may not appear all at once. They can manifest themselves in decades to come in future generations, in the form of cancer, genetic mutations, etc. For this reason, we have developed in this paper a microcontroller-based radiation monitoring system. The system determines an accumulated radiation dose for a certain period, as well as gives alarm signals when the rate of the equivalent dose exceeds. The high reliability of this system is ensured by a rapid response to emergency situations: excess of the allowable power of the equivalent radiation dose and the accumulator charge control. Further, we have composed a microcontroller electronic circuit for the monitoring radiation system. Additionally, an operation algorithm, as well as software for the ATmega328P microcontroller of the Arduino Uno board, have been developed.

Keywords: background radiation monitoring system; Atmel AVR ATmega328 microcontroller (MC); Geiger-Mueller counter; Petri net model

---

1. Introduction

Radiation is dangerous due to its high damageability as well as the fact that it is not perceived by human sense organs. None of the human sense organs is able to detect the presence of the nearby radioactive source. This makes it possible to hide the information about accidents at nuclear power plants (NPPs) and their consequences. Even after the events at the Chernobyl nuclear power plant, the radioactive consequences contaminated the countries of Europe, and for several years in the USSR the devices for the determination of the level of radioactive pollution were forbidden for the civilian population.

The appearance of radiometers and dosimeters on free sale made it possible to provide better radiation safety for citizens. The effects of radiation may not appear all at once. The effects of radiation can manifest themselves in decades to come in future generations, in the form of cancer, genetic mutations, etc. It is enough to mention the consequences of the atomic bombings of the Japanese cities.

As the level of radiation in the human body increases, disorders occur that can lead to death in a matter of days or hours. Having established the increased level of radiation, it is necessary to immediately determine the source of radioactivity. In order to do this a radiometer-dosimeter is needed.

Ionizing radiation is any radiation that causes an ionizing medium. Ionizing radiation belongs to cosmic radiation, and its natural sources on Earth are radioactive substances distributed in the geosphere. Artificial sources of ionizing radiation are nuclear reactors, artificial radioactive isotopes, nuclear explosions, X-ray equipment, etc. (OSS-2000 [1] basic sanitary rules, NRBU-97 [2] norms of radioactive safety in Ukraine).

Nowadays, a significant proportion of the world's electricity is generated by nuclear power plants. Most countries around the world attempt to reduce this share by replacing it with wind, solar stations, and others. At the same time, despite the very large investments in renewable and green energy, the process of replacing nuclear power plants is much slower than the world community expects. Therefore, in the near future we are forced to use nuclear power plants. On the other hand, we observe large-scale changes in the Earth's climate with extreme conditions that threaten nuclear safety in many parts of the globe. In such conditions inexpensive and mobile tools for monitoring ionizing radiation of environment are necessary. Therefore, the development of devices and systems for radiation monitoring is a topical task.

2. Related Work

There are various scientific and technical articles which are devoted to the development of background radiation monitoring systems. In particular, in [3] the authors have developed a personal portable effective dose dosimeter with a radio frequency data transmission channel. The construction of the microcontroller X-ray detector has been developed. The sensor element of the device is implemented on the basis of a radiochromic film, Gafchromic EBT3, which changes its transparency under the influence of ionizing radiation.

The schematic diagram of the dosimeter and its electronic circuit have been designed and implemented. The developed dosimeter evaluates the change in the effective dose according to the degree of the sensory element transparency. The dosimeter has a radio frequency (RF) data transmission system. The active mode of the dosimeter operation performs only at the moments of data reading using external RF devices. The lifetime of a personal device for measuring an effective dose of ionizing radiation, due to the absence of its own power source, is limited by the maximum radiation dose and the property of the photodetector to record the optical stream that passes through the film sensor.

There is no aforementioned shortcoming in the developed microcontroller systems [4,5]. The implemented radiation monitoring system [4] is intended for monitoring of subatomic high-frequency particles in the gamma-radiation zone. The device is built on the PIC16F84 microcontroller and the SBM-19 Geiger–Mueller counter. Unlike other works, in experiments of which personal computers for processing and remote data transmission were used, in this work, the independent autonomous microcontroller system, which includes standard Internet protocols, has been developed.

The measurement data, as well as warning signals, are sent to the decision-making system through communication channels (e.g., the Internet, mobile phone, or radio amateur strap). The system is not portable and requires an Internet or mobile network for remote data transmission. The device for registration and measurement of radioactive radiation [5] was developed with the help of sensors and the transmission of the collected data is performed wirelessly. At the same time, the registration and measurement system of radiation is expensive, as well as not enough compact and mobile.

An example of a wireless mobile system for measuring gamma radiation using a commercial portable instrument (dosimeter-radiometer) GAMMA-SCOUT is given in [6]. Since the GAMMA-SCOUT device does not provide built-in wireless connectivity, an additional hardware module was developed. The prototype of the module is implemented on the MIC PIC16F887 by Microchip and RF-Telit LE70-868 transmitters. The paper presents the structure and algorithm of module work. At the same time, the system is commercial and contains closed software that limits its area of use.

In [7], a digital search dosimeter for measuring low-level gamma radiation was developed. The dosimeter uses a scintillation detector (scintillation counter) as a radiation detector (radioactive radiation) and the PIC microcontroller PIC16F876 to control the functions of the developed system. The microcontroller generates a frequency of rectangular shape with a defined pulse width for forming and regulating high voltage order of +1200 V. High voltage is required to bring the scintillation detector into operation.

An amplifier and signal amplifier were designed for further processing by the microcontroller. MK records the pulses from the output of the amplifier, programmatically treats them, and outputs the result. The software for the microcontroller is created in C using the PCWH compiler. There are no means of wireless data transmission and information about the parameters of the radioactive radiation detector used, which limits the scope of use of the developed device.

Worthy of attention is the development of devices for monitoring the radiation background from air and under water [8,9]. In particular, in [8] the development of an unmanned aeronautical complex for remote monitoring of background radiation is considered. The research was carried out using the developed aircraft in the exclusion zone of the Chernobyl Nuclear Power Plant. The descriptions of the created dosimeter and experimental data of the radiation background measurement with automatic recording of the longitude, latitude and height of the measuring point are given. As a drawback, it can be noted that the dosimeter does not belong to the class of portable measuring devices.

An example of a compact wireless radiation monitoring system in real-time in an underwater environment is given in [9]. The system is marked by a high selling price.

Accordingly, the best approaches and technical solutions among the above-described solutions for the range of tasks involved include the methods described in [4,7,8]. The research methods were based on conducting experiments that included measurements of radiation, and their processing and analysis with the help of self-developed hardware and software systems. The developed systems are closed and have a high price, which does not allow us to expand their functionality and modification to the corresponding operational needs.

Consequently, the analysis of the existing approaches and technical solutions makes it possible to state the necessity of developing a qualitatively new, inexpensive, open-source mobile microcontroller radiation background monitoring system. Such a system should be built on affordable and inexpensive components, and open source software with the ability to modify or extend its functionality in accordance with the user and application requirements.

The goal of the work is to develop and research a microcontroller based system for background radiation monitoring, which has low cost and a wide range of functional capabilities.

In order to achieve the goal, the following tasks have to be solved:

- Develop the block diagram and operation algorithm of the microcontroller system for monitoring radiation which is based on the modular principle.
- Develop the physical model of the low-cost microcontroller-based radiation monitoring system.
- Develop the embedded software of the microcontroller-based radiation monitoring system, which makes it possible to implement a wide range of functionality.

3. Development of Structure and Operation Algorithms of Microcontroller Based System for Background Radiation Monitoring

The developed microcontroller system measures the level of radiation, analyzes the received data and transmits them to the PC via serial interface. It sends the alarm signal when the level of radiation exceeds. In Figure 1, the block diagram of the microcontroller-based radiation monitoring system is shown. The system is built on the Arduino Uno board based on the ATmega328P microcontroller [10–12].

The developed structure includes three main subsystems, such as a subsystem for collecting the background radiation information, a data processing subsystem, and a subsystem for displaying output results. The developed structure implies the use of the modular principle in the process of the system implementation.

The algorithm of the microcontroller system operation for monitoring background radiation is shown in Figure 2. When the microcontroller software (firmware) starts, it performs its initial setup and initialization of the parameters of the dosimeter-radiometer, LCD, serial interface, and I/O ports of the Arduino Uno board [13].

In the next step, the system checks the battery voltage supply. If the voltage is normally checked for the anode voltage of the tube, the SBM-20 Geiger–Mueller counter [14] and the timer1 start. The system counts the number of impulses per 1 min with the SBM-20. Then, the system processes the measurement results and displays them on the display and via the serial interface on PC. On the liquid crystal display (LCD), the system outputs the value of the power of radioactive radiation in µSv/h and the number of impulses per minute.

Each minute the timer is reset and restarted for the next measurement. When the permissible power of radiation is exceeded, the system having a sound alarm, turns on the alarm LED. If the "Start/Stop" button is pressed for the first time, the system enters the dosimeter mode. In the dosimeter mode, timer2 starts to measure the received radiation dose. The LCD displays the irradiation dose value. If the "Start/Stop" button is pressed again, the dosimeter mode will be disabled and the system returns to the radiometer mode.

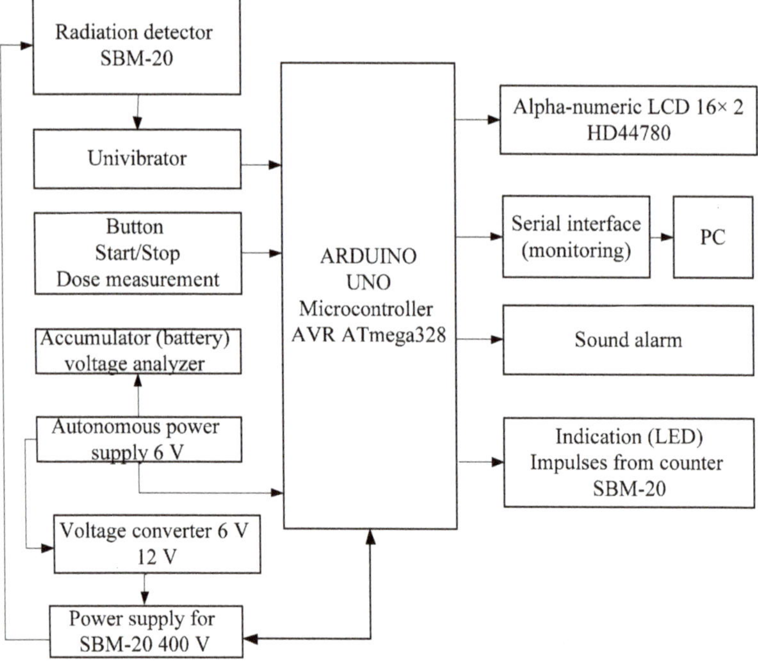

Figure 1. The block diagram of the microcontroller-based radiation monitoring built on the Arduino Uno board [15].

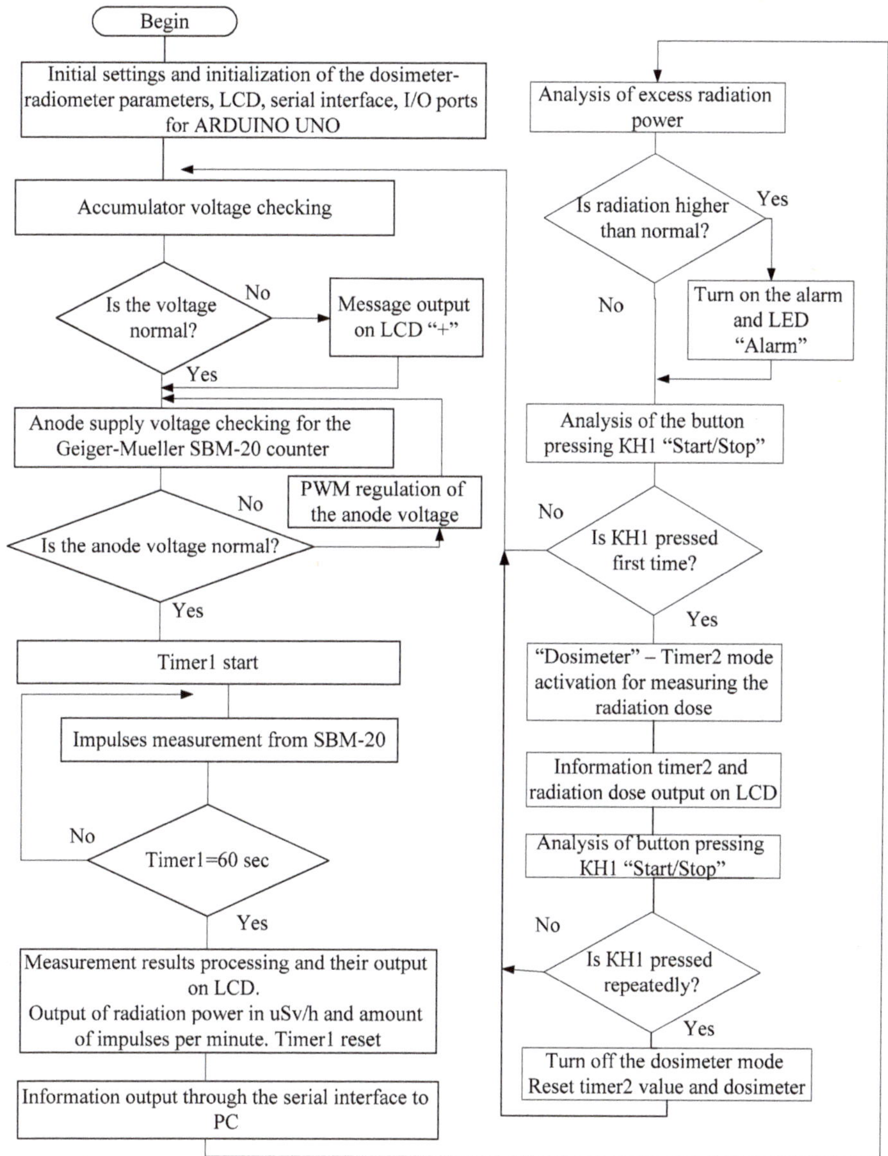

Figure 2. Operation algorithm of the microcontroller-based system for background radiation monitoring [15].

The measurement accuracy depends on a number of settings for the components of the designed system. In particular, in Figure 3, we see a block diagram of the adjustment algorithm of the anode voltage for the Geiger–Mueller counter which is one of the main components of the system.

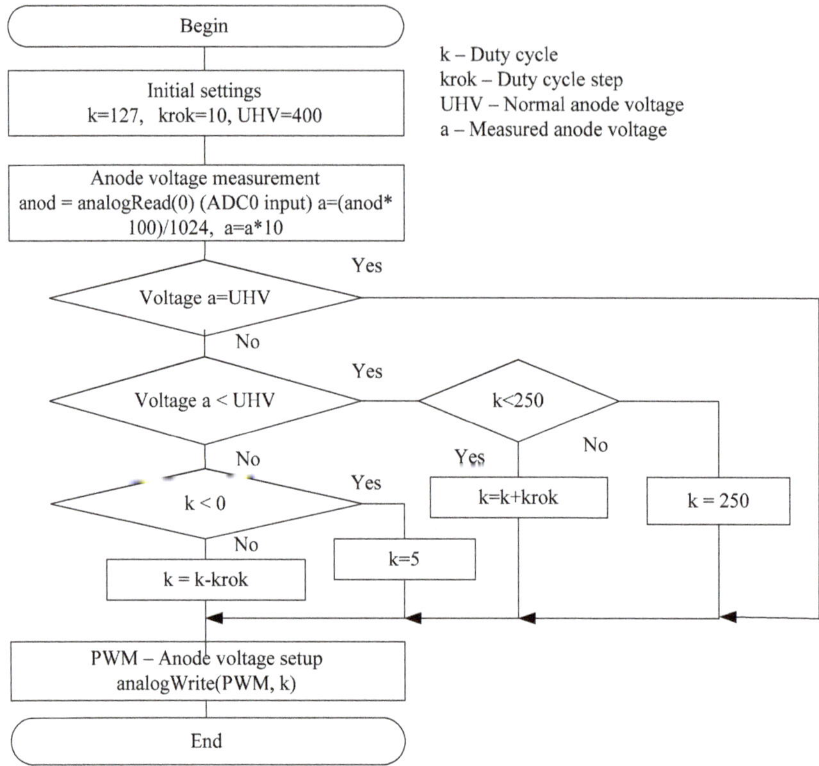

Figure 3. Algorithm of the anode voltage adjustment for the Geiger–Mueller counter.

At the stage of developing the structure of the microcontroller system, the models based on theory of Petri nets have been developed [16–18]. The models are based on the use of the expression:

$$N_{model} = \{P, T, F, M_0\}, P = \{P_1, P_2, \ldots, P_n\}, T = \{t_1, t_2, \ldots, t_m\}, \tag{1}$$

where P is a set of positions (states); T is a set of transitions; F is a set of arcs, which includes two subsets of input and output arcs in relation to the transition; M_0 is a set of the initial marking of the Petri net; n is a number of positions; m is a number of conversions.

The obtained models made it possible to construct a reachability graph of states [19] in which the system may be located and to investigate the dynamics of the operation of the developed system. In particular, an example of the schematic representation of the model based on a simple Petri net for the process of the anode voltage adjustment for the Geiger–Mueller counter is shown in Figure 4. The reachability graph of states is shown in Figure 5 [20]. The developed graph allows us to obtain the information about the states in which the researched system can be:

$$St = \{S_0, S_1, \ldots, S_l\}, \tag{2}$$

where St is a the set of states, and l is a number of states. The obtained results allow us to state that the system is alive, there are no dead ends, and the system operates correctly.

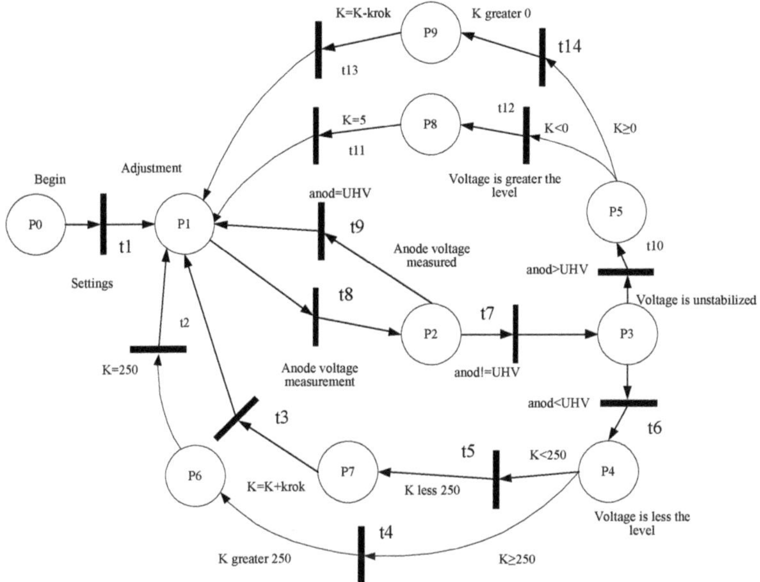

Figure 4. Schematic view of the model based on Petri nets for researching the process dynamics of the anode voltage adjustment for the Geiger–Mueller counter.

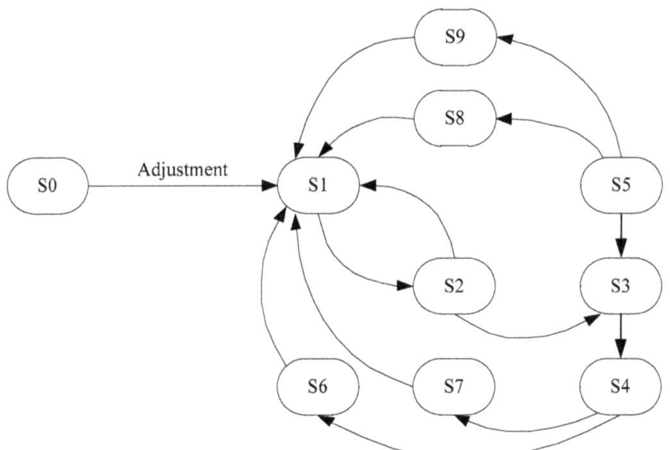

Figure 5. The reachability graph of the states for studying the process dynamics of the anode voltage adjustment for the Geiger–Mueller counter.

Consequently, the developed model based on the theory of Petri nets allows us to research the dynamics of the designed system.

As a result of the performed research using the developed models, based on the theory of Petri nets, it can be argued that the created Petri nets are alive, there are no dead ends, and certain states are achievable. Accordingly, it can be concluded that the designed system at the system level operates correctly and all the technical requirements are met.

4. Development of the Physical Model of the Microcontroller-Based System for Background Radiation Monitoring

The developed system includes hardware and software. The developed physical model of the system is shown in Figure 1. It consists of the data processing module, components of the collection of environmental data and control, and components for the processed data displaying. The data collection is implemented using a radiation detector, a gas-discharge Geiger–Mueller SBM-20 meter [14,21,22].

The control of the operating modes of the microcontroller system is performed using the key (button) (start-stop). The received data are processed by using the Arduino Uno microcontroller [10]. The processed data are displayed on the liquid-crystal display module as well as using the sound and light indication module. In addition, the physical model includes a number of modules that have been developed to ensure the functioning of the main components, such as a module for charging the battery, a DC 6 V–DC 12 V converter, a module of the anode voltage generator 400 V for power supply of SBM-20, and a univibrator circuit.

The SBM-20 counter detects beta- and gamma radiation in the range of dose rates up to 40 μR/s (0.4 μSv/h). The nominal operating voltage is 400 V. The operating voltage range is from 350 to 475 V. The voltage range is from 260 to 320 V. The range of registered capacities of exposure doses of gamma radiation is 0.004 ... 40 μR/s (0.014 ... 144 μR/h) or 0.00004 ... 0.4 μSv/h (0.00014 ... 1.44 μSv/h). Sensitivity to gamma radiation from radium (226Ra) is 29 cps/mR/h (3 cps/μSv/h), and from cobalt (60Co) is 22 cps/mR/h (2 cps/μSv/h) (60 ÷ 75 cps/μR/s = 600 ÷ 750 cps/μSv/s). The plate-counting characteristic length is at least 100 V. The slope of the plate-counting characteristic is 0.1% per 1 V. Dead time at the voltage supply of 400 V is 64 R/μs. Calculation speed (counting speed) at P = 4 μR/s (0.04 μSv/s) from the source 137Cs is from 240 to 280 im/s (280 cps). The unit's own background is no more than 1 im/s.

The dosimeter-radiometer operates on 6 V 1.2 Ah accumulator (rechargeable battery). In order to charge the accumulator (rechargeable battery), the special module is used. In order to provide the required voltage for the high-voltage converter, the 12 V DC voltage is used which is received from the DC-DC converter. The DC-DC converter converts the input voltage from the accumulator (5.5–6.8 V rechargeable battery) to 12 V. The DC-DC converter is built with an MC34063 chip, The MC34063 chip operates in the circuit as the step-up converter.

The transistor Q1 is used to increase the power of the converter to 3 A. Using the potentiometer RV1 the output voltage can be adjusted. In order to operate the Geiger counters need a high voltage source. Usually, such sources are autonomous. One of the classic converters is a 12 V to 400 V converter.

The circuit of the pulse-width modulator (PWM oscillator) is built using a 555 timer IC. With the potentiometer RV1 the oscillator frequency is adjusted, and with the potentiometer RV2, the oscillator duty cycle. The impulses of the required frequency and duty cycle arrive at the gate of the transistor Q1. The transistor Q1 increases the voltage on the coil L1 to 400 V. The accurate voltage value is adjusted with the potentiometer RV2. From the capacitor C4, the high voltage is applied to the Geiger counter (tube). The frequency of the oscillator impulses is 4–14 kHz. The high voltage oscillator can be implemented on the microcontroller which provides PWM functions. From the output PE3 (OC3A) to the input Q1 the PWM sequence arrives that sets the voltage to 400 V.

For adjusting the output voltage in the required range we apply the divider R6-RV3 to the microcontroller ADC input ADC0. Depending on the voltage value at the output PE3 (OC3A) the necessary PWM sequence is generated.

The Arduino Uno board based on the ATmega328P microcontroller is the main module for processing data from the Geiger–Mueller counter (tube), displaying the needed information on the LCD module, sending data to the PC through the serial port [15,23].

In Figure 6, the circuit of the microcontroller-based radiation monitoring system is shown. The system monitors the power of radiation, displays information and transmits the output data through the serial port to the PC.

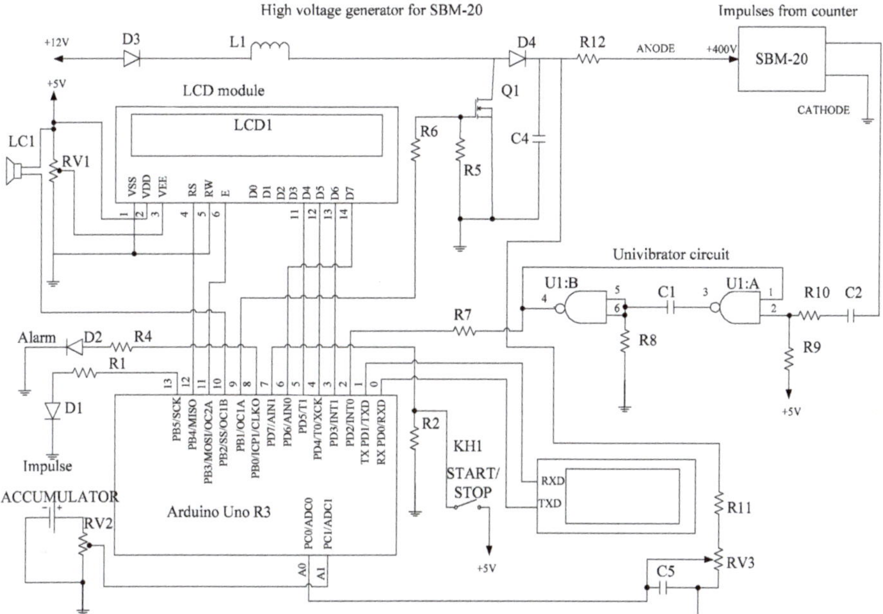

Figure 6. Circuit diagram of the Arduino microcontroller based system for radiation monitoring [15].

As the radiation power increases more than the permissible norms (0.3 µSv/h), the sound (LS1 buzzer) and the light (the red LED "Alarm" blinks) alerts are switched on.

The indication of the intensity of the radiation impulses occurs with the blue LED1. In the LCD module the information about the radiation power in µSv/h and CPM (counts per minute) is continually displayed.

By pressing the key KH1 "START/STOP" the timer starts and additionally the obtained radiation dose during the period (µSv) is displayed on the LCD module.

For the correct operation of the system, additionally the accumulator (rechargeable battery) charge status and the supply voltage of the SBM-20 counter are measured (the ADC inputs of the Arduino Uno board A1, A0, respectively). The high voltage of the counter is generated with the pulse width modulator at the output 9 of the Arduino Uno board. It is controlled by the ADC input A0. The high voltage generator consists of the coil L1, transistor Q1, diode D4, resistors R6, R5, R11, RV3.

The short impulses from the SBM-20 counter arrive using the single-vibrator (univibrator) (the U1 chip, elements C1, C2, R10, R9, R8, R7) to the input INT0 of the Arduino Uno board for detection and analysis.

The developed hardware of the microcontroller-based radiation monitoring system has a low cost and the modular structure [15,22–25] that makes it possible to rapidly improve the system.

5. Features of Specialized Software Development for the Microcontroller-Based Background Radiation Monitoring System

The system software is written in Arduino IDE [26,27] with the maximum use of libraries of this environment and tools for downloading compiled code into the board. In particular, as an example, the software implementation of the algorithm for adjusting the anode voltage of the Geiger–Mueller radiation counter according to Figure 3 is shown in Appendix A.

The developed software has been tested [28]. It has a modular structure that enables the system to be upgraded quickly and efficiently [29]. In addition, the specialized software makes it possible to extend the functionality of the system when needed.

6. Simulation and Analysis of the Operation of the Microcontroller-Based System for Background Radiation Monitoring

The compiled program in Arduino Software (IDE) [26,27] for the Arduino Uno board (based on the ATmega328P microcontroller) is a hex file. The hex file is flashed into the microcontroller of the Arduino board. In the Figures 7–9, the results of the operation simulation of the developed system are shown.

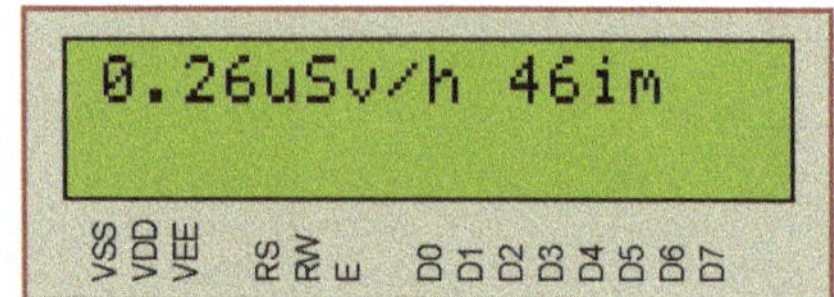

Figure 7. Simulation of the microcontroller-based system for background radiation monitoring (radiometer operation mode).

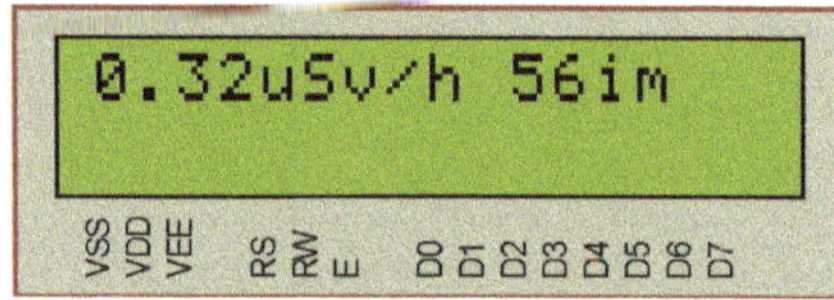

Figure 8. Simulation of the situation of exceeding the allowed radiation dose.

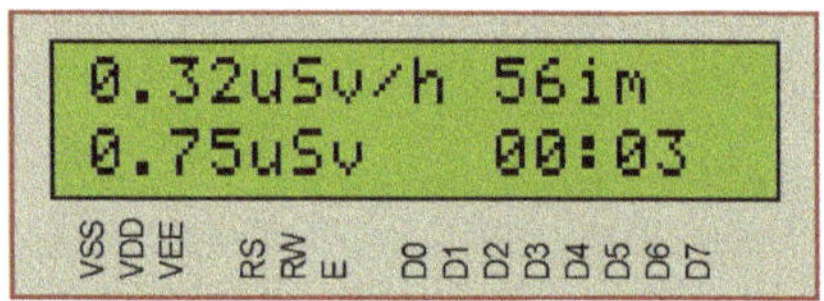

Figure 9. Simulation of the system operation in dosimeter mode.

In the radiometer mode, Figure 7, the radiation power in μSv/h and the number of impulses per minute are displayed in the LCD module. The blue LED D1 "Impulse" blinks when the incoming impulse is detected. Simultaneously, the information is displayed in the LCD and sent to the PC via the serial interface (Table 1). Information update occurs every minute.

Table 1. Output of radiation dose in uSv/h and counts per minute.

cpm	uSv/h
0	0.00
46	0.26
42	0.24
45	0.26
43	0.25

In Figure 8 the simulation results of the radiation dose exceeding are shown. The critical radiation dose is set during programming and is equal to 0.3 μSv/h. When the value of this dose exceeds, the red "Alarm" LED and alarm sound (LS1) are switched on. If the radiation is within the set tolerances, then the "Alarm" signals are off.

In Figure 9 the results of the system simulation in dosimeter mode are shown. When the KH1 button ("Start/Stop") is pressed, the timer switches on and the system displays on the LCD screen the value of the accumulated radiation dose during the timer operation. The maximum timer value is 24 h.

then the account is executed from scratch. The timer displays minutes and hours. The accumulated radiation dose in μSv. Pressing the KH1 key again stops the dosimeter mode and resets the LCD.

The obtained results of the device operation testing can be confirmed by the following data from the official technical documentation on the Geiger–Mueller counter SBM-20 [29–33], its radium calibrated sensitivity (Ra-226) is 29 cps/mR/h, cobalt calibrated sensitivity (Co-60) 22 cps/mR/h. The formula for converting the number of impulses per minute into the accumulated radiation dose in μSv/h is as follows:

$$[cpm]CF = [\mu Sv/h]. \tag{3}$$

The conversion factor (CF) (conversion factor) from cpm into μSv/h for SBM-20:

$$29 \text{ cps/mR/h} - 29 \times 60 \text{ cpm/mR/h} = 1740/8.77 \text{ (1 mR/h } \sim= 8.77 \text{ μSv/h)} \sim = 198 \text{ cpm/μSv/h,}$$
$$1 \text{ cpm} \sim 1/198 = 0.005040 \text{ μSv/h.}$$

Therefore, CF = 0.00504 (Ra-226).

$$22 \text{ cps/mR/h} - 22 \times 60 \text{ cpm/mR/h} = 1320/8.77 \sim= 150 \text{ cpm/μSv/h,}$$
$$1 \text{ cpm} \sim 1/150 = 0.006644 \text{ μSv/h.}$$

Thus, CF = 0.00664 (Co-60).

For SBM-20 the conversion factors are: CF = 0.00504 (Ra-226) and CF = 0.00664 (Co-60). In the software of the developed system, the average value of the coefficient CF = 0.0058 is taken, therefore, 56 imp. corresponds to a radiation dose of 0.32 μSv/h (μSv/h).

Consequently, the obtained results of the simulation and testing of the developed device and their coincidence with the theoretical calculations allow us to assert that the implemented system operates correctly.

7. Discussion of the Obtained Results

The obtained results of testing the developed physical model of the system in different modes coincide with the theoretical grounding and allow us to assert that the device is designed and implemented correctly and performs all previously defined functions. The system makes it possible to measure the equivalent dose of radiation, and the accumulated dose for a given period, outputs the received information on the serial port to the PC, and also gives warning signals in excess of the equivalent dose rate. Output information is displayed on the liquid crystal display.

A distinctive feature of the developed radiation background monitoring system is its multifunctionality, since inexpensive components are used, and the functionality of the device can be changed by improving the software of the microcontroller. In addition, the feature of the developed system is the ability to operate autonomously on accumulators and permanently from the 220 V network.

The limits of the developed system regarding the value and accuracy of the dose of radiation are determined by the parameters of the Geiger–Muller counter [21,29–33], namely, the upper limit of the radiation dose range can be adjusted using the built-in software.

However, it should be noted that the technical solution using the Arduino Uno platform is associated with a number of problems for the mobile version of the radiation background monitoring system. Accordingly, for such an option one must use industrial microcontrollers, which, in turn, significantly increases the cost of the system.

The further improvement of the developed device is possible by means of using the STM family of microcontrollers [34–36], which will increase the reliability [24,25,28]. In addition, the use of microcontrollers enables one to increase the functionality of the background radiation monitoring system.

8. Conclusions

In this paper, we have developed a background radiation monitoring system using the Arduino Uno board (based on the ATmega328P microcontroller) and the SBM-20 Geiger counter tube. The block diagram and operation algorithm for such a system have been composed. In addition, to research the operation dynamics of the designed system, we have built a model based on the theory of Petri nets. Next, we have built a low-cost physical model of the system. Appropriate software has also been developed. It allows us to implement wide-range functionality. The developed microcontroller-based system for background radiation monitoring can operate in both radiometer and dosimeter mode. The microcontroller-based radiation monitoring system measures the intensity of radiation in the radiometer mode. In the dosimeter mode, the microcontroller device determines an accumulated radiation dose over a period of time (max. 24 h), as well as gives alarm signals when the intensity of radiation is exceeded. The system can operate autonomously on accumulators and in stationary situations from the 220 V electricity network. The high reliability of the microcontroller-based system for background radiation monitoring is ensured by the rapid system response to the emergency situations: excess of the allowable intensity of equivalent radiation dose and the accumulator charge control. To conclude, our radiation monitoring system can be used in stationary premises as well as in mobile measurements.

Author Contributions: Formal analysis: A.H. and V.T.; investigation: V.T. and N.K.; algorithms: A.H. and V.T.; models: A.H., A.K., and V.T.; resources: N.K. and A.K.; software: A.H. and A.K.; hardware: A.H. and V.T.; validation: V.T. and N.K.; writing—original draft preparation: A.H. and V.T.; writing—review and editing: V.T. and A.H.; visualization: V.T. and A.K.; supervision: V.T.; data curation: V.T. and N.K. All authors have read and agreed to the published version of the manuscript.

Funding: This research received no external funding.

Conflicts of Interest: The authors declare no conflict of interest.

Appendix A. Software Implementation of the Algorithm for Adjusting the Anode Voltage of the Geiger–Mueller Counter

```
void HV(void) // function of generation and control of the counter anode voltage
{
anod = analogRead(0); // analog Read – pin A0 – anode voltage
k = 127; // duty = 50%
a = (anod * 100)/1024; // anode voltage * 10
a = a * 10;
if (a < UHV)
{k = k + krok; // duty voltage increasing
if (k > 250)
{k = 250;};
}
else
{k = k – krok; // duty voltage decreasing
if (k < 0)
{k = 5;};
};
analogWrite(PWM, k); // PWM – anode voltage setting
}
```

References

1. Osnovni Sanitarni Pravyla Zabezpechennia Radiatsijnoyi Bezpeky Ukrayiny. Pryjniattia vid 02.02.2015. Nabrannia Chynnosti vid 31.05.2015. Baza Danyh "Zakonodavstvo Ukrayiny". Available online: https://zakon.rada.gov.ua/laws/show/z0552-05 (accessed on 1 November 2020). (In Ukrainian)
2. Normy Radiatsijnoyi Bezpeky Ukrayiny (НРБУ-97). Radiatsijnyy Zahyst vid Dzherel Potentsijnoho Oprominennia. Zatverdzheni Ministerstvom Ohorony Zdorovja Ukrayiny 12 Lypnia 2000 Roku. Baza Danyh "Zakonodavstvo Ukrayiny". Available online: https://zakon.rada.gov.ua/rada/show/v0062282-97 (accessed on 1 November 2020). (In Ukrainian)
3. Novikov, S.G.; Berintsev, A.V.; Alekseyev, A.S.; Somov, A.I.; Prikhodko, V.V.; Guskov, P.A.; Svetukhin, V.V. Development of a personal portable dosimeter of an effective dose with the RFID data channel. *Radiopromyshlennost* **2018**, *28*, 78–85. [CrossRef]
4. Buruiană, V.; Oprea, M. A Microcontroller-Based Radiation Monitoring and Warning System. *IFIP Int. Conf. Artif. Intell. Appl. Innov.* **2012**, 380–389. [CrossRef]
5. Priya, P.; Gowdami, J.; Roshini, V.; Sinthanai Selvi, G. Detection and Measurement of Nuclear Radiations. *Proc. Mater.* **2017**, *4*, 4213–4218. [CrossRef]
6. Petrušić, Z.; Jovanović, I.; Jovanović, U.; Mančić, D. Wireless System for Measurement of Natural Background Gamma Radiation. *Ser. Work. Living Environ. Prot.* **2014**, *11*, 177–184.
7. Mashud, M.A.A.; Hossain, S.M.S.; Razzaque, M.A.; Islam, M.S. A First Response Microcontroller Based Digital Radiation Survey Meter Using Scintillation Detector. *Int. J. Cybern. Inform. (IJCI)* **2014**, *3*, 11–18.
8. Babak, S. Radiation Monitoring of Environment using Unmanned Aerial Complex. *Adv. Sci. J.* **2014**, *12*, 41–44. [CrossRef]
9. Kim, J.H.; Park, K.H.; Joo, K.S. Development of low-cost, compact, real-time, and wireless radiation monitoring system in underwater environment. *Nucl. Eng. Tech.* **2018**, *50*, 801–805. [CrossRef]
10. Banzi, M. *Getting Started with Arduino*, 2nd ed.; O'Really Media Inc.: Sebastopol, CA, USA, 2011; p. 130.
11. Cobo, A.; Villalba-Mora, E.; Hayn, D.; Ferre, X.; Pérez-Rodríguez, R.; Sánchez-Sánchez, A.; Bernabé-Espiga, R.; Sánchez-Sánchez, J.-L.; López-Diez-Picazo, A.; Moral, C.; et al. Portable Ultrasound-Based Device for Detecting Older Adults' Sit-to-Stand Transitions in Unsupervised 30-Second Chair–Stand Tests. *Sensors* **2020**, *20*, 1975. [CrossRef] [PubMed]
12. Jidin, A.Z.; Yusof, N.M.; Sutikno, T. Arduino based paperless queue management system. *Telecommun. Comput. Electron. Control* **2016**, *14*, 839–845. [CrossRef]
13. Parkhomenko, A.; Gladkova, O.; Sokolyanskii, A.; Shepelenko, V.; Zalubovskiy, Y. Implementation of reusable solutions for remote laboratory development. *iJOE* **2016**, *12*, 24–29. [CrossRef]
14. Shampo, M.A.; Kyle, R.A.; Steensma, D.P. Hans Geiger-German physicist and the Geiger counter. *Mayo Clin. Proc.* **2011**, *86*, e54. [CrossRef] [PubMed]
15. Holovatyy, A.; Teslyuk, V.; Lobur, M.; Sokolovskyy, Y.; Pobereyko, S. Development of Background Radiation Monitoring System Based on Arduino Platform. In Proceedings of the 2018 IEEE 13th International Scientific and Technical Conference on Computer Sciences and Information Technologies (CSIT), Lviv, Ukraine, 11–14 September 2018; pp. 139–142.
16. Diaz, M. *Petri Nets: Fundamental Models, Verification and Applications*; John Wiley & Sons: Hoboken, NJ, USA, 2010; p. 768.
17. Grobelna, I.; Wisniewski, R.; Grobelny, M.; Wisniewska, M. Design and verification of real-life processes with application of petri nets. *IEEE Trans. Syst. Man Cybern. Syst.* **2017**, *47*, 2856–2869. [CrossRef]
18. Wisniewski, R.; Wisniewska, M.; Jarnut, M. C-exact hypergraphs in concurrency and sequentiality analyses of cyber-physical systems specified by safe petri nets. *IEEE Access* **2019**, *7*, 13510–13522. [CrossRef]
19. Romansky, R. An Approach for Mathematical Modeling and Investigation of Computer Processes at a Macro Level. *Mathematics* **2020**, *8*, 1838. [CrossRef]
20. Goran, F. An Introduction to Hybrid Automata, Numerical Simulation and Reachability Analysis. In *Formal Modeling and Verification of Cyber-Physical Systems*; Drechsler, R., Kühne, U., Eds.; Springer Vieweg: Wiesbaden, Germany, 2015; pp. 50–81.
21. Lichylnyk, G.-M. Istoria Stvorennia, Pryntsypy Roboty i Pryznachennia. Available online: https://www.quarta-rad.ru/useful/statii-o-dozimetrax-radone/schetchik-geygera-myullera/ (accessed on 1 November 2020). (In Ukrainian)

22. Momose, Y.; Sakurai, T.; Nakayama, K. Thermal Analysis of Photoelectron Emission (PE) and X-ray Photoelectron Spectroscopy (XPS) Data for Iron Surfaces Scratched in Air, Water, and Liquid Organics. *Appl. Sci.* **2020**, *10*, 2111. [CrossRef]

23. Holovatyy, A.; Teslyuk, V.; Iwaniec, M.; Mashevska, M. Development of a system for monitoring vibration accelerations based on the raspberry pi microcomputer and the adxl345 accelerometer. *East. -Eur. J. Enterp. Technol.* **2017**, *6*, 52–62. [CrossRef]

24. Stefanovych, T.; Shcherbovskykh, S.; Droździel, P. The reliability model for failure cause analysis of pressure vessel protective fittings with taking into account load-sharing effect between valves. *Diagnostyka* **2015**, *16*, 17–24.

25. Sydor, A.R.; Teslyuk, V.M.; Denysyuk, P.Y. Recurrent expressions for reliability indicators of compound electropower systems. *Tech. Electrodyn.* **2014**, *4*, 47–49.

26. Huang, B.; Runberg, D. *The Arduino Inventor's Guide*; SparkFun Electronics: Boulder, CO, USA, 2017; 472p.

27. Schwartz, M. *Internet of Things with Arduino Cookbook*; Packt Publishing: Birmingham, UK, 2016; 187p.

28. Bobalo, Y.; Seniv, M.; Yakovyna, V.; Symets, I. Method of Reliability Block Diagram Visualization and Automated Construction of Technical System Operability Condition. *Adv. Intell. Syst. Comput. III* **2019**, *871*, 599–610.

29. Batyuk, A.; Voityshyn, V.; Verhun, V. Software Architecture Design of the Real-Time Processes Monitoring Platform. In Proceedings of the Second International Conference on Data Stream Mining & Processing (DSMP), Lviv, Ukraine, 21–25 August 2018; pp. 98–101.

30. SBM-20. Parameters and Characteristics. Available online: https://www.gstube.com/data/2398/ (accessed on 1 November 2020).

31. NEW SBM-20/SBM20/СБМ-20 (an STS-5, SI22G) Geiger Muller Tube Counter TESTED! Available online: https://www.ebay.com/itm/NEW-SBM-20-SBM20-20-an-STS-5-SI22G-Geiger-Muller-Tube-Counter-TESTED-/191831193846 (accessed on 1 November 2020).

32. SBM-20. Available online: http://sselec.sub.jp/pdf/SBM-20_general_specifications.pdf (accessed on 1 November 2020).

33. Sapporo Radiation Levels. Conversion Factor. Available online: https://sapporohibaku.wordpress.com/2011/10/15/conversion-factor/ (accessed on 1 November 2020).

34. Geiger Counter—Radiation Sensor Board for Arduino and Raspberry Pi. Available online: https://www.cooking-hacks.com/documentation/tutorials/geiger-counter-radiation-sensor-board-arduino-raspberry-pi-tutorial (accessed on 1 November 2020).

35. Chudzik, S. Zastosowanie tanich czujników inercyjnych w układzie regulacji kąta pochylenia pojazdu balansującego. *Przegląd Elektrotechniczny* **2015**, *6*, 177–180. [CrossRef]

36. Ma, X.; Li, Y.; Yang, L. Design of environment parameter adjustment system for greenhouse based on STM32. In Proceedings of the 13-th IEEE Conference on Industrial Electronics and Applications (ICIEA'2018), Wuhan, China, 31 May–2 June 2018; pp. 719–723.

Publisher's Note: MDPI stays neutral with regard to jurisdictional claims in published maps and institutional affiliations.

© 2020 by the authors. Licensee MDPI, Basel, Switzerland. This article is an open access article distributed under the terms and conditions of the Creative Commons Attribution (CC BY) license (http://creativecommons.org/licenses/by/4.0/).

Article

Magnetic Bioreactor for Magneto-, Mechano- and Electroactive Tissue Engineering Strategies

Nelson Castro [1,*], Margarida M. Fernandes [2,3], Clarisse Ribeiro [2,3], Vítor Correia [4], Rikardo Minguez [5,*] and Senentxu Lanceros-Méndez [1,6]

[1] BCMaterials, Basque Centre for Materials, Applications and Nanostructures, UPV/EHU Science Park, E-48940 Leioa, Spain; senentxu.lanceros@bcmaterials.net
[2] Centre of Physics, University of Minho, Campus de Gualtar, 4710-057 Braga, Portugal; margaridafernandes@fisica.uminho.pt (M.M.F.); cribeiro@fisica.uminho.pt (C.R.)
[3] Centre of Biological Engineering, University of Minho, Campus de Gualtar, 4710-057 Braga, Portugal
[4] Algoritmi Research Centre, University of Minho, Campus de Azurém, 4800-058 Guimarães, Portugal; vcorreia@dei.uminho.pt
[5] Department of Graphic Design and Engineering Projects, University of the Basque Country, E-48013 Bilbao, Spain
[6] IKERBASQUE, Basque Foundation for Science, E-48013 Bilbao, Spain
*   Correspondence: nelson.castro@bcmaterials.net (N.C.); rikardo.minguez@ehu.eus (R.M.)

Received: 12 May 2020; Accepted: 9 June 2020; Published: 12 June 2020

Abstract: Biomimetic bioreactor systems are increasingly being developed for tissue engineering applications, due to their ability to recreate the native cell/tissue microenvironment. Regarding bone-related diseases and considering the piezoelectric nature of bone, piezoelectric scaffolds electromechanically stimulated by a bioreactor, providing the stimuli to the cells, allows a biomimetic approach and thus, mimicking the required microenvironment for effective growth and differentiation of bone cells. In this work, a bioreactor has been designed and built allowing to magnetically stimulate magnetoelectric scaffolds and therefore provide mechanical and electrical stimuli to the cells through magnetomechanical or magnetoelectrical effects, depending on the piezoelectric nature of the scaffold. While mechanical bioreactors need direct application of the stimuli on the scaffolds, the herein proposed magnetic bioreactors allow for a remote stimulation without direct contact with the material. Thus, the stimuli application (23 mT at a frequency of 0.3 Hz) to cells seeded on the magnetoelectric, leads to an increase in cell viability of almost 30% with respect to cell culture under static conditions. This could be valuable to mimic what occurs in the human body and for application in immobilized patients. Thus, special emphasis has been placed on the control, design and modeling parameters governing the bioreactor as well as its functional mechanism.

Keywords: magnetic bioreactor; magnetoactive scaffolds; tissue engineering; magnetic actuator; magnetoelectric stimulation

---

1. Introduction

Fundamental biological studies and therapeutic applications rely on tissue engineering (TE) techniques, which aim to mimic the physicochemical and bioactive characteristics of natural cellular matrices [1,2], in order to achieve the replacement and/or regeneration of damaged tissues or organs [3,4]. When building a new tissue culture, three tools are mainly used: cells, scaffolds and stimuli. The cells are the building blocks for tissue culture as they contain the pre-programmed information that allows tissue regeneration. Cells are thus placed in a scaffold, which acts as the cell culture support, where the necessary environment is present, mainly in terms of biochemical stimuli, through the inclusion of growth factors and biophysical stimuli, by using a bioreactor [5]. Bioreactors allow us to introduce different chemical or physical stimuli on tissue culture, depending on both bioreactor technology and on how scaffold structures respond to those stimuli, in order to create a synergistic environment, thus stimulating cell response. Biomaterials used as scaffolds can be tailored, allowing them to be passively tolerated by the organism or actively providing the most appropriate and specific cell responses [6,7].

In the context of physically active TE by using bioreactors, different approaches have been implemented, including mechanoactive and electroactive scaffolds, among others [1,8]. In particular, electroactive materials are gaining increased attention due to the possibility of regulating different cell functions by providing electrical signals to the tissue culture [9–11]. Examples of such materials are piezoelectric scaffolds, which provide the necessary stimuli for the effective regeneration of bone [7,9,12], neural tissue [13,14], muscle [15–17], among others. The underlying mechanism relies on the possibility for mechanoelectrical transduction from materials to the cells [12], but lack of appropriate bioreactors able to stimulate those materials and take full advantages of their smart and multifunctional nature. In this work, a bioreactor is presented able to apply magnetic, mechanical and electrical stimuli to the cells in culture, based on the application of a magnetic field to a magnetically responsive scaffold containing magnetostrictive nano/microparticles embedded in a specific matrix [18,19] which can be electrically responsive (e.g., piezoelectric [20]) or not. These stimuli can be important for different types of tissue and, in particular, for bone TE due to the piezoelectric characteristics of bone [21], allowing to address novel TE strategies [22]. In particular, the performance of bone tissue along with its development and functional characteristics is strongly influenced by the voltage variation generated by mechanical stress to which the bone is subjected and, therefore, piezoelectric stimuli must be considered for proper regeneration strategies [23].

In fact, the electric sensitivity of osteoblasts has been regarded as an important tool for enhancing the ossification and healing through electric stimulation, as proven by piezoelectric scaffolds stimulated by a mechanical bioreactor, thus providing a proper electroactive environment to the cells [24,25]. In a similar approach, conductive composites have been proven to deliver exogenous electric currents to cells and increase their function [26], while evidence of electric stimulation influence on the ossification has been indeed observed [27].

In a novel approach, magnetoelectric (ME) scaffolds were used in order to provide support for cell seeding and proliferation while taking advantage of the scaffolds material, which enables electromechanical stimuli to the cells, generated by a magnetic field [19,28,29]. These composites are composed of magnetostrictive and piezoelectric layers working synergistically to produce the ME effect [30]. The ME effect can be described as a transduction from a magnetic field to an electrical field once the vibration of the magnetostrictive phase generated by an alternated magnetic field results in an electrical charge variation at the piezoelectric phase terminals at room temperature [31,32]. Further, if the support is not piezoelectric, just a mechanical vibration will be induced, providing such stimuli to the neighboring cells [33].

The flexibility, versatility and biocompatibility of these materials [8,34] can take advantage of in-vitro dynamic cultures through the support of a remote magnetic field [33]. Thus, materials with ME properties are therefore regarded as breakthrough platforms for TE applications that allow for remote generation of these physical stimuli, resulting in a controled influence on the surrounding tissue [9]. This effect has already been proven to induce a magnetomechanical [20] or a local magnetomechanical and magnetoelectrical effect [19], on the cells thus triggering improved cell proliferation and differentiation effect.

It is important to notice that the application of the magnetic field by itself, without further magnetomechanical or magnetoelectrical transduction, is also interesting for biomedical applications. This enables stimulation of cellular functions and cell manipulation to create cellular clusters, enabling more complex tissue structures than conventional strategies based on static scaffolds [35]. As a recent example, the use of superparamagnetic iron oxide nanoparticles proved to be a promising bioactive additive for scaffold fabrication [36], the scaffold enhancing the performance of human dental pulp stem cells yielding a higher count of phosphatase activity, higher osteogenic marker gene expression and improved cell-synthesized bone minerals. Other methods include the marking of C2C12 cells with magnetite cationic liposomes, mixed in a collagen solution, and seeded in a cell culture space of a hollow-fiber bioreactor [37]. The results demonstrated that high cell-density and viable tissue constructs containing myotubes were successfully obtained. Magnetic stimuli through permanent magnetic displacement were also proposed [38]. Rotation of permanent magnets was also employed in order to induce cellular growth proving that the variation of the magnetic field between 7 and 10 Hz increased the growth of neurite on chromaffin cells [39]. These devices can thus give an important contribution to the field, in order to overcome the issues related to the traditional cell culture conditions, improving the cellular distribution and accelerating cellular growth [40].

Besides biocompatibility and sterility, the design of mechanical bioreactors requires accurate control of the applied stimulus in order to get accurate data and to replicate results within the same parameters. In order to comply with these requirements and build a user-friendly device, which enables the user to apply controlled stimuli while providing control over temperature, culture stimuli active and resting time schedule, as well as total time, a modular magnetic bioreactor with an interchangeable magnets table was designed and developed. The interchangeable magnets table module has been designed to be used with 24-multiwell standard plates that can be easily operated and calibrated by the user. Magnetic stimulation has been previously reported [19,20,41], however, the present work reports on the development and validation of a novel bioreactor for magnetic stimulation of cells and/or scaffolds, comprehending cells-scaffold-stimuli relation, electromechanical study for actuation, control user interface and cell culture validation.

In terms of practical applications, the herein developed bioreactor can be a valuable tool for novel and more efficient tissue engineering strategies, including (i) to perform cell culture assays in vitro to validate the use of magneto-active materials for tissue engineering, thus avoiding extensive animal testing; (ii) to grow in vitro cellular tissue to be further implanted in a patient, after detaching the tissue from the surface of a magneto-active material and (iii) to be used in vitro to grow cellular tissue that is further implanted in a patient with the magneto-active material, which would allow for a remote stimulation of the material in the body, an important tool for immobilized patients.

2. Bioreactor Design

A bioreactor is an important tool for TE purposes since it allows us to mimic essential elements of the tissue environment and thus evaluate the influence of the stimuli on cell proliferation and differentiation. In order to study the influence of the electromechanical stimulation in bone tissue cells remotely [19,20], magnetoelectric scaffolds have been developed to provide the required stimuli [19], where an alternated magnetic field was applied to supply the necessary magnetic stimulation to the scaffolds. In this way, the herein developed device was designed to fulfill these characteristics and to meet experimental requirements to which the bioreactor will be subjected, which includes an incubation chamber with controlled temperature and humidity (37 °C and 95%, respectively). Thus, increasing the scaffolds temperature while applying the magnetic field is a critical design parameter, excluding the use of electromagnets, which requires high current flow, consequently resulting in radial heat from windings [42]. On the other hand, the displacement of permanent magnets as magnetic actuators on magnetoelectric scaffolds were used instead, to avoid a bulkier system and further heat. A schematic representation of the designed system is presented in Figure 1. The actuation system is composed of a permanent magnetic table that is displaced at a controlled frequency until certain limits, in order to get the required alternated magnetic field at the culture plate. For that, a mechanical structure comprising a motor in a ball screw assembly was installed to obtain an electromechanical actuation system. For mechanical protection, limit switches and precision sensors were applied to obtain electronic control of table position through a linear sensor and magnetic encoder for speed, as main operation components.

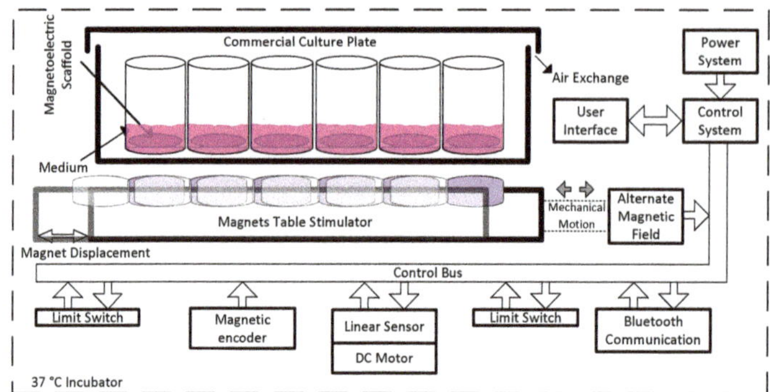

Figure 1. Magnetoelectric bioreactor operating principle through the use of electrical and mechanical controls to produce an alternated magnetic field and thus stimulate the magnetoelectric scaffolds and, consequently, the cells.

Furthermore, the system required a stable power supply in order to handle all digital and power electronic components, as well as remote control through wireless communication such as Bluetooth®® and respective user-machine interface through buttons and light and/or display feedback. To facilitate the use of this bioreactor in TE, the design included the application of a commercially available culture plate on the top of the system, where the ME scaffolds are easily placed and tested for cell culture as represented in Figure 1. The starting position of the magnets is user-defined as well as its displacement and motion frequency. It is important to note that the displacement will influence the magnetic variation over time in a specific place of the culture wells. The magnetic field peak values are directly dependent on magnets grade and distance to the scaffold, which can be calibrated mechanically by adding extra layers below the magnet table. The required magnetic field for the stimulation of the ME samples must reach values within a range of 20 to 50 mT [20]. The selected magnets are nickel-plated neodymium disks S–15–03-N52N from Supermagnete. The distance between the culture bottom and the N52 grade neodymium permanent magnets influences the magnetic field intensity, as analyzed through simulation with ANSYS®® Software. The magnetic field generated by a permanent magnet is easily stronger at a given z distance from the source, by comparison with fair current amplitude within a reasonable size coil for this system. It can be calculated for a cylinder type of permanent magnet, using Equation (1).

$$B = \frac{B_r}{2}\left(\frac{D+z}{\sqrt{R^2 + (D+z)^2}} - \frac{z}{\sqrt{R^2 + z^2}} \right). \tag{1}$$

This way, magnetic flux density can be calculated at a certain distance, where B_r is the remnant field, independent of the magnet's geometry, z the distance from a pole face on the symmetrical axis, D the thickness (or height) of the cylinder and R the semi-diameter (radius) of the cylinder [43]. Figure 2 displays the more appropriate distance according to each culture wells plates setup, which is different according to the number of magnets and radius used, resulting in a distance of 10 mm, where a field intensity of 30 mT was achieved at culture plate bottom (Figure 2a). In Figure 2b, it is possible to observe the side views with a larger range of the magnetic field variation, due to the substantial higher field at the N52 permanent magnets core. This fact enables higher magnetic fields at the culture bottom by reducing distance through magnetic table mechanical elevation in the same mechanism represented in Figure 2c. Furthermore, tailoring/exchanging the permanent magnets table with different magnetic grades, sizes and geometries, enables the mechanics to fit with the number of wells and geometries that a culture plate may present.

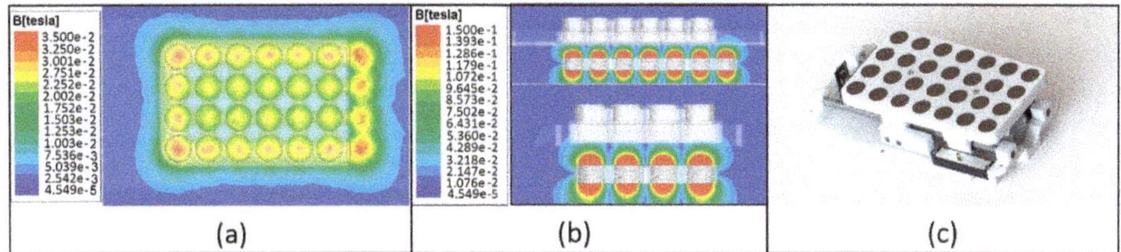

Figure 2. (**a**) Magnetic field intensity distribution at the bottom of 24-wells cell culture plates, (**b**) magnetic field force lines simulation in frontal and side planes and (**c**) rendered model of the mechanical permanent magnets table ball-screw assembly.

As schematized in Figure 1, the DC motor is controlled by a custom electrical system built and designed for the purpose of this application (control system). For this role, the system takes advantage of the sensors (limit switches, encoder and linear) in order to perform a close loop control of the magnets table positioning and displacement frequency. Furthermore, the system handles a user firmware interface that allows us to select the culture cycles for active and resting times. Regarding the firmware tasks, they are divided into two main control units: one unit controls the user inputs while the other controls the magnets positioning, sensors and display feedback. The local user interface is composed of an ILI9341 LCD providing 240 × 320 resolution with 262 k color and a side capacitive touch wheel with one button and a sliding circular panel for selection and option confirmation. The setup allows the user to locally stop, start or change the culture control parameters. A remote interface was also installed by Bluetooth, which allows us to monitor the culture status and the sensor reads from outside the incubator, using a mobile terminal. Since cell culture experiments require aseptic environments, requiring a sterilization process before each cell culture experiment, the creation of a waterproof enclosure was necessary. Nevertheless, such enclosure should withstand the temperature without overheating the bioreactor thus damaging the cell culture. The three-dimensional project of the device complying with such requirements is presented in Figure 3.

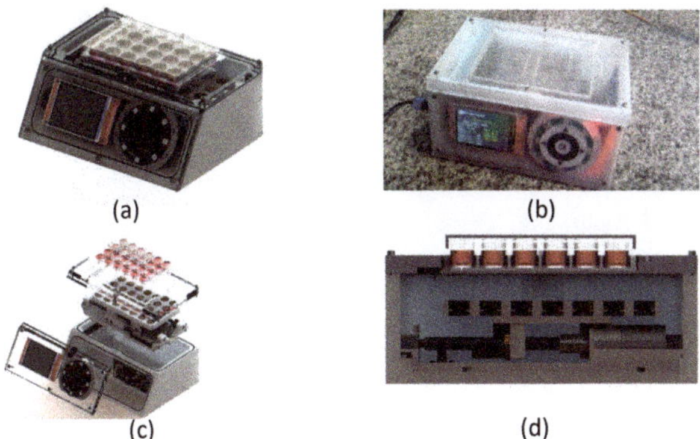

Figure 3. (**a**) Representation of the bioreactor assembled with a cell culture plate; (**b**) Bioreactor prototype built mechanism with every component; and schematic representation of (**c**) all disassembled main electric and mechanical components and (**d**) of the mechanical component represented as a transversal cut.

The magnetostriction of the scaffolds is obtained as a consequence of the magnetic field application, achieved by the movement of the magnets along the horizontal axis of the magnets table, below the cell culture plate. The selected construction material was nylon in order to avoid magnetic field interference. The system was designed for 24 and 48 well culture plates, although several types of commercial plates can be used with a different number of culture wells. Further, it was designed to be modular, thus the permanent magnets table can be replaced with a higher or lower number of magnets to fit the culture plate and apply an even magnetic field to all scaffolds (Figure 3). The result of the assembled system is simple, compact and a sealed design (Figure 3a). It is worth noting that the herein used permanent magnets were selected according to the magnetic field level required for inducing an electroactive environment on the ME scaffolds. Thus, when subjected to a magnetic field, a magneto-mechanical and magneto-electric stimulation is induced on the scaffold due to the incorporated magnetostrictive particles [19,44]. The magnetostrictive particles deform and generate the mechanical stimulus to the piezoelectric polymer within the scaffold which, in turn, develops an electrical charge that passes to the cells. The movement of magnets along the horizontal axis is achieved by using two side supports and a central motor shaft coupled with a DC motor, a component that can be observed in the device sectional detail in Figure 3d. The mechanical setup is implemented with a 25GA370 DC motor with 400 rpm and is controlled through an H-Bridge at 20 kHz pulse width modulation (PWM) pulses. This was selected over a stepper motor due to the possibility of applying lower currents and avoid heating in order to keep the system temperature low to protect the cell culture. The magnet table position is controlled by a linear sensor 9615R5.1KL2.0 (from BEI Sensors, Attleboro, MA, USA), together with an ADC resolution of 12 bits, which results in a linear resolution of 0.01 mm. The bioreactor was designed with an IP68 waterproof rating system to withstand the sterilization process, through waterproof power connector, rubber protection in the joints and capacitive touch avoiding mechanical buttons and leaks. Figure 3 displays a photo of the prototype device after the design and development phase.

2.1. Power and Control Circuitry

The main system communication and control between modules is illustrated in the block diagram of Figure 1. Thus, an electrical system was developed according to the mechanical design and user interface operational requirements. Further, the designed circuits were implemented using commercial electronic components. The electrical circuits designed are represented in the schematic of Figure 4. These circuits can be divided between power conversions (A), user interface (B), sensors (C), main control of operation (D), actuation (E) and system wireless communications (F). Power conversion is required in order to comply with three different voltage levels, whereas the motor operates with 12 V, the LED lighting and sensors operate with 5 V and logical CMOS level of 3.3 V. The user interface was designed in order to be intuitive and waterproof with no leaking fissures to the interior of the device. Thus, the capacitive interface was selected with a dedicated microcontroller STM32F091CBT6 (IC2) ARM®® 32-bit Cortex®® M0 CPU frequency up to 48 MHz for both RGB led lighting (D1~D8) and capacitive detection. The main control of the operation was performed by STM32F303RET6 (IC1) ARM®® Cortex®® M4 32-bit CPU with 72 MHz FPU, which controlled the overall firmware architecture. Therefore, IC1 was able to control the operation of the motor with the aid of an H-Bridge DRV8872-Q1 (IC3) with a high range of operation up to 3.6 A and 45 V. In addition, it made the interface communication with IC2 and Bluetooth HM−10 (IC7) through UART, LCD screen design through SPI, sensors input through ADC channels and system-timings for culture operation. In order to measure the temperature, an LMT85 sensor (IC4) and magnetic field (AD22151) (IC5) in the most adequate position an extra PCB was designed in order to house both sensors and wire communicate with the main board for a closed-loop system response. A linear position sensor 9610R3.4KL2.0 (IC6) was used in order to access magnets platform position to control permanent magnets positioning and resulting magnetic field, was easily integrated with an ADC channel input. All integrated circuits were designed with their respective decoupling capacitors in order to avoid high-frequency power transitions to dwell in the rest of the circuit power lines. The 5 V regulator LM2596 (IC9) was a switching step-down with 80% efficiency, being important for this application by comparison with linear regulators, during long periods of use the amount of heat produced is considerably less for this level of power required. In order to power the logic circuits at 3.3 V, an LD1117 voltage regulator (IC10) was employed.

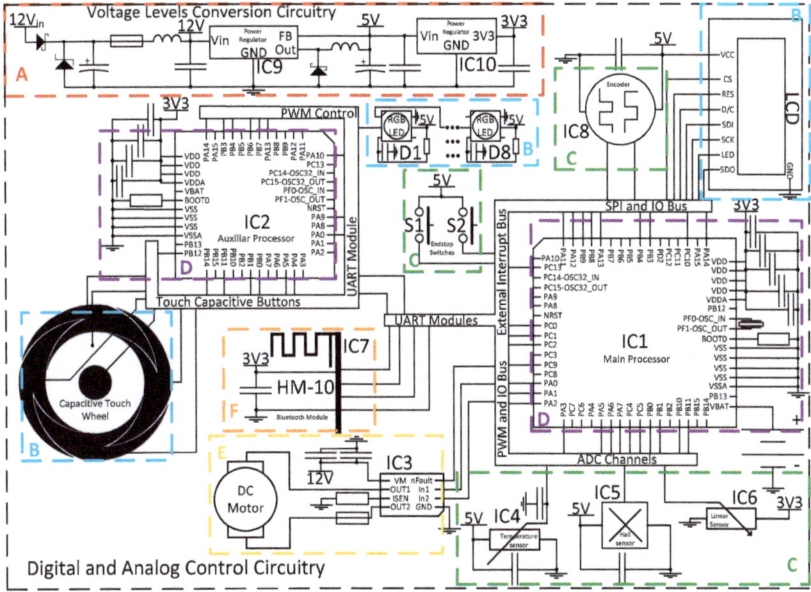

Figure 4. Main circuits used for the power conversion (**A**), user interface (**B**), sensors (**C**), system control (**D**), actuators (**E**) and wireless communications (**F**).

2.2. Firmware and Interface Design

The developed circuit was based on microcontrollers and digital communication with every component on the control boards. Regarding magnetic output, it must be noted that the peak amplitude was topped by the permanent magnets magnetization grade, thus the electromechanical component will control solely how much amplitude will reach the culture plate through the Hall sensor, as well as the displacement and frequency. Thus, the resulting mechanism control followed an approach by calculated displacement steps that will fit a distance at a given displacement frequency in order to output at the culture plate an approximate alternated magnetic field wave. However, motor sliding was a problem, using firmware breaking mechanisms with the aid of the IC3 H-Bridge, together with linear sensor for displacement error, more precise control was enabled over the mechanical response, and consequently the magnetic field. The interface, communication and each machine state (Figure 5) was controlled by firmware developed for each microprocessor according to their own respective tasks. The different machine states allowed the user to interface with each system functionality and take advantage of its calibration, variable settings and running processes.

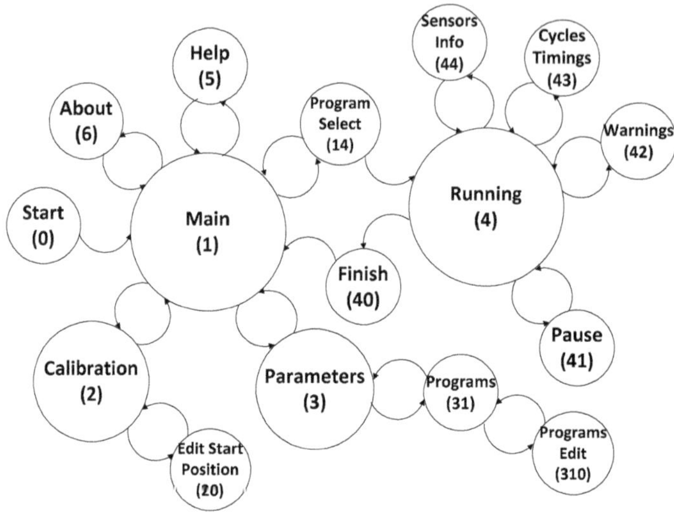

Figure 5. State machine control nodes.

The device works through four main states: (i) a menu state where every state returns to and consists of the core control of the device; (ii) a calibration state where the user sets the starting position; (iii) a program calibration state where the user sets the culture variables; (iv) a running state where the machine performs the programmed culture by the user. Through the capacitive interface, the developed system allows us to navigate a menu to adjust variables (Table 1) such as whole culture duration; active and resting cycles duration which work through two temporal levels (short and long cycles); table position calibration and setting starting point. It is possible to store up to three culture programs through emulated electric erasable programmable read-only memory (EEPROM) in the program memory. Mechanical characteristics of the developed system allows us to move the magnets up to 25 mm at a speed of up to 2.5 mm per second. The firmware provides these limitations in order to protect the cells and hardware from human error by calculating hardware limits according to distance and operation frequency. The user can define various stimulation parameters such as critical temperature, displacement between 1 and 25 mm, resulting in operating frequencies between 0.1 and 2 Hz and different stimuli cycle timings to adapt the culture to the cell's native environments.

Table 1. Bioreactor user control variables and respective ranges to be set in the programs menu.

User Control Variables	Description	Ranges
Displacement	Distance traveled by the respective permanent magnet platforms.	5–25 mm
Frequency	Frequency of stretch or magnetic field stimuli signal to be applied.	0.1–2 Hz
Runtime	Culture total running time.	1–180 d
Cycle 1 active time	Active time of sublayer cycle included in the main layer active time.	1–360 min
Cycle 1 resting time	Resting time of sublayer cycle included in the main layer active time.	1–360 min
Cycle 2 active time	Active time of the main layer cycle.	1–24 h
Cycle 2 resting time	Resting time of the main layer cycle.	1–24 h
Shutdown temperature	Temperature value, which shuts down stimuli until culture temperature lowers to safety values again.	30–40 (°C)

3. Bioreactor Evaluation

The performance of the herein developed bioreactor was evaluated using magneto-active materials based on composites comprising the piezoelectric poly(vinylidene fluoride) (PVDF) and the magnetostrictive Terfenol-D (TD) particles (Etrema Products) with approximately 1 μm diameter, as described in References [20,45]. This material was selected due to its magnetoelectrical properties, i.e., actively responding to the magnetic field provided by the magnetic bioreactor. Due to their magnetostrictive component (TD), the material senses the magnetic field, inducing a mechanical stimulation on PVDF, which due to its piezoelectric properties further induce an electrical polarization variation, creating the electrically active microenvironment that is translated to the cells [46].

MC3T3-E1 pre-osteoblast cells (Riken Bank) were used for the cell proliferation assays, as a proof of concept for bone regeneration studies. Previous to the cell culture studies, the cells were maintained in Dulbecco's modified Eagle's medium (DMEM from Gibco, ThermoFisher, Loughborough, UK) containing 1 g L^{-1} glucose, 10% fetal bovine serum (FBS from Biochrom, Cambridge, UK), and 1% penicillin/streptomycin (P/S, Biochrom) in a controlled atmosphere at 37 °C and 5% CO_2. The culture medium was replaced every 2 days, and at pre-confluence, cells were harvested using trypsin–ethylenediaminetetraacetic acid (EDTA)(Biochrom). Non-poled (non-charged) films were used to study the effect of the mechanical stimuli provided by the magnetostrictive particles in cell proliferation while poled films were used to study the influence of the mechano-electrical stimulus provided by the combination of electroactive PVDF and TD particles. ME films with a diameter of 1.3 cm were sterilized using UV for 1 h each side and placed in a 24-well tissue culture polystyrene plate. Then, 3×10^4 cells·mL^{-1} in DMEM were seeded on each well and incubated for 24 h. For this, a drop method was used, in which approximately thirty-five microliter of DMEM containing 15,000 cells was first placed on the surface of the material for 30 min to allow the cell adhesion, and then 250 μL DMEM was added to the well. After 24 h incubation time, one plate was used for the static cell culture (without any stimulation) and the other was transferred onto the bioreactor for 48 h at 37 °C in a 95% humidified air containing 5% CO_2, totalizing two cycles of magnetic stimulation. The dynamic stimuli provided by the magnetic bioreactor was achieved through the following procedure: an active time of 16 h under magnetic stimuli, which was divided into intervals of 5 min active stimuli and 25 min of resting followed by a period of complete inactivity of 8 h (Figure 6a). After 48 h, the 3-(4,5-dimethylthiazol-2-yl)-5-(3-carboxymethoxyphenyl)-2-(4-sulfophenyl)-2 H-tetrazolium (MTS, Promega) assay was used in order to determine the cell viability at the defined time-points. MTS assay is a coloring method that allows determining the cell viability and is based on the NADPH or NADP-assisted bioreduction in living cells. For this assay, the samples were transferred to a new 48-well plate and further incubated with an MTS solution (in a 1:5 ratio) at 37 °C and 5% CO_2. After 2 h, 100 μL of each well was transferred to a 96-well plate, and the optical density (OD) of each well was measured at 490 nm using a spectrophotometric plate reader (Synergy HT from BioTek, Colmar Cedex, France).

The selected conditions were employed in order to resemble the human body's daily mechanical conditions divided by 16 h of activity and 8 h of sleep, also considering the fact that bone is piezoelectric itself [21] and the magnetoelectrical scaffold is able to mimic the electroactive microenvironment upon magnetic stimulation. Those short bursts of stimuli for a duration of 5 min were performed by displacing 25 mm the magnetic table at a frequency of 0.3 Hz, resulting in a magnetic field variation of up to 23 mT within the cell culture wells. For every studied condition, three samples were assayed, and growing cell viability was determined through the MTS assay. For this assay, three main variables were considered: (i) under static conditions, i.e., without magnetic stimulus and bearing in mind the single effect of the different morphologies related to pore size differences, (ii) under dynamic conditions considering magnetic stimuli effect, and (iii) the relative effect between the material surface charge and magnetic stimuli.

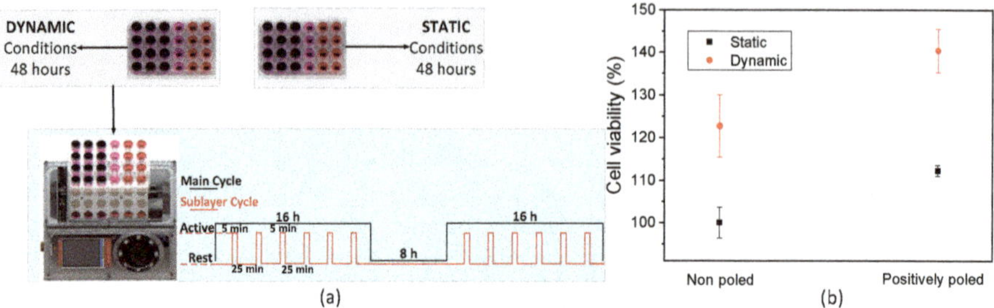

Figure 6. (**a**) Stimuli schedule timing programmed in the bioreactor for pre-osteoblast tissue culture assays using either static or dynamic conditions and (**b**) cell viability after 48 h of cell culture on TD/PVDF films with and without magnetic stimuli. The cell viability was calculated regarding the cells growing on the non-poled ME film at static conditions presented as % of growth. In each study, three samples were assayed per studied condition.

The bioreactor system was found to be completely biocompatible and suitable for cell culture. After 48 h of cell culture, MTT results show that cells grow and proliferate independently of the condition applied, showing more than 100% of cell growth under all conditions. All the components are thus biocompatible and the system working properly to avoid the increase in the temperature, one of the main concerns related to this device.

Different conditions applied to the scaffolds further induce different effects in terms of cell proliferation. The application of magnetic stimuli brings an unequivocal increase of proliferation rate in all samples, indicating a clear response of the ME films to the magnetic field, thus demonstrating that the bioreactor provides a suitable microenvironment to the pre-osteoblast cells especially in positively charge TD/PVDF film (Figure 6b). The clear increase in cell viability upon application of the stimuli indicates that a mechanoelectrical effect occurs on non-poled samples while a magnetoelectric effect occurs on poled samples, being the later more beneficial for cell growth. On all tested magneto-responsive materials, statistically significant differences in proliferation rate were observed on the growing cells.

4. Conclusions

There is an ever-increasing need for more efficient strategies in TE applications. This fact is becoming a driving force in the R&D efforts to develop a new class of materials, smart materials that respond to stimuli that further triggers appropriate cellular response through the creation of a proper microenvironment. The complexity of these microenvironments where cells are able to optimize cell growth and control cellular functions, make it difficult to recapitulate in vitro. Thereby, the need for these materials and devices capable of recreating in vivo conditions are key elements for the next developments in TE applications.

The magnetic bioreactor developed in this study represents an advance on the state of the art in TE and provides, together with specific magnetoelectric scaffolds, the electrically active microenvironments necessary for cell tissue regeneration. The bioreactor was designed and constructed taking into consideration the envisaged operating principles, by using biocompatible materials, conventional mechanics and digital electronics. It also has the potential to integrate other electronic modules that support digital communication for synchronization. This study further proved that tissue cultures may be performed with this system since a boosting effect on the proliferation rate was observed upon application of the stimuli and no signs of toxicity were found. Simultaneously, this experiment demonstrates the suitability of magneto-responsive scaffolds for adhesion and proliferation of pre-osteoblasts, availing itself from the mechanical and electrical microenvironment conceived in the material. It was possible to conclude that the magnetic module of this bioreactor was able to provide an important contribution to building the proper microenvironment as a device, whereas a scaffold which provides the proper cues to cells with the physical environment, mechanical and electrical stimuli that can be used synergistically with the system.

Therefore, this work has provided the development of a novel bioreactor based on magnetic stimulation that has proven that the developed bioreactor is biocompatible and that may be used for advanced tissue engineering applications, allowing for advanced tissue engineering strategies. It will certainly act as a valuable tool for mimicking in vitro the human stimulations provided by the electrically active tissues that are present in the body. It could also be important for growing well-formed cellular tissues in vitro in a more effective and rapid way, which could be further implanted in the human body without the material. In the case that magnetoactive materials to be implanted in the human body, it would provide a suitable platform to evaluate the remote stimulation and thus for effective growth and differentiation of cells in immobilized patients. In fact, mimicking cell microenvironments is thus a key issue to recapitulate in vitro what occurs in vivo and this bioreactor holds great promise to fulfill such requirements.

Author Contributions: Conceptualization, S.L.-M., V.C. and C.R.; methodology V.C. and C.R.; hardware N.C. and V.C.; software N.C.; validation M.M.F., C.R. and N.C.; formal analysis N.C., V.C., R.M., S.L.-M., C.R. and M.M.F.; investigation C.R. and M.M.F.; writing—original draft preparation N.C.; writing—review and editing S.L.-M., R.M., M.M.F. and N.C.; supervision, S.L.-M. and R.M.; project administration, S.L.-M.; funding acquisition, S.L.-M. All authors have read and agreed to the published version of the manuscript.

Funding: FCT—Fundação para a Ciência e Tecnologia: UID/FIS/04650/2020; PTDC/BTM-MAT/28237/2017; PTDC/EMD-EMD/28159/2017 and SFRH/BPD/121464/2016. Spanish Ministry of Economy and Competitiveness (MINECO): MAT2016–76039-C4-3-R (AEI/FEDER, UE). Basque Government Industry and Education Department: ELKARTEK, PIB and PIBA (PIBA−2018–06) programs, respectively.

Acknowledgments: The authors thank the FCT—Fundação para a Ciência e Tecnologia—for financial support under Strategic Funding UID/FIS/04650/2020 and projects PTDC/BTM-MAT/28237/2017 and PTDC/EMD-EMD/28159/2017. MMF thank FCT for the SFR H/BPD/121464/2016 grant. The authors acknowledge funding by the Spanish Ministry of Economy and Competitiveness (MINECO) through the project MAT2016–76039-C4-3-R (AEI/FEDER, UE) and from the Basque Government Industry and Education Department under the ELKARTEK, HAZITEK and PIBA (PIBA−2018–06) programs, respectively.

Conflicts of Interest: The authors declare no conflict of interest.

References

1. Chung, S.; King, M.W. Design concepts and strategies for tissue engineering scaffolds. *Biotechnol. Appl. Biochem.* **2011**, *58*, 423–438. [CrossRef]

2. Lutolf, M.; Hubbell, J. Synthetic biomaterials as instructive extracellular microenvironments for morphogenesis in tissue engineering. *Nat. Biotechnol.* **2005**, *23*, 47. [CrossRef]

3. Einhorn, T.A.; Gerstenfeld, L.C. Fracture healing: Mechanisms and interventions. *Nat. Rev. Rheumatol.* **2015**, *11*, 45. [CrossRef]

4. Quinlan, E.; López-Noriega, A.; Thompson, E.; Kelly, H.M.; Cryan, S.A.; O'brien, F.J. Development of collagen–hydroxyapatite scaffolds incorporating PLGA and alginate microparticles for the controlled delivery of rhBMP−2 for bone tissue engineering. *J. Control. Release* **2015**, *198*, 71–79. [CrossRef] [PubMed]

5. Place, E.S.; George, J.H.; Williams, C.K.; Stevens, M.M. Synthetic polymer scaffolds for tissue engineering. *Chem. Soc. Rev.* **2009**, *38*, 1139–1151. [CrossRef] [PubMed]

6. Bacakova, L.; Filova, E.; Parizek, M.; Ruml, T.; Svorcik, V. Modulation of cell adhesion, proliferation and differentiation on materials designed for body implants. *Biotechnol. Adv.* **2011**, *29*, 739–767. [CrossRef] [PubMed]

7. Ribeiro, C.; Correia, D.M.; Ribeiro, S.; Sencadas, V.; Botelho, G.; Lanceros-Méndez, S. Piezoelectric poly(vinylidene fluoride) microstructure and poling state in active tissue engineering. *Eng. Life Sci.* **2015**, *15*, 351–356. [CrossRef]

8. O'brien, F.J. Biomaterials & scaffolds for tissue engineering. *Mater. Today* **2011**, *14*, 88–95.

9. Ribeiro, C.; Correia, D.M.; Ribeiro, S.; Fernandes, M.M.; Lanceros-Mendez, S. Piezo- and Magnetoelectric Polymers as Biomaterials for Novel Tissue Engineering Strategies. *MRS Adv.* **2018**, *3*, 1671–1676. [CrossRef]

10. Cardoso, V.F.; Correia, D.M.; Ribeiro, C.; Fernandes, M.M.; Lanceros-Méndez, S. Fluorinated Polymers as Smart Materials for Advanced Biomedical Applications. *Polymers* **2018**, *10*, 161. [CrossRef]

11. Qazi, T.H.; Rai, R.; Boccaccini, A.R. Tissue engineering of electrically responsive tissues using polyaniline based polymers: A review. *Biomaterials* **2014**, *35*, 9068–9086. [CrossRef] [PubMed]

12. Ribeiro, C.; Sencadas, V.; Correia, D.M.; Lanceros-Méndez, S. Piezoelectric polymers as biomaterials for tissue engineering applications. *Colloids Surf. B* **2015**, *136*, 46–55. [CrossRef] [PubMed]

13. Ghasemi-Mobarakeh, L.; Prabhakaran, M.P.; Morshed, M.; Nasr-Esfahani, M.H.; Baharvand, H.; Kiani, S.; Al-Deyab, S.S.; Ramakrishna, S. Application of conductive polymers, scaffolds and electrical stimulation for nerve tissue engineering. *J. Tissue Eng. Regen. M* **2011**, *5*, e17–e35. [CrossRef] [PubMed]

14. Seil, J.T.; Webster, T.J. Electrically active nanomaterials as improved neural tissue regeneration scaffolds. *Wiley Interdiscip. Rev. Nanomed. Nanobiotechnol.* **2010**, *2*, 635–647. [CrossRef] [PubMed]

15. Martins, P.M.; Ribeiro, S.; Ribeiro, C.; Sencadas, V.; Gomes, A.C.; Gama, F.M.; Lanceros-Méndez, S. Effect of poling state and morphology of piezoelectric poly(vinylidene fluoride) membranes for skeletal muscle tissue engineering. *RSC Adv.* **2013**, *3*, 17938–17944. [CrossRef]

16. Ribeiro, S.; Puckert, C.; Ribeiro, C.; Gomes, A.C.; Higgins, M.J.; Lanceros-Méndez, S. Surface Charge-Mediated Cell–Surface Interaction on Piezoelectric Materials. *ACS Appl. Mater. Interfaces* **2020**, *12*, 191–199. [CrossRef]

17. Marques-Almeida, T.; Cardoso, V.F.; Ribeiro, S.; Gama, F.M.; Ribeiro, C.; Lanceros-Mendez, S. Tuning Myoblast and Preosteoblast Cell Adhesion Site, Orientation, and Elongation through Electroactive Micropatterned Scaffolds. *ACS Appl. Biol. Mater.* **2019**, *2*, 1591–1602. [CrossRef]

18. Hermenegildo, B.; Ribeiro, C.; Pérez-Álvarez, L.; Vilas, J.L.; Learmonth, D.A.; Sousa, R.A.; Martins, P.; Lanceros-Méndez, S. Hydrogel-based magnetoelectric microenvironments for tissue stimulation. *Colloids Surf. B* **2019**, *181*, 1041–1047. [CrossRef]

19. Fernandes, M.M.; Correia, D.M.; Ribeiro, C.; Castro, N.; Correia, V.; Lanceros-Mendez, S. Bioinspired three-dimensional magneto-active scaffolds for bone tissue engineering. *ACS Appl. Mater. Interfaces* **2019**. [CrossRef]

20. Ribeiro, C.; Correia, V.; Martins, P.; Gama, F.M.; Lanceros-Mendez, S. Proving the suitability of magnetoelectric stimuli for tissue engineering applications. *Colloids Surf. B* **2016**, *140*, 430–436. [CrossRef]

21. Williams, W.S.; Breger, L. Piezoelectricity in tendon and bone. *J. Biomech.* **1975**, *8*, 407–413. [CrossRef]

22. Costa-Pinto, A.R.; Reis, R.L.; Neves, N.M. Scaffolds based bone tissue engineering: The role of chitosan. *Tissue Eng. Part B Rev.* **2011**, *17*, 331–347. [CrossRef] [PubMed]

23. Dimitriou, R.; Jones, E.; McGonagle, D.; Giannoudis, P.V. Bone regeneration: Current concepts and future directions. *BMC Med.* **2011**, *9*, 66. [CrossRef] [PubMed]

24. Ribeiro, C.; Pärssinen, J.; Sencadas, V.; Correia, V.; Miettinen, S.; Hytönen, V.P.; Lanceros-Méndez, S. Dynamic piezoelectric stimulation enhances osteogenic differentiation of human adipose stem cells. *J. Biomed. Mater. Res. Part A* **2015**, *103*, 2172–2175. [CrossRef]

25. Ribeiro, C.; Moreira, S.; Correia, V.; Sencadas, V.; Rocha, J.G.; Gama, F.M.; Ribelles, J.L.G.; Lanceros-Mendez, S. Enhanced proliferation of pre-osteoblastic cells by dynamic piezoelectric stimulation. *RSC Adv.* **2012**, *2*, 11504–11509. [CrossRef]

26. Bodhak, S.; Bose, S.; Kinsel, W.C.; Bandyopadhyay, A. Investigation of In Vitro Bone Cell Adhesion and Proliferation on Ti Using Direct Current Stimulation. *Mater. Sci. Eng. C Mater. Biol. Appl.* **2012**, *32*, 2163–2168. [CrossRef]

27. Sollazzo, V.; Palmieri, A.; Pezzetti, F.; Massari, L.; Carinci, F. Effects of pulsed electromagnetic fields on human osteoblastlike cells (MG–63): A pilot study. *Clin. Orthop. Relat.* **2010**, *468*, 2260–2277. [CrossRef]

28. Guduru, R.; Khizroev, S. Magnetic field-controlled release of paclitaxel drug from functionalized magnetoelectric nanoparticles. *Part. Part. Syst. Char.* **2014**, *31*, 605–611. [CrossRef]

29. Lodi, M.B.; Fanti, A.; Casu, S. Analysis of superparamagnetic scaffolds: For. bone tissue engineering in static magnetic and dynamic fields. In Proceedings of the 2016 Loughborough Antennas & Propagation Conference (LAPC), Loughborough, Leicestershire, UK, 14–15 November 2016.

30. Evans, D.; Schilling, A.; Kumar, A.; Sanchez, D.; Ortega, N.; Arredondo, M.; Katiyar, R.; Gregg, J.; Scott, J. Magnetic switching of ferroelectric domains at room temperature in multiferroic PZTFT. *Nat. Commun.* **2013**, *4*, 1534. [CrossRef]

31. Gonçalves, R.; Martins, P.; Correia, D.M.; Sencadas, V.; Vilas, J.; León, L.; Botelho, G.; Lanceros-Méndez, S. Development of magnetoelectric CoFe 2 O 4/poly (vinylidene fluoride) microspheres. *RSC Adv.* **2015**, *5*, 35852–35857. [CrossRef]

32. Li, Y.; Wang, Z.; Yao, J.; Yang, T.; Wang, Z.; Hu, J.-M.; Chen, C.; Sun, R.; Tian, Z.; Li, J.; et al. Magnetoelectric quasi-(0–3) nanocomposite heterostructures. *Nat. Commun.* **2015**, *6*, 6680. [CrossRef] [PubMed]

33. Martins, P.; LancerosMéndez, S. Polymerbased magnetoelectric materials. *Adv. Funct. Mater.* **2013**, *23*, 3371–3385. [CrossRef]

34. Mistry, A.S.; Mikos, A.G. Tissue engineering strategies for bone regeneration. In *Regenerative Medicine II*; Springer: Berlin, UK, 2005; pp. 1–22.

35. Lee, E.A.; Yim, H.; Heo, J.; Kim, H.; Jung, G.; Hwang, N.S. Application of magnetic nanoparticle for controlled tissue assembly and tissue engineering. *Arch. Pharmacal Res.* **2014**, *37*, 120–128. [CrossRef]

36. Xia, Y.; Chen, H.; Zhao, Y.; Zhang, F.; Li, X.; Wang, L.; Weir, M.D.; Ma, J.; Reynolds, M.A.; Gu, N. Novel magnetic calcium phosphate-stem cell construct with magnetic field enhances osteogenic differentiation and bone tissue engineering. *Mater. Sci. Eng. C* **2019**, *98*, 30–41. [CrossRef]

37. Yamamoto, Y.; Ito, A.; Jitsunobu, H.; Yamaguchi, K.; Kawabe, Y.; Mizumoto, H.; Kamihira, M. Hollow fiber bioreactor perfusion culture system for magnetic force-based skeletal muscle tissue engineering. *J. Chem. Eng. Jpn.* **2012**, *45*, 348–354. [CrossRef]

38. El Haj, A.J.H.; Dobson, J.P. Culturing Tissue Using Magnetically Generated Mechanical Stresses. U.S. Patent 7,553,662, 30 June 2009.

39. Hernández-Hernández, H.; Cruces-Solis, H.; Elías-Viñas, D.; Verdugo-Díaz, L. Neurite outgrowth on chromaffin cells applying extremely low frequency magnetic fields by permanent magnets. *Arch. Med. Res.* **2009**, *40*, 545–550. [CrossRef]

40. Freed, L.E.; Guilak, F.; Guo, X.E.; Gray, M.L.; Tranquillo, R.; Holmes, J.W.; Radisic, M.; Sefton, M.V.; Kaplan, D.; Vunjak-Novakovic, G. Advanced tools for tissue engineering: Scaffolds, bioreactors, and signaling. *Tissue Eng.* **2006**, *12*, 3285–3305. [CrossRef]

41. Reizabal, A.; Brito-Pereira, R.; Fernandes, M.M.; Castro, N.; Correia, V.; Ribeiro, C.; Costa, C.M.; Perez, L.; Vilas, J.L.; Lanceros-Méndez, S. Silk Fibroin magnetoactive nanocomposite films and membranes for dynamic bone tissue engineering strategies. *Materialia* **2020**, 100709. [CrossRef]

42. Lee, A.-C.; Hsiao, F.-Z.; Ko, D. Analysis and testing of magnetic bearing with permanent magnets for bias. *JSME Int. J. Ser. C Dyn. Control Robot. Des. Manuf.* **1994**, *37*, 774–782. [CrossRef]

43. Camacho, J.M.; Sosa, V. Alternative method to calculate the magnetic field of permanent magnets with azimuthal symmetry. *Rev. Mex. Fis. E* **2013**, *59*, 8–17.

44. Brito-Pereira, R.; Correia, D.M.; Ribeiro, C.; Francesko, A.; Etxebarria, I.; Pérez-Álvarez, L.; Vilas, J.L.; Martins, P.; Lanceros-Mendez, S. Silk fibroin-magnetic hybrid composite electrospun fibers for tissue engineering applications. *Compos. Part B* **2018**, *141*, 70–75. [CrossRef]

45. Ribeiro, C.; Costa, C.M.; Correia, D.M.; Nunes-Pereira, J.; Oliveira, J.; Martins, P.; Gonçalves, R.; Cardoso, V.F.; Lanceros-Méndez, S. Electroactive poly (vinylidene fluoride)-based structures for advanced applications. *Nat. Protoc.* **2018**, *13*, 681. [CrossRef] [PubMed]

46. Fernandes, M.M.; Carvalho, E.O.; Lanceros-Mendez, S. Electroactive Smart Materials: Novel Tools for Tailoring Bacteria Behavior and Fight Antimicrobial Resistance. *Front. Bioeng. Biotechnol.* **2019**, *7*, 277. [CrossRef] [PubMed]

© 2020 by the authors. Licensee MDPI, Basel, Switzerland. This article is an open access article distributed under the terms and conditions of the Creative Commons Attribution (CC BY) license (http://creativecommons.org/licenses/by/4.0/).

Letter

Component Analysis of Gas Mixture Based on One-Dimensional Convolutional Neural Network

Canjian Zhan, Jiafeng He *, Mingjin Pan and Dehan Luo

School of Information Engineering, Guangdong University of Technology, Guangzhou 510006, China; zcj@mail2.gdut.edu.cn (C.Z.); illusionnol66@gmail.com (M.P.); dehanluo@gdut.edu.cn (D.L.)
* Correspondence: jfhe@gdut.edu.cn; Tel.:+86-1814-891-8377

Abstract: Indoor harmful gases are a considerable threat to the health of residents. In order to improve the accuracy of indoor harmful gas component identification, we propose an indoor toxic gas component analysis method that is based on the combination of bionic olfactory and convolutional neural network. This method uses the convolutional neural network's ability to extract nonlinear features and identify each component of bionic olfactory response signal. A comparison with the results of other methods verifies the improvement of recognition rate while with the same level of time cost, which proved the effectiveness of the proposed model. The experimental results showed that the recognition rate of different types and concentrations of harmful gas components reached 90.96% and it solved the problem of mutual interference between gases.

Keywords: electronic nose; convolutional neural network; component analysis

Citation: Zhan, C.; He, J.; Pan, M.; Luo, D. Component Analysis of Gas Mixture Based on One-Dimensional Convolutional Neural Network. *Sensors* **2021**, *21*, 347. https://doi.org/10.3390/s21020347

Received: 16 November 2020
Accepted: 2 January 2021
Published: 6 January 2021

Publisher's Note: MDPI stays neutral with regard to jurisdictional claims in published maps and institutional affiliations.

Copyright: © 2021 by the authors. Licensee MDPI, Basel, Switzerland. This article is an open access article distributed under the terms and conditions of the Creative Commons Attribution (CC BY) license (https://creativecommons.org/licenses/by/4.0/).

1. Introduction

Because low-concentration indoor harmful gases are invisible and tasteless, they are difficult for people to distinguish. We can detect low-concentration indoor toxic gases through physical and chemical identification methods. Still, they are cumbersome and complicated operations, and to use the instrument needs to be professionally trained. It is difficult to promote in the market.

Many different methods are applied in indoor air environment monitoring for the quantitative analysis of harmful gases, including the non-dispersive infrared method [1], gas chromatography [2], nessler's reagent colorimetry [3], and ion-selective electrode method [4]. The methods, as mentioned above, are relatively complicated and they cannot perform real-time on-site air quality testing. With the rapid development of information science and sensor technology, the bionic olfactory system has been applied in medical, food processing, and environmental detection fields, with its advantages of simplicity and economy.

However, when machine olfactory technology is used in the quantitative analysis of substance odor in an open environment, it is easily affected by interfering gases and environmental temperature and humidity, which causes the problem of reduced recognition accuracy.

The machine learning algorithms and their optimization methods were applied to the quantitative analysis of machine olfactory rapidly. For example, Xianjiang Li et al. proposed an optimization model for mixed gas quantitative detection that combines an adaptive genetic algorithm and a traditional BP neural network. This algorithm can overcome the shortcomings of the slow search rate and easily fall into local minimum. The BP neural network can obtain better initial weights and early stage thresholds through the adaptive genetic algorithm. Subsequently, the experiment uses this algorithm to quantitatively analyze the machine olfactory odor data of a set of five-element gas mixtures. The experiment shows that the accuracy of the algorithm for identifying the gas concentration of the gas mixture is higher than that of the traditional BP neural network [5]. Shurui

Fan et al. used principal component analysis and random forest modeling, which can qualitatively identify methane and ethylene gases. Subsequently, the optimal regression model is constructed through support vector regression and particle swarm optimization for quantitative analysis of the two types of gases. The experimental results show that the average recognition rate of principal component analysis, combined with random forest, is 97% higher than logistic regression and the support vector machine. The fitting effect of support vector regression is optimized by the particle swarm algorithm, and better fitting results are obtained than support vector regression [6].

In practical applications, when using machine olfaction to quantitatively analyze the odor of substances, it is often accompanied by the influence of interfering gas. Therefore, nonlinear or linear inseparable phenomenon, or even gas shielding, will often appear in the machine olfactory system. The phenomenon of shielding between gases means that the response of the electronic nose sensor when measuring mixed gas is lower than that of measuring pure gas of the same concentration [7]. However, the above-mentioned literature does not fully consider the influence of these phenomena on the quantitative analysis of machine olfaction [8], so some improved algorithms are proposed.

Yu Lu et al. combined the artificial neural network with the basic concepts of analytical chemistry, designed an error function based on analytical chemistry, and applied the error function to the neural network. This method can be used to control alcohol, petroleum gas, and water. The experimental results show that the method predicts a gas concentration error less than 10% [9]. Tang K TZ et al. proposed a Locally Weighted Nearest Neighbor (LWNN) algorithm that is based on the K-Nearest Neighbor classifier (KNN) algorithm to determine the odor components, and then combined with the types of odor components, while using the weighted and constrained least squares (Weighted and Constrained Least-Squares, WCLS) gas concentration estimation method that is based on the least squares method measures the target gas concentration in the mixed gas [10].

The above analysis shows that, in addition to investigate the feasibility of applying machine odor perception to quantitative gas analysis, great progress has been made in the component analysis methods of machine odor perception; however, there are still various shortcomings. The methods are not universally applicable and they do not take the mutual influence between different gas concentrations and the reversal of the response curves into consideration. Therefore, none of the above methods can be used for the component analysis of indoor pollutant gases while using the odor perception engine. It is necessary to find appropriate methods that are based on the properties of the pollutant gas odor data for indoor spaces in order to extract the component properties of the target gas (for example, formaldehyde) among other indoor gases. Based on the background values of indoor air pollution, this paper proposes a component analysis method for indoor pollutant gases when considering the influence of interfering gases.

2. Gas Data Preprocessing

According to the requirement of the research objectives and content of this article, a machine olfactory system is required for collecting the odor information of indoor harmful gases. This article contains the odor data of multiple indoor toxic gas samples through the PEN3 electronic nose system in order to ensure the reliability of the data. The PEN3 electronic nose is the third generation of the PEN series developed by AIRSENSE, Germany. It is built with 10 cross-sensitive metal oxide gas sensors. Table 1 shows the characteristics of the sensors in PEN3 electronic nose.

Table 1. The characteristics of the sensors in PEN3 electronic nose.

The Number of Sensors	The Name of Sensors	Sensitive Substances (Primary Response Components)
S1	W1C	aromatic substances such as toluene
S2	W5S	nitrogen dioxide and other nitrogen oxides
S3	W3C	ammonia and aromatic substances
S4	W6S	hydrogen
S5	W5C	alkanes and aromatic substances
S6	W1S	methane
S7	W1W	sulfide
S8	W2S	ethanol
S9	W2W	aromatic substances and organic sulfur substances
S10	W3S	alkanes

This article selected the four most common pollution gases in daily life based on the needs of the research objectives and content of this article. They are formaldehyde, ammonia, benzene, and methanol, produced by Henan Testing Center. Additionally, the multi-channel gas mixing system performs the ratio of indoor harmful gases in order to ensure the objectivity and reliability of the data. Appropriate experimental materials and exhaust gas treatment equipment should be selected when designing the practical plan in order to ensure the safety of the experiment. Figure 1 shows the process of harmful indoor gas collection based on machine olfactory. In order to obtain accurate gas concentration sample data, the MT-500X dynamic gas mixing system (gas mixing instrument) is used in this paper to carry out the precise ratio of indoor harmful gas samples, to ensure the objectivity and accuracy of the data that were collected in this paper. The system is equipped with a high-precision mass flow controller, which can meet the requirements of stable, reliable, and high-precision gas distribution.

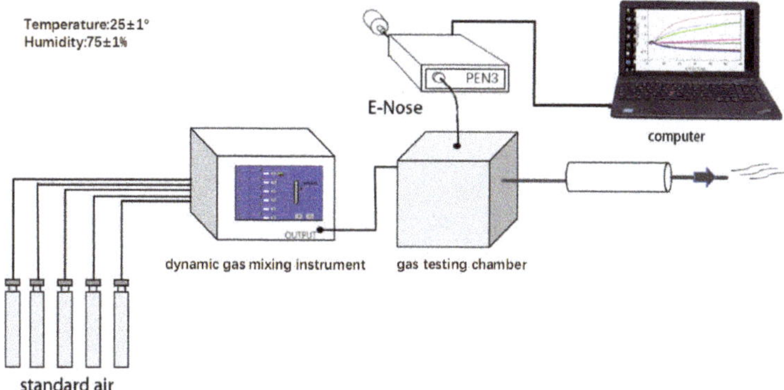

Figure 1. The experimental procedure of indoor hazardous gas collection based on machine olfaction.

Before starting the experiment, the indoor air conditioner and humidifier should be turned on, so that the temperature should be controlled within the range of $25 \pm 1\ ^\circ$C and the humidity should be controlled at $75 \pm 1\%$. Table 2 shows the other experimental parameters. First, we prepare the gas samples that are required for the experiment with the standard gas through the gas mixing instrument, transport the prepared gas samples to the gas testing chamber, and then collect the odor data through the PEN3 electronic nose. Finally, the residual test gas in the gas test box is passed through the tail gas treatment device for harmless treatment.

Table 2. The parameters of the gas data acquisition experiment.

Experimental Parameters	Parameter Setting
collect temperature	$25 \pm 1\,^{\circ}\mathrm{C}$
collect humidity	$75 \pm 1\%$
gas sample volume	4 L
gas testing chamber volume	4 L
sample standing time	30 min
sensor array automatic cleaning time	120 s
sampling time	60 s
sample interval	1 s
gas-flow rate	60 mL/min

According to different interference groups, the response curve of the electronic nose to different indoor harmful gases in the same interference group can be drawn through the data set. The concentrations of formaldehyde gas containing 0.01 mg/m^3, 0.05 mg/m^3, 0.09 mg/m^3, 0.13 mg/m^3, 0.17 mg/m^3, and 0.21 mg/m^3 are drawn, respectively, as shown in Figures 2 and 3.

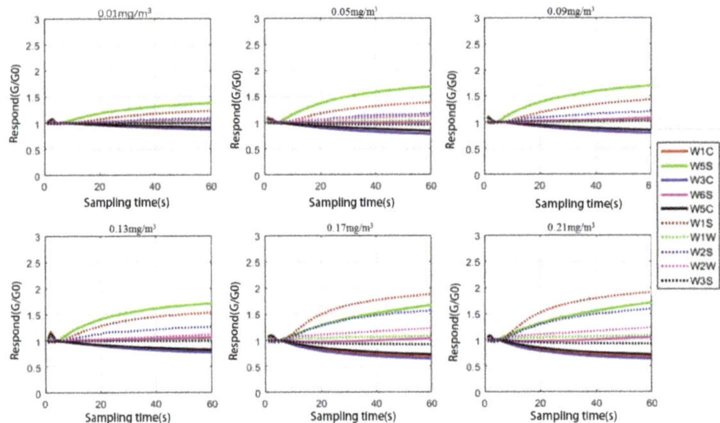

Figure 2. The response curve of weak interference group.

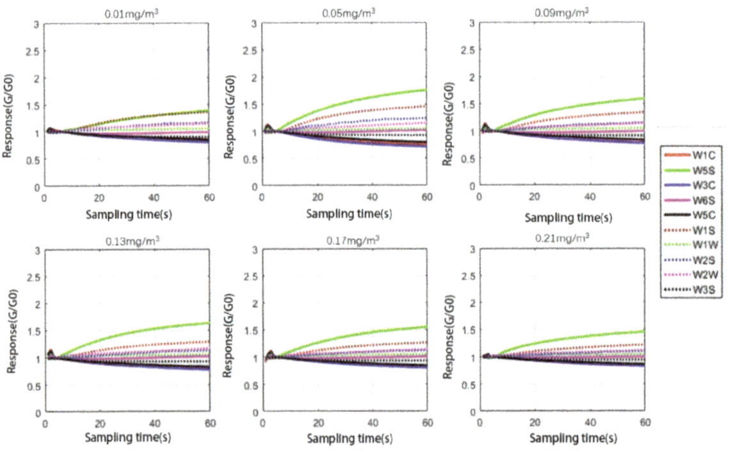

Figure 3. The response curve of moderate interference group.

By observing Figures 2 and 3, it can be found that, in the case of the same interference group, each response curve of the sensor is very similar in the overall listing, and it is still necessary to use the radar chart to supplement the observation. The radar chart data still uses the data with a sampling interval of 30 s to 40 s in order to calculate the average value, and then the average value is converted into a radar chart, as shown in Figure 4.

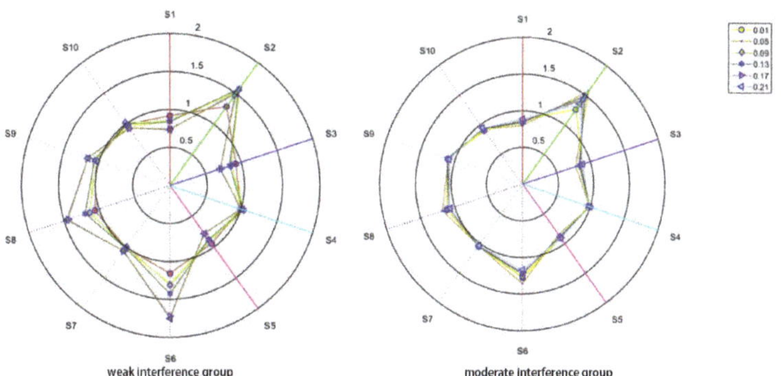

Figure 4. Radar diagram for each set of data.

The response values of sensors numbered S2, S6, and S8 are quite different from other sensors, especially for S2. The response of the sample is more sensitive. However, with the exception of the S2, S6, and S8 sensors, the difference in the response values of other sensors is small enough to determine that "high dimensionality, redundant information, and non-linearity" are the characteristics of indoor harmful gas data that are collected by machine olfactory.

The above data show that we can infer that indoor harmful gas samples containing different formaldehyde gas have differences in data that are based on the difference between the response curve of the sensor and the radar chart. According to their differences, suitable identification methods can be selected in order to train the computer to identify the level of formaldehyde pollution in indoor harmful gases.

In this experiment, Table 3 shows the 60 × 10 odor data matrix generated by PEN3. Each column represents a different sensor, and each row represents the response of the same sensor at different sampling times.

Table 3. Raw data format.

sensor1	sensor2	......	sensor10
1.0040	0.9695	......	1.0023
1.0224	1.0341	......	1.0065
....	....	......	......
1.3567	1.1501	......	1.2354

According to the original data format of the PEN3 electronic nose in Table 3, the data matrix of 60 × 10 is first transposed to the data matrix of 10 × 60, and then the data matrix is represented by a brand new row vector. The specific method is as follows. Based on the corresponding sensor that generates the response, the original matrix is divided into 10 row vectors of 60, and is then sequentially connected by the sensor number to form a new row vector, and stored in the .csv format. If the sampling time of the acquisition experiment is 60 s to collect m samples, then the converted data set of the electronic nose data file is a .csv file with m 600-dimensional features. In this paper, 1040 gas samples are collected through experiments, so the data in this paper are concentrated.

In this paper, a total of 1040 data samples were collected through the collection experiment. According to the indoor air quality standard (GB/T18883-2002), the data samples are divided into three types of data with different pollution levels. Among them, 320 data samples containing formaldehyde gas from the concentration of 0.01 mg/m^3 to 0.08 mg/m^3 (without 0.08 mg/m^3) are qualified (Normal), and 320 data samples with the formaldehyde concentration from 0.08 mg/m^3 to 0.16 mg/m^3 were classified as mildly polluted (Mild), and 400 data samples containing formaldehyde concentrations that were greater than 0.16 mg/m^3 were classified as severely polluted (Serious). Select 70% of the data samples of each type of pollution level as the training set and the remaining 30% data samples as the test set of the experiment. The data set used for model training has 728 data samples, and the data set used for testing has 312 data samples.

Therefore, we construct an indoor hazardous gas odor information collection system. A number of indoor harmful gas samples with different concentrations were prepared by a dynamic mixing gas distributor, and the PEN3 electronic noses was used in order to collect odor data from these samples, thereby obtaining a data set of the response of the indoor harmful gases. The exhaust gas of the experiment was treated in a harmless manner in order to eliminate the impact of harmful gases on the health of the experimenters. According to the characteristics of the electronic nose response data, a component analysis method of indoor harmful gases based on machine olfaction and global average pooling convolutional neural network model is proposed.

3. One-Dimensional GAP-CNN

CNN is a neural network with deep structure and a classic algorithm that is widely used in deep learning [11]. Nowadays, many typical and widely used CNN models have been proposed, such as LeNet-5 [12], AlexNet [13], and GoogleLeNet [14]. They have been successfully appled in face detection [15], role recognition [16], pedestrian detection [17], and robot navigation area [18]. Because CNN has the structural characteristics of local connection, weight sharing, and down-sampling, the model is sample-invariant to translation, scaling, and distortion, and, thus, has strong robustness [19]. This feature makes convolutional neural networks a great success in the field of image processing. The main difference between CNN and the traditional BP (Back Propagation, BP) neural network lies in the two aspects of weight sharing and local connection. Weight sharing makes the convolutional neural network more suitable for the structure of biological neural networks. The local connection of convolutional neural network is not like a traditional neural network. Each neuron in the first layer is connected to all neurons in the first layer, but the neurons in the first layer are partially connected to the neurons in the first layer. The role of these two characteristics makes the model have lower model complexity and fewer weights than traditional BP neural networks.

The convolution layer performs convolution processing on the input data through multiple convolution kernels and extracts the convolutional features, which is the feature map. A corresponding type of feature is extracted through a convolution kernel. Because the operation of the same convolution kernel has the characteristics of local connection, parameter sharing, and multiple convolution kernels, when compared with the fully connected layer, the convolution layer can propose more features with fewer parameters when extracting data features. Because the convolution structure is not affected by the input dimensions and the training depth structure is simple, it can effectively extract features from complex and high-latitude inputs. The convolution formula of the convolution layer is:

$$g(i) = \sum_{x=1}^{m} \sum_{y=1}^{n} \sum_{z=1}^{p} \alpha_{x,y,z} \times w_{x,y,z}^{i} + \beta^{i}, i = 1, 2, \ldots, q \tag{1}$$

where: i is the i-th convolution kernel, $g(i)$ is the feature map that is extracted by the i-th convolution kernel; α is the input data; β is the bias of the convolution kernel; x, y, z represent three different dimensions of data. After completing the convolution of the data, it is necessary to use a nonlinear activation function in order to perform nonlinear conversion on the data. The commonly used activation function in CNN is generally ReLU, and its formula is:

$$y(i) = f(g(i)) = \max\{0, g(i)\}, i = 1, 2, \ldots, q \tag{2}$$

$$p^{l(i,j)} = \max_{(j-1)w < t < jw} \left\{ \alpha^{l(i,t)} \right\}, j = 1, 2, \ldots, q \tag{3}$$

$$p^{l(i,j)} = \underset{(j-1)w < t < jw}{\mathrm{avg}} \left\{ \alpha^{l(i,t)} \right\}, j = 1, 2, \ldots, q \tag{4}$$

where: $\alpha^{l(i,t)}$ is the t-th neuron of the i-th feature map in the l-th layer, w is the width of the convolution kernel, and j is the j-th pooling kernel [20].

In the convolutional neural network, the convolutional layer and the pooling layer both perform feature extraction on the data, and the operation of data classification is performed in the fully connected layer based on the feature. The fully connected layer can integrate features through the feature maps output by the convolutional layer, thereby obtaining classification information with high-level meaning, and then classify and output according to these. As the output of the CNN model, the output of the fully connected layer is a fixed-length feature vector that is obtained by transforming the feature map input from the layer. The feature information of all combinations of the original data will be integrated by this feature vector. Although this fixed-length feature vector does not have the location information of the original data, the feature information used to effectively complete the data classification task has been fully extracted [21].

In the traditional convolutional neural network, the parameter ratio of the fully connected layer is almost 80% of the entire neural network model [22]. In order to achieve the purpose of reducing network parameters, there are generally two common methods. The first method is to replace the fully connected layer in the model with a convolutional layer. If the convolution kernel of the same size as the feature map in the full connection layer is used as the input of the output layer, this often makes the model output result far inferior to using the fully connected layer as the input of the output layer accurate. The second approach is to reduce the feature output dimensions of all the layers in the network, but the disadvantage of this approach is that the neural network lacks sufficient features for network training, which results in the final output feature information that contains less feature information than the original dimension, which reduces the model accuracy of recognition.

In this article, we refer to the idea of using global pooling in GoogleNet,and use the global pooling layer to replace the fully connected layer in order to extract features [23]. There are generally two methods for global pooling, global average pooling (GAP) and global maxing pooling (GMP). This article takes the global average pooling as an example to illustrate the difference between global pooling layer and full connection. The global average pooling layer, like the fully connected layer, has the function of connecting global feature information, so as to ensure the accuracy of the output result of the recognition model. However, because the global pooling layer does not contain any parameters, the phenomenon where the fully connected layer is easy to overfit is avoided during the model training process.

4. Result

In order to verify that GAP-CNN can reduce the weight parameters while ensuring the accuracy of identifying formaldehyde pollution in indoor harmful gases, a CNN model with the same depth and the same convolution kernel size as GAP-CNN will be used as the control group. Both network models use the same convolution layer and convolution kernel with the same parameter settings. However, in the CNN model, the number of parameters that need to be trained is 14,963, which is much higher than the number of 4259 parameters shown in the GAP-CNN model. The reason is that in the CNN model, the number of parameters in the first dense layer, which is the fully connected layer, accounts for 70% of the total network model parameters. This also proves that global pooling can effectively compress the volume of the convolutional neural network model. The parameters of the prediction models of the two convolutional neural networks are set as: the number of training data is epoch = 400, the size of the training set data and the test set data is randomly selected each time batchsize = 32, and the learning rate α is 0.01. Figures 5 and 6 show the training curves of the two models, respectively.

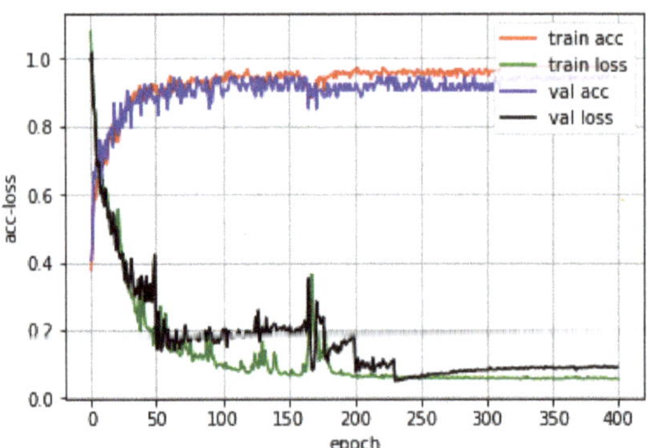

Figure 5. Training curve of global average pooling-CNN (GAP-CNN) model.

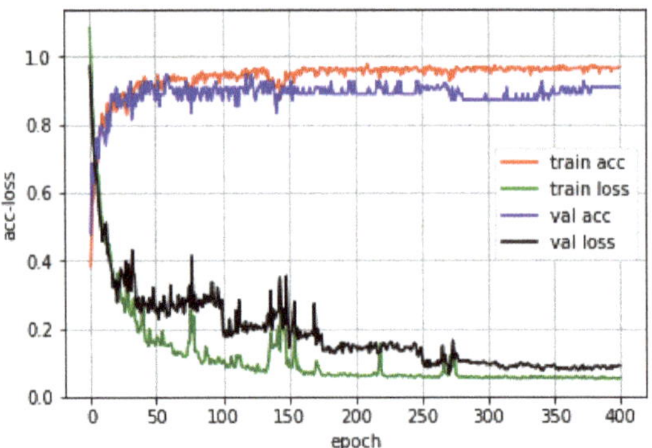

Figure 6. Training curve of CNN model.

Among them, the abscissa represents the number of epochs of model training, the ordinate represents the accuracy and the size of the loss function value, train acc represents the recognition rate of the model to the training set, train loss represents the value of the loss function of the model to the training set, the val acc model represents the recognition rate of the set, and val loss represents the value of the loss function of the model to the training set. In Figure 5, the values of train loss and train acc of GAP-CNN reach a stable state after the number of training is 260. The GAP-CNN model has been trained. The value of train acc of the model is 0.9684, and the value of train loss is 0.0954. The correct recognition rate val acc of the model on the training set is 0.9140. From Figure 6, the train loss and train acc values of the CNN model reach a stable state after 230 training times. The train acc value of the CNN model is 0.9615 and the train loss value is 0.1083. The correct recognition rate val acc of the CNN model on the training set is 0.9071.

When comparing the training curves of the two models, we can find that, under the same training parameters, although the convergence speed of the GAP-CNN network model is slightly slower than that of the classic CNN, the recognition of GAP-CNN on the test set of indoor harmful gases. The accuracy is slightly better than the classic CNN model, which also proves that the GAP-CNN network model can reduce the model parameters while reducing the occurrence of overfitting, and it can also ensure a certain model accuracy. By saving the GAP-CNN model to identify the test set again, it is found that there are three main reasons why the classification accuracy rate cannot reach 100%. The first is that GAP-CNN will make a small part of the formaldehyde concentration in the strong interference group close to 0.08 mg/m. The Normal class is incorrectly identified as the Mild class. The second reason is that GAP-CNN will incorrectly classify a small part of the Mild and Serious, especially gas samples whose concentration is at the classification boundary. For example, gas samples containing Mild's 0.14 mg/m^3 formaldehyde will be classified as Serious. The Serious class' 0.18 mg/m^3 will be classified into the Mild class. The third reason is that there is a high concentration of interfering gas, indoor harmful gas samples containing low concentration of formaldehyde gas, the model can easily identify it as a higher level of formaldehyde pollution level.

In order to further verify the effectiveness of the GAP-CNN algorithm, in Table 4, PLS model, PCA+LDA model, CNN model, t-SNE+RF model, and GAP-CNN model are used in order to calculate the average of the classification results while using five-fold cross-validation.

Table 4. Cross validation results of 5 models.

Algorithm	PLS	PCA+LDA	CNN	t-SNE+RF	GAP-CNN
Accuracy (%)	71.16	41.54	90.58	85.58	90.96

Through the analysis and research of the above experimental results, it can be known that the traditional PCA+LDA machine olfactory recognition method cannot effectively extract the effective features for the classification of formaldehyde pollution in indoor harmful gases. The PLS model can extract the effective features of formaldehyde pollution classification in indoor harmful gas to a certain extent, but, in the presence of other indoor harmful gas interference, the accuracy of algorithm recognition is low. The t-SNE+RF algorithm is less accurate than the classification that is based on the CNN neural network. In the CNN and GAP-CNN models, although the classification accuracy of the Normal level of formaldehyde pollution is not as good as t-SNE+RF, as compared with the t-SNE+RF algorithm, it is found that the advantage of the CNN neural network formaldehyde pollution classification is that CNN When the neural network recognizes the level of formaldehyde pollution of Mild and Serious, it has higher classification accuracy. At the same time, GAP-CNN has better generalization performance than t-SNE+RF.

By observing the results of cross-validation experiments, it can be found that CNN, t-SNE+RF, and GAP-CNN have an ideal effect on identifying the level of formaldehyde pollution in indoor harmful gases. Therefore, this paper uses the aboved algorithm to compare and analyze the total computing time of data preprocessing, feature extraction, model training, and output results in order to further compare the characteristics and performance of the three machine learning algorithms to identify machine olfactory odor data. The operation time presented in Table 5 shows that GAP-CNN algorithm needs to go through multi-layer convolution operation, and the convergence speed is slow, which leads to greater time consumption of GAP-CNN algorithm.

Table 5. Computational time comparison of discrimination algorithm.

Algorithm	Computing Time (s)
CNN	19.3968
t-SNE+RF	12.1842
GAP-CNN	25.7133

5. Discussion

This paper proposes an indoor harmful gas component analysis algorithm that is based on the combination of CNN and bionic olfactory. This method uses one dimensional convolutional neural network weight sharing and adding a global maximum pooling layer, so that the neural network has a higher recognition rate for indoor harmful gases when the number of training parameters is small. The research of this algorithm is of great significance to the solution of the subsequent concentration estimation problem of the bionic olfactory system. The algorithm that is proposed in this paper has not been well verified in the concentration regression experiment. In the process of the experiment, the influence of external factors on the experimental results has not been considered. This will be the direction of follow-up research.

6. Conclusions

In this work, a novel component analysis strategy was proposed for formaldehyde pollution in harmful indoor gases based on machine olfaction. While using a portable electronic nose system, indoor harmful gas samples' odor information is collected and processed to achieve an real-time monitoring. By converting the smell information of the machine olfaction into a one-dimensional time series, the machine olfactory analysis method GAP-CNN that is based on global average pooling and one-dimensional convolutional neural network is innovatively proposed. Additionally, through the comparison of the experimental results, in the case of compressing the model volume, the recognition accuracy of the model is guaranteed, which proves the performance of the GAP-CNN model. At the same time, this paper uses the GAP-CNN model and the classic CNN model to build an indoor hazardous gas formaldehyde pollution classification model. Through comparing the experimental results, it is found that the accuracy of the two classification models is greater than 90%, and the weight parameters of the GAP-CNN model are much lower.

Author Contributions: The work described in this article is the collaborative development of all authors. C.Z. and J.H. contributed to the idea of data processing and designed the algorithm. D.L. and M.P. made contributions to data measurement and analysis. C.Z. and M.P. participated in the writing of the paper. All authors have read and agreed to the published version of the manuscript.

Funding: The authors acknowledge the financial support of the General Program of the National Natural Science Fund (Grant No. 61571140), Guangdong Provincial Project (Grant No. 2016A020226018), Guangdong Provincial Administration of traditional Chinese Medicine Project (Grant No. 20161152), Guangdong University Project (Grant No. 51348000).

Data Availability Statement: All data generated or appeared in this study are available upon request by contact with the corresponding author. Furthermore, the models and code used during the study cannot be shared at this time as the data also forms part of an ongoing study.

Conflicts of Interest: The authors declare no conflict of interest.

Abbreviations

The following abbreviations are used in this manuscript:

CNN	Convolutional Neural Network
KNN	K-Nearest Neighbor
WCLS	Weighted and Constrained Least-Squares
KNN	K-Nearest Neighbor
GMP	Global Maxing Pooling
PCA	Principal Component Analysis
LDA	Linear Discriminant Analysis
GAP	Global Average Pooling

References

1. Tury, E.L.; Kaste, K.; Johnson, R.E.; Danielson, D.O. Non-Dispersive Infrared Gas Analyzer System. U.S. Patent 5,060,505, 29 October 1991.
2. Kmita, A.; Fischer, C.; Hodor, K.; Holtzer, M.; Roczniak, A. Thermal decomposition of foundry resins: A determination of organic products by thermogravimetry–gas chromatography–mass spectrometry (TG–GC–MS). *Arab. J. Chem.* **2018**, *11*, 380–387. [CrossRef]
3. Makino, A.; Mae, T.; Ohira, K. Colorimetric measurement of protein stained with Coomassie Brilliant Blue R on sodium dodecyl sulfate-polyacrylamide gel electrophoresis by eluting with formamide. *Agric. Biol. Chem.* **1986**, *50*, 1911–1912.
4. Pungor, E.; Tóth, K. Ion-selective membrane electrodes. A review. *Analyst* **1970**, *95*, 625–648. [CrossRef]
5. Yang, X.; Yuan, L.; Wang, Y. Quantitative detection for gas mixtures based on the adaptive genetic algorithm and BP network. In Proceedings of the 2012 International Conference on Industrial Control and Electronics Engineering, Xi'an, China, 23–25 August 2012; pp. 1341–1344.
6. Fan, S.; Li, Z.; Xia, K.; Hao, D. Quantitative and qualitative analysis of multicomponent gas using sensor array. *Sensors* **2019**, *19*, 3917. [CrossRef] [PubMed]
7. Hua, Z.; Tian, C.; Qiu, Z.; Li, Y.; Tian, X.; Wang, M.; Li, E. An investigation on NO_2 sensing mechanism and shielding behavior of WO_3 nanosheets. *Sens. Actuators B Chem.* **2018**, *259*, 250–257. [CrossRef]
8. Kim, K.H. Experimental demonstration of masking phenomena between competing odorants via an air dilution sensory test. *Sensors* **2010**, *10*, 7287–7302. [CrossRef] [PubMed]
9. Lu, Y.; Bian, L.; Yang, P. Quantitative artificial neural network for electronic noses. *Anal. Chim. Acta* **2000**, *417*, 101–110. [CrossRef]
10. Tang, K.T.; Lin, Y.S.; Shyu, J.M. A local weighted nearest neighbor algorithm and a weighted and constrained least-squared method for mixed odor analysis by electronic nose systems. *Sensors* **2010**, *10*, 10467–10483. [CrossRef] [PubMed]
11. Goodfellow, I.; Bengio, Y.; Courville, A.; Bengio, Y. *Deep Learning*; MIT Press Cambridge: Cambridge, UK, 2016; Volume 1.
12. LeCun, Y.; Bottou, L.; Bengio, Y.; Haffner, P. Gradient-based learning applied to document recognition. *Proc. IEEE* **1998**, *86*, 2278–2324. [CrossRef]
13. Krizhevsky, A.; Sutskever, I.; Hinton, G.E. Imagenet classification with deep convolutional neural networks. In Proceedings of the Advances in Neural Information Processing Systems, Lake Tahoe, NV, USA, 3–6 December 2012; pp. 1097–1105.
14. Szegedy, C.; Liu, W.; Jia, Y.; Sermanet, P.; Reed, S.; Anguelov, D.; Erhan, D.; Vanhoucke, V.; Rabinovich, A. Going deeper with convolutions. In Proceedings of the IEEE Conference on Computer Vision and Pattern Recognition, Boston, MA, USA 7–12 June 2015; pp. 1–9.
15. Lawrence, S.; Giles, C.L.; Tsoi, A.C.; Back, A.D. Face recognition: A convolutional neural-network approach. *IEEE Trans. Neural Netw.* **1997**, *8*, 98–113. [CrossRef] [PubMed]
16. Tivive, F.H.C.; Bouzerdoum, A. An eye feature detector based on convolutional neural network. In Proceedings of the Eighth International Symposium on Signal Processing and Its Applications, Sydney, Australia, 28–31 August 2005.
17. Szarvas, M.; Yoshizawa, A.; Yamamoto, M.; Ogata, J. Pedestrian detection with convolutional neural networks. In Proceedings of the IEEE Proceedings. Intelligent Vehicles Symposium, Las Vegas, NV, USA, 6–8 June 2005; pp. 224–229.
18. Sabes, P.N.; Jordan, M.I. Advances in neural information processing systems. In *Advances in Neural Information Processing Systems*; Tesauro, G.D., Touretzky, T.L., Eds.; Citeseer: Cambridge, MA, USA, 1995.
19. Park, K.; Kim, D.H. Accelerating image classification using feature map similarity in convolutional neural networks. *Appl. Sci.* **2019**, *9*, 108. [CrossRef]
20. Cui, Y.; Zhou, F.; Wang, J.; Liu, X.; Lin, Y.; Belongie, S. Kernel pooling for convolutional neural networks. In Proceedings of the IEEE Conference on Computer Vision and Pattern Recognition, Honolulu, HI, USA, 21–26 July 2017; pp. 2921–2930.
21. Alaskar, H.; Hussain, A.; Al-Aseem, N.; Liatsis, P.; Al-Jumeily, D. Application of Convolutional Neural Networks for Automated Ulcer Detection in Wireless Capsule Endoscopy Images. *Sensors* **2019**, *19*, 1265. [CrossRef] [PubMed]
22. Sze, V.; Chen, Y.H.; Yang, T.J.; Emer, J.S. Efficient processing of deep neural networks. *Synth. Lect. Comput. Archit.* **2020**, *15*, 1–341. [CrossRef]
23. Lin, M.; Chen, Q.; Yan, S. Network in network. *arXiv* **2013**, arXiv:1312.4400.

Letter

0.5 V Fifth-Order Butterworth Low-Pass Filter Using Multiple-Input OTA for ECG Applications

Montree Kumngern [1], Nattharinee Aupithak [1], Fabian Khateb [2,3,*] and Tomasz Kulej [4]

[1] Department of Telecommunications Engineering, Faculty of Engineering, King Mongkut's Institute of Technology Ladkrabang, Bangkok 10520, Thailand; montree.ku@kmitl.ac.th (M.K.); yoknattharinee@gmail.com (N.A.)

[2] Department of Microelectronics, Brno University of Technology, Technicka 3058/10, 612 00 Brno, Czech Republic

[3] Faculty of Biomedical Engineering, Czech Technical University in Prague, nám. Sítná 3105, 272 01 Kladno, Czech Republic

[4] Department of Electrical Engineering, Czestochowa University of Technology, 42-201 Czestochowa, Poland; kulej@el.pcz.czest.pl

* Correspondence: khateb@feec.vutbr.cz; Tel.: +420-54114-6128

Received: 15 November 2020; Accepted: 14 December 2020; Published: 21 December 2020

Abstract: This paper presents a 0.5 V fifth-order Butterworth low-pass filter based on multiple-input operational transconductance amplifiers (OTA). The filter is designed for electrocardiogram (ECG) acquisition systems and operates in the subthreshold region with nano-watt power consumption. The used multiple-input technique simplifies the overall structure of the OTA and reduces the number of active elements needed to realize the filter. The filter was designed and simulated in the Cadence environment using a 0.18 μm Complementary Metal Oxide Semiconductor (CMOS) process from Taiwan Semiconductor Manufacturing Company (TSMC). Simulation results show that the filter has a bandwidth of 250 Hz, a power consumption of 34.65 nW, a dynamic range of 63.24 dB, attaining a figure-of-merit of 0.0191 pJ. The corner (process, voltage, temperature: PVT) and Monte Carlo (MC) analyses are included to prove the robustness of the filter.

Keywords: fifth-order low-pass filter; operational transconductance amplifier; multiple-input bulk-driven technique; subthreshold region; nanopower

1. Introduction

Continuous-time filters are widely used in biomedical systems devoted to applications in electroencephalographic (EEG), electromyographic (EMG), and electrocardiographic (ECG) systems. The biological signals processed in these systems typically occupy the frequency range of 0.05–250 Hz, with an amplitude of 15 μV–5 mV [1]. In more detail, the frequency/amplitude ranges for EEG, EMG, and ECG signals are 0.05–60 Hz/15–100 μV, 10–200 Hz/0.1–5 mV, and 0.05–250 Hz/100 μV–5 mV, respectively.   Figure 1 shows a typical data acquisition system for ECG signal processing. The pre-amplifier stage amplifies a low-amplitude ECG signal, then the low-pass filter selects the frequency range and eliminates out-of-band noise. The filtered analog signal is converted into digital form by an analog-to-digital converter (ADC) and then it is further processed by a digital signal processing (DSP) block. This work focused on the design of a low-pass filter with the cutoff frequency of 250 Hz. The analog low-pass filters for ECG acquisition systems should be designed to meet specific requirements, such as high dynamic range, low-power consumption, and small chip area. There are many low-pass filters for ECG acquisition systems described in the literature [2–10]. The Butterworth approximation is usually used because it provides a better linear phase and flat response within each bandwidth. Considering the analog filters in [2–10], one can distinguish two main techniques that have been used to realize the low-pass Butterworth filters: the cascade approach [2–6] and the ladder simulation approach [7–10]. The cascade structure can be obtained by cascading several biquad filters, which leads to a simple and easy-to-tune realization.

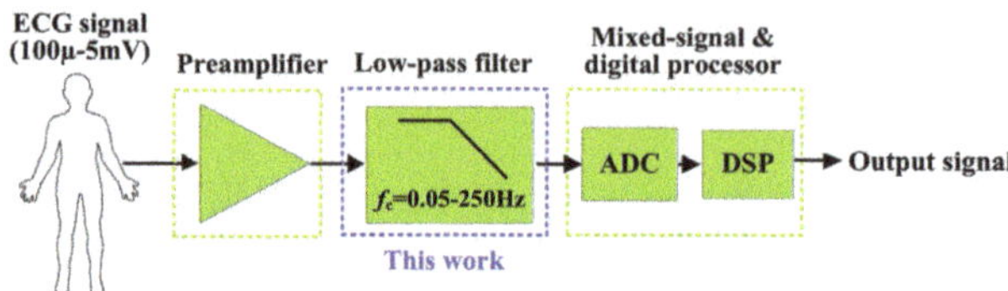

Figure 1. Electrocardiogram acquisition system.

The present work focused on the second approach, i.e., the ladder simulation of a prototype filter. In particular, we designed a fifth-order low-pass Butterworth filter based on the RLC prototype shown in Figure 2. As it is widely known, the high-order filters based on the RLC prototypes have lower pass-band sensitivity to the variation of passive elements, compared with that of the cascade approach.

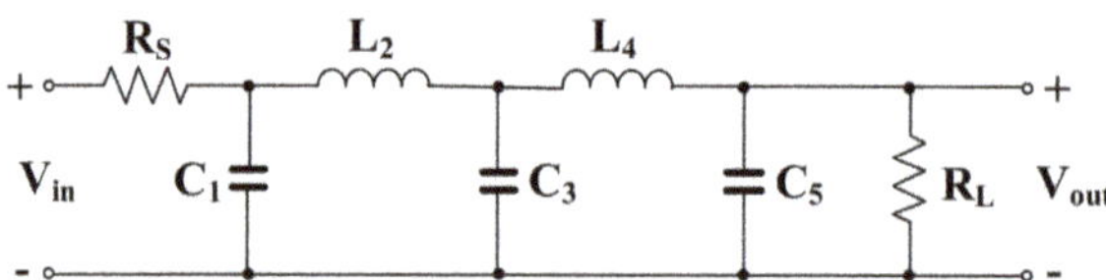

Figure 2. Prototype of a fifth-order low-pass filter.

The fifth-order low-pass Butterworth filters derived from the LC ladder-type filter were reported in [7–10]. The fifth-order Butterworth low-pass filter using fully differential operational transconductance amplifiers (FD-OTAs) is shown in Figure 3a [7]. The floating inductors L_2 and L_4 are simulated using OTA-based gyrators. The resistors R_S and R_L are simulated using OTAs as well. It should be noted that the filter in [7] employs eleven FD-OTAs and consumes 453 nW of power. The number of active devices that are used to realize this fifth-order Butterworth filter can be reduced by using multiple-output fully differential OTA (MOFD-OTA) as shown in Figure 3b [8,9], or fully differential-difference transconductance (FDDA) (a multiple-input active device) as shown in Figure 3c [10]. The structures in [8,9] employ six MOFD-OTA while the structure in [10] employs five FDDAs and one OTA. The filter in [8] consumes 350 nW of power and offers a 49.9 dB dynamic range while the filter in [9] consumes 41 nW of power and offers a 61.2 dB dynamic range. The filter in [10] consumes 453 nW of power and offers a 50 dB dynamic range.

This paper proposes a fifth-order Butterworth low-pass filter based on multiple-input operational transconductance amplifiers. It is clearly shown that the number of active devices needed to realize the fifth-order low-pass filter can be reduced by using the multiple-input OTA and results in reducing the power consumption and the active chip area. A novel technique with a multiple-input gate-driven (MIGD) transistor is used to realize multiple-input OTA with an internal CMOS structure as simple as a conventional OTA, hence, no additional current branches or cascade connections of multiple OTAs is needed. Unlike the floating-gate technique, the multiple-input technique does not require any additional processing steps to eliminate the trapped charge effect on the isolated gate nor any auxiliary circuit. Another advantage is that the multiple-input gate-driven P-orN-MOS transistors can be realized with any CMOS process. It is worth noting that the results presented in this work are based on pre-layout simulation and this work does not include the physical realization of the filter, nor the experimental testing in the context of ECG applications. However, the principle of multiple-input transistors, as multiple-input bulk-driven and multiple-input bulk-driven quasi-floating-gates, have been confirmed experimentally by Khateb et al. in previous works [11–13]. The paper is organized as follows: Section 2 shows the principle of multiple-input gate-driven OTA and the filter design based on it, Section 3 the simulation results, and finally Section 4 the conclusion.

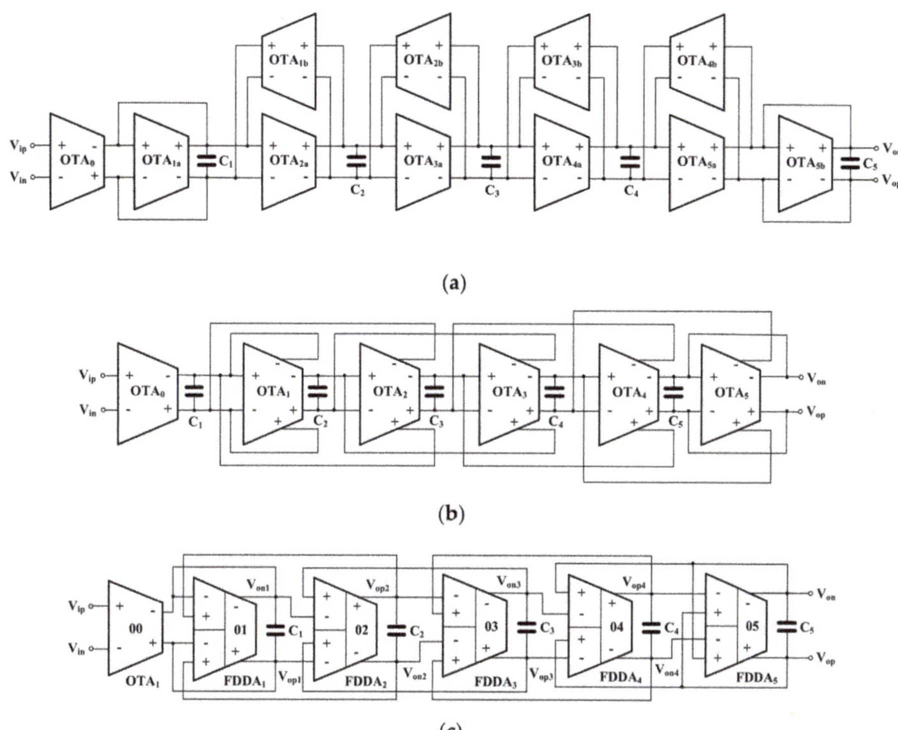

(a)

(b)

(c)

Figure 3. Fifth-order Butterworth low-pass filters, (a) FD-OTA-C filter [7], (b) MOFD-OTA-C filter [8,9], (c) FDDA-based filter [10].

2. Fifth-Order Butterworth Low Pass Filter

2.1. Multiple-Input Gate-Driven OTA

The active filter proposed in this work exploits multiple-input OTAs, which allows for simplifying its overall structure [14]. The multiple-input OTA is realized using a concept of a multiple-input MOS transistor. The symbol and CMOS realization of this element are shown in Figure 4a,b, respectively. As it is seen in Figure 4b, the multiple-input MOS can be seen as a connection of an "internal" MOS transistor and a voltage divider/analog summing circuit, composed of capacitances C_{Gi} ($i = 1 \ldots N$). The capacitors C_{Gi} are shunted by the large resistances R_{Li}, which ensures proper biasing of the gate terminal of the internal MOS for DC. The large resistances can be realized using an anti-parallel connection of two minimum-size MOS transistors operating in a cutoff region, as shown in Figure 4b. The small-signal equivalent circuit of the resulting multiple-input MOS is shown in Figure 4c. Assuming $1/\omega C_{Gi} \ll R_{Li}$, the gate potential V_G is given by

$$V_G = \sum_{i=1}^{N} \frac{C_{Gi}}{C_\Sigma} V_{ini}$$

(1)

where C_Σ is the sum of the capacitances C_{Gi} and the input capacitance of an internal MOS seen from its gate terminal C_{in}:

$$C_\Sigma = C_{in} + \sum_{i=1}^{N} C_{Gi} \tag{2}$$

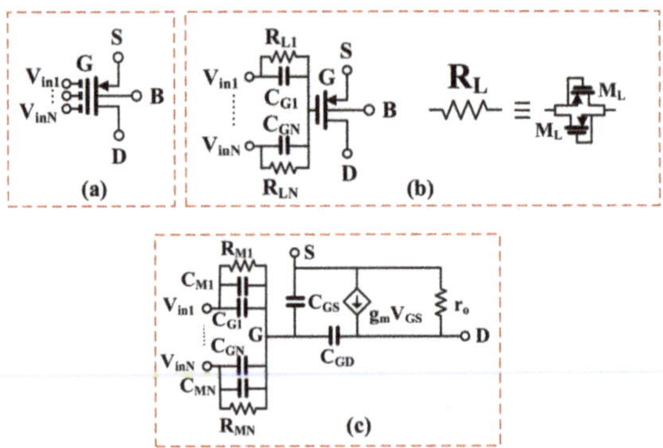

Figure 4. MIGD MOS transistor, (**a**) symbol, (**b**) realization, (**c**) small-signal model.

Since the AC signal at the gate of the internal MOS transistor is attenuated by the capacitive divider, the transconductance of the multiple-input device seen from its *i*-th input, and operating in the subthreshold region, can be expressed as:

$$g_{mi} = \frac{I_D}{n_p U_T} \cdot \frac{C_{Gi}}{C_\Sigma} \tag{3}$$

where I_D is the DC drain current, n_p. is the subthreshold slope, and U_T. is the thermal potential. As it is seen from (3), the transconductance seen from the *i*-th input is equal to the transconductance of the internal MOS, multiplied by the voltage gain of the capacitive voltage divider.

The lower input transconductance g_{mi} entails a lower intrinsic voltage gain of the multiple-input MOS, as well as an increased input-referred noise. Both parameters are degraded by the factor of C_Σ/C_{Gi}. However, it is worth noting that the linear range for such a device is also increased by the factor of C_Σ/C_{Gi}, therefore, its dynamic range (DR) remains the same as that of the internal MOS.

The multiple-input MOS transistors were used to design a multiple-input OTA. The symbol and CMOS realization of the circuit are shown in Figures 5 and 6, respectively. The multiple-input MOS transistors M_1 and M_2 were used to create a multiple-input differential pair, biased by the self-cascode current sources $M_{7,7c}$ and $M_{8,8c}$. The drain currents of the input differential pair are transferred to the outputs (I_{o+} and I_{o-}) through the current mirrors composed of the self-cascode transistors $M_{3/3c}$-$M_{4/4c}$ and $M_{5,5c}$-$M_{6,6c}$. The current mirrors are loaded with the self-cascode current sources $M_{10,10c}$ and $M_{9,9c}$. Note that the tail node that supplies the differential pair in Figure 6 is drawn with two branches for esthetic reasons. The application of self-cascode connections in this design allows for an increase in the output resistance of the OTA, which entails increasing the DC voltage gain of this circuit. The transistors M_{9c}-M_{11c} form a simple common-mode feedback circuit (CMFB) circuit, which forces the output common-mode level to be equal to the reference potential V_{CM}. All the transistors operate in a subthreshold triode region. If the common-mode level is increasing/decreasing, the channel resistances of $M_{10C1,c2}$ are increasing/decreasing as well, thus lowering the currents flowing through M_{10} and M_9, and consequently, decreasing/increasing the common-mode level to the desired value. The transistors M_{9c} and M_{10c} are divided into two devices, which makes the circuit insensitive to the output differential signals of the OTA, at least for small amplitudes of the signal. For larger amplitudes of the output signals, one can observe nonlinear components of the drain currents I_{D9} and I_{D10}, caused by the differential output voltage of the OTA. However, this nonlinear effect is not apparent at the differential output of OTA, since variation of I_{D9} and I_{D10} are identical. This effect, however, causes variation of the output common-mode level. Figure 7 illustrates the large signal transfer characteristics and the common-mode level variation for unloaded OTA in Figure 6 controlled with differential signals. Note, moderate nonlinear effects are caused by the nonlinear output conductance of the OTA rather than that of the CMFB. Variations of the common-mode output voltage are maintained at an acceptable level.

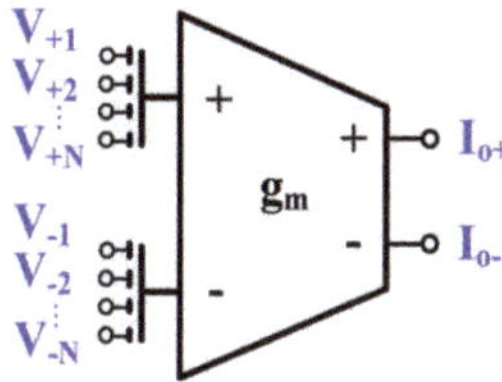

Figure 5. Symbol of a multiple-input operational transconductance amplifier (OTA).

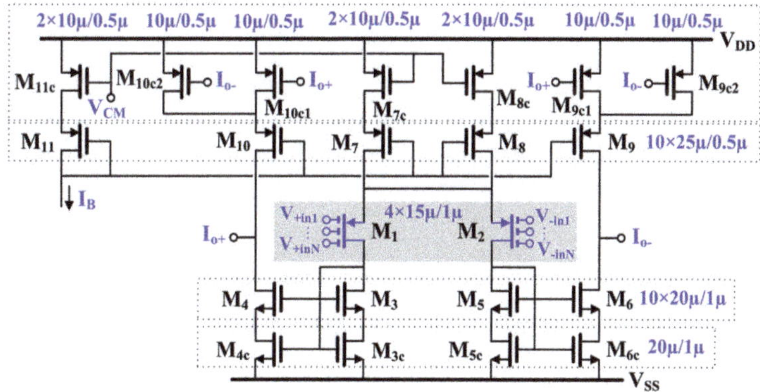

Figure 6. CMOS implementation for an MIGD OTA.

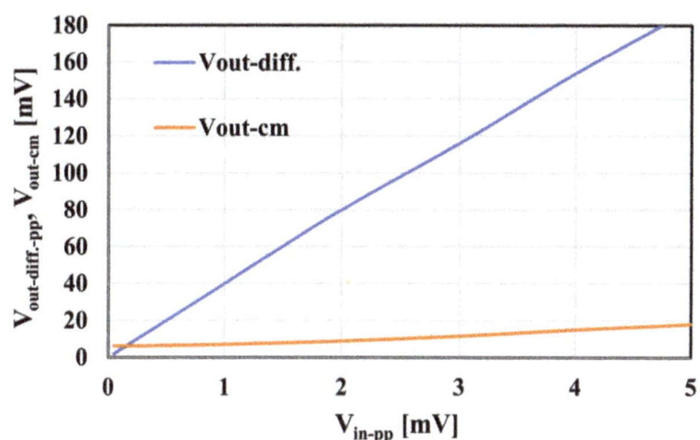

Figure 7. Output differential voltage and common-mode level versus input differential voltage for unloaded OTA in Figure 6.

One can say that the applied CMFB has a simple structure and does not consume additional power from supply rails. On the other hand, it slightly limits the maximum output voltage swing due to nonzero voltage drops across transistors M_{9c}–M_{11c} and variations of the output common-mode level caused by differential signals. However, the negative effects can be maintained at an acceptable level.

Assuming $1/\omega C_{Gi} \ll R_{Li}$, the differential output current of the OTA can be expressed as:

$$I_{o+} - I_{o-} = I_B \tanh\left(\sum_{i=1}^{N} \frac{V_{+ini} - V_{-ini}}{n_p U_T} \cdot \frac{C_{Gi}}{C_\Sigma}\right) \tag{4}$$

where I_B is the biasing current (it was assumed that $I_{D7} = I_{D8} = I_{D11}$). From (4), the small-signal transconductance from i-th input is given by:

$$g_{mi} = \frac{I_D}{n_p U_T} \cdot \frac{C_{Gi}}{C_\Sigma} \tag{5}$$

The DC voltage gain of the OTA from the i-th input can be expressed as:

$$A_{vd} = g_{mi} r_{out} \tag{6}$$

where r_{out} is the output resistance of the OTA, given by:

$$r_{out} \cong g_{m4,6} r_{ds4,6} r_{ds4c,6c} \big\| g_{m9,10} r_{ds9,10} \left(r_{ds9,10c}/2\right) \tag{7}$$

Thanks to the self cascode connections, the voltage gain of the OTA can be at an acceptable level, despite the lower transconductance of the input differential pair.

From (4), the third order harmonic distortion of the OTA for a sinusoidal signal applied to one pair of input terminals, while the other pairs are shorted to ground the AC signals, can be expressed as:

$$HD_3 = \frac{1}{48}\left(\frac{V_{+i} - V_{-i}}{n_p U_T} \cdot \frac{C_{Gi}}{C_\Sigma}\right)^2 \tag{8}$$

Thus, as it is seen from (8), the input linear range is increased by the factor of C_Σ/C_{Gi}, i.e., the voltage attenuation factor introduced by the input capacitive divider.

The input referred noise of the OTA, including both thermal and flicker noise components, can be expressed as:

$$\overline{v_{nti}^2} = 2\left(\frac{U_T}{I_B}\right)^2 \left(\frac{C_\Sigma}{C_{Gi}}\right)^2 \left[\overline{i_{1,2}^2} + 2\overline{i_{3-6}^2} + 2\overline{i_{9,10}^2}\right] \tag{9}$$

where:

$$\overline{i_{1,2}^2} = 2qI_B + \frac{K_{Fp}\left(\frac{I_B}{U_T}\right)^2}{f C_{OX}(WL)_{1,2}} \tag{10}$$

$$\overline{i_{3-6}^2} = 4kT g_{ds3-6c}\left(1 + \frac{2}{3}\frac{g_{ds3-6c}}{g_{m3-6}}\right)\frac{(g_{m3-6} r_{ds3-6c})^2}{(1 + g_{m3-6} r_{ds3-6c})^2} + \frac{1}{4} \cdot \frac{K_{Fn}\left(\frac{I_B}{U_T}\right)^2}{f C_{OX}\left[(WL)_{3-6eff}\right]} \tag{11}$$

$$\overline{i_{10-14}^2} = 4kT g_{ds9-10c}\left(1 + \frac{2}{3}\frac{g_{ds9-10c}}{g_{m9-10}}\right)\frac{(g_{m9-10} r_{ds9-10c})^2}{(1 + g_{m9-10} r_{ds9-10c})^2} + \frac{1}{4} \cdot \frac{K_{Fp}\left(\frac{I_B}{U_T}\right)^2}{f C_{OX}\left[(WL)_{9,10eff}\right]} \tag{12}$$

where $g_{ds9-10c} = g_{ds9-10c1}//g_{ds9-10c1}$, $WL_{ieff} = (WL_i \cdot WL_{ic})/(WL_i + WL_{ic})$, $i = 3 \ldots 10$, $WL_{9,10c} = WL_{9,10c1} + WL_{9-10c2}$, K_{Fn} and K_{Fp} are the flicker noise constants for n- and p-channel transistors, respectively, and C_{OX} is the oxide capacitance per unit area.

As it is easy to note from (9), the input referred noise is increased by the factor of C_Σ/C_{Gi}, as compared with the input noise of a single-input OTA biased with the same current. However, if the multiple input OTA is realized with N identical OTAs, each biased with the current of I_B/N, then the input transconductance from each input and the input referred noise would be the same as that for the proposed realization (see the Appendix A). Since the linear range in the proposed design is increased C_Σ/C_{Gi} times, then the DR of the proposed solution is also increased in the same proportion. The improved DR can be considered as the most important advantage of the proposed approach. Note that a similar capacitive attenuation approach that increase the dynamic range of OTAs has been presented before [15].

2.2. Proposed Filter

The proposed fifth-order Butterworth low-pass filter is shown in Figure 8a. It was developed from the LC-ladder filter based on the OTA-C topology. Its signal flow graph is shown in Figure 8b, where $\tau_1 = C_1/g_{m1}$, $\tau_2 = C_2/g_{m2}$, $\tau_3 = C_3/g_{m3}$, $\tau_4 = C_4/g_{m4}$, and $\tau_5 = C_5/g_{m5}$. The filter comprises five MIGD OTAs and five capacitors. The number of active devices is reduced from 6 to 5, as compared with [8–10], which allows for the reduction of the active area and power.

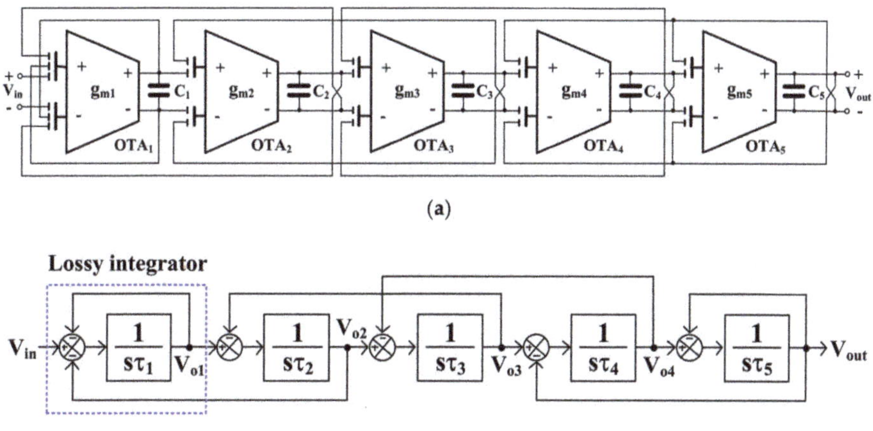

(a)

(b)

Figure 8. (a) Proposed fifth-order Butterworth low-pass filter, (b) signal flow graph.

Considering OTA_0, OTA_1 in Figure 3b and OTA_0, $FDDA_1$ in Figure 3c, it can be noted that these devices are used to realize a floating resistor [9]. In this work these components together with the capacitor C_1 create a lossy integrator as shown in Figure 9a [8], Figure 9b [10]. The ideal transfer function of these circuits can be expressed as:

$$\frac{V_{op1} - V_{on1}}{V_{ip} - V_{in}} = \frac{g_{mo}/g_{m1}}{(sC_1/g_{m1}) + 1} \tag{13}$$

It is evident that the circuits work as lossy integrators, where the voltage gain can be controlled by g_{mo}. Usually, all transconductances are set to be equal for easy tuning. Figure 9c shows the lossy integrator based on the three-input OTA that is proposed in this paper. The ideal transfer function of the circuit in Figure 9c can be expressed as:

$$\frac{V_{op1} - V_{on1}}{V_{ip} - V_{in}} = \frac{1}{(sC_1/g_{m1}) + 1} \tag{14}$$

Thus, the circuit works as a lossy integrator with unity gain. Assuming that $g_{mo} = g_{m1}$, Equations (13) and (14) will be identical. Thus, it can be concluded that the OTA_0 in Figure 3b,c can be removed by using multiple-input OTA. This application can only be realized using multiple-input OTA and it is not possible by using conventional OTA. It should be noted that only the parts mentioned above in Figure 9a of [8], Figure 9b of [10] are modified, the other parts (OTA$_{2-5}$ or FDDA$_{2-5}$) are not changed and the feedback connection is still similar to the filters in [8,10].

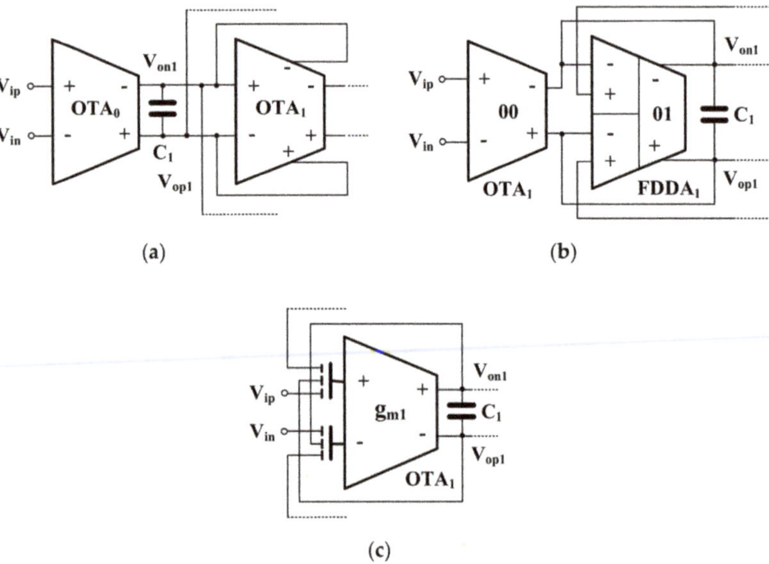

Figure 9. Lossy integrator, (**a**) circuit in [8], (**b**) circuit in [10], (**c**) proposed circuit.

3. Results and Discussion

The circuit was designed in the Cadence environment using a TSMC 0.18 µm CMOS process with a metal-insulator-metal (MIM) capacitor. The OTA with bias current I_B = 3.3 nA consumes 8.25 nW under a 0.5 V supply voltage. The isolation between OTA inputs is assured by the large value resistance of the MOS transistor operating in a cutoff region. The input currents are well below 100 pA for input range rail-to rail.

The RLC filter in Figure 2 was designed for the cut-off frequency of 250 Hz. The prototype element values were chosen as follows: R_S = R_L = 1 Ω, C_1 = C_5 = 393.4 µF, C_3 = 1.27 mF, and L_2 = L_4 = 1.03 mH. For the OTA-C filter C_1 = C_5 = 5.43 pF, C_2 = C_4 = 14.2 pF, C_3 = 17.57 pF, and the bias current for each OTA was I_B = 3.3 nA. Note that the bias current circuit serves to bias all OTAs hence the maximum power consumption of the filter is 34.65 nW. Figure 10 shows the frequency responses of the RLC and the proposed filter. The gain magnitude at low frequency was −6 dB and −6.4 dB and the cut-off frequency (f_c) was 250.2 Hz and 250.4 for the RLC and OTA filters, respectively. Both curves are in good agreement up to −70 dB. Figure 11 shows the frequency response of the filter with different bias currents ranging from 0.1 nA to 3.3 nA while the f_c was in the range of 17.11 Hz to 250.4 Hz. The tuning capability and the linear relation between f_c and I_B are demonstrated in Figure 12. The transient response of the filter for the input sine wave of V_{inpp} = 100 mV and 10-Hz frequency are illustrated in in Figure 13. The total harmonic distortion (THD) was 1%.

To check the influence of the process, voltage, and temperature (PVT) variations on the filter performance, the corner analysis was performed. The MOS transistor corners (ss, sf, fs, ff), MIM capacitor corners (ss, ff), voltage supply corners (490 mV, 510 mV), and temperature corners (0 °C, 60 °C) were used. The variation of the gain was in the range of −7.2 dB to −6.13 dB while the variation of the cut-off frequency was in the range of 100.6 Hz to 326.7 Hz, as shown in Figure 14. Note that the temperature corner has the most effect of the variation of the frequency response since the circuit operates in a subthreshold region. However, since the circuit is proposed for biomedical applications it is expected that the temperature variation will be less than the chosen temperature corners. Although the variation of the cut-off frequency is large, the needed value can be simply re-adjusted by the bias current. Note that the amplitudes of the bumps at low bias currents in Figure 11 and at higher frequencies in Figure 14 do not exceed 1.6 dB and do not affect stability of the circuit in a significant manner.

The Monte Carlo analysis with 200 runs was performed for the filter gain and cut-off frequency as shown in Figures 15 and 16, respectively. The mean value of the gain was −6.23 dB with standard deviation of 0.14 dB, while the mean value of the cut-off frequency was 251.7 Hz with standard deviation of 4.9 Hz. Figure 17 shows the output referred noise density of the filter. The integrated in-band noise between 0.1 Hz to 250 Hz shows that the output referred noise is 77 μV_{rms}. Figure 18 shows the performance of the proposed filter in processing the ECG signal where (a) depicts the ECG signal with a distortion signal (5 mV/500 Hz) that was applied at the input of the filter and (b) depicts the filtered output signal.

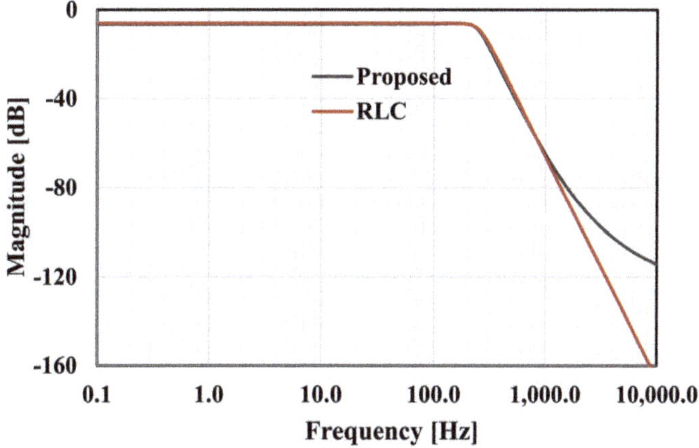

Figure 10. The frequency response of the RLC and the proposed filter.

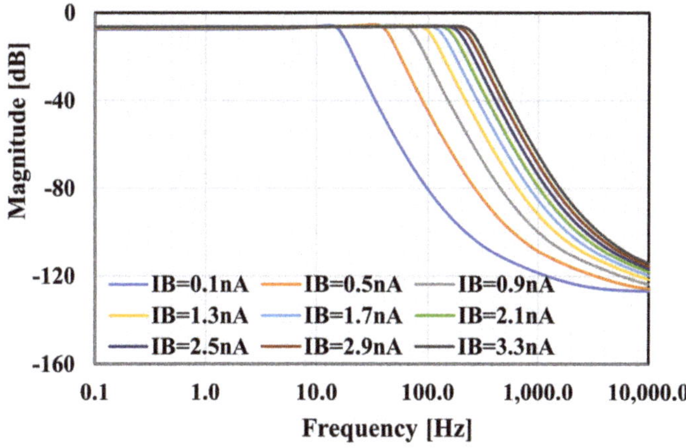

Figure 11. The frequency response of the proposed filter with different bias currents.

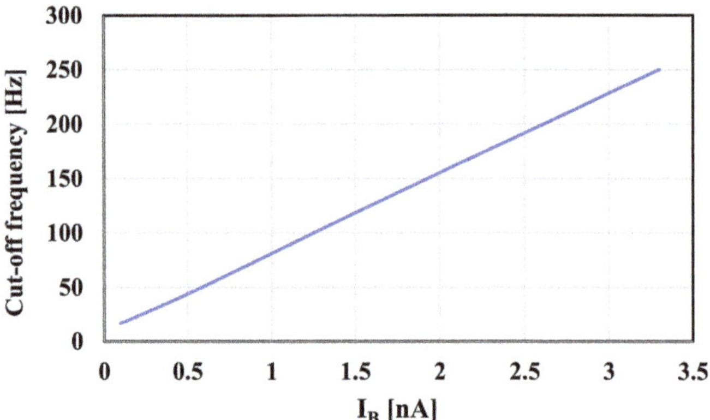

Figure 12. The cut-off frequency versus the bias current.

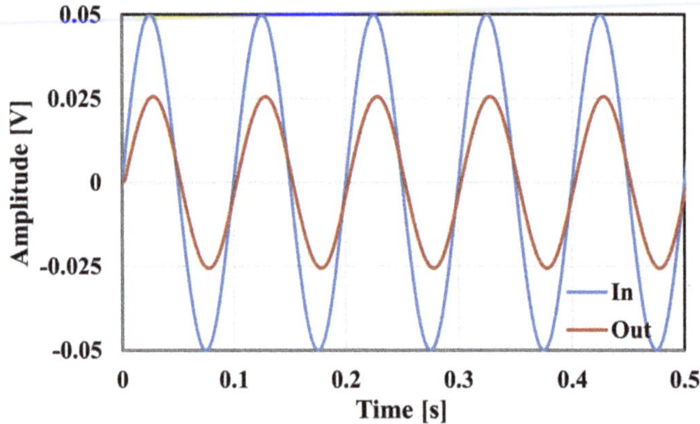

Figure 13. The transient response of the filter for input sine wave with $V_{inpp} = 100$ mV and 10 Hz.

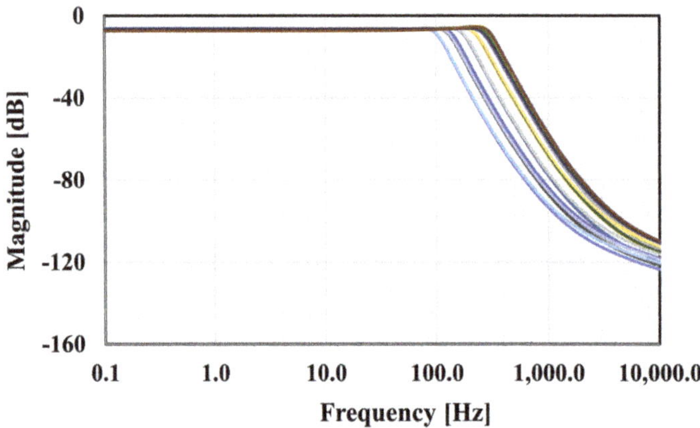

Figure 14. The frequency response of the proposed filter under process, voltage and temperature (PVT) corners.

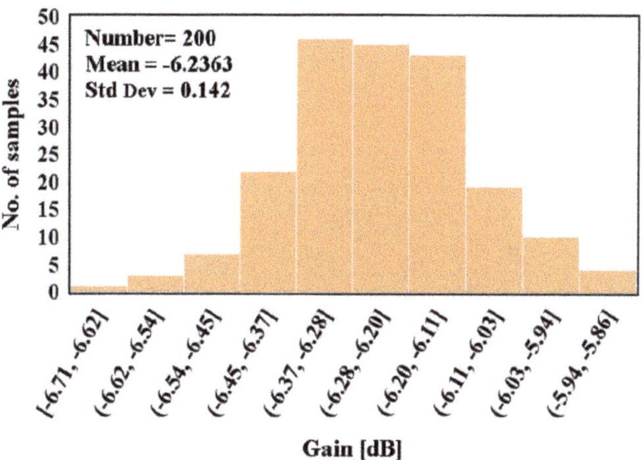

Figure 15. Monte Carlo simulation of the voltage gain.

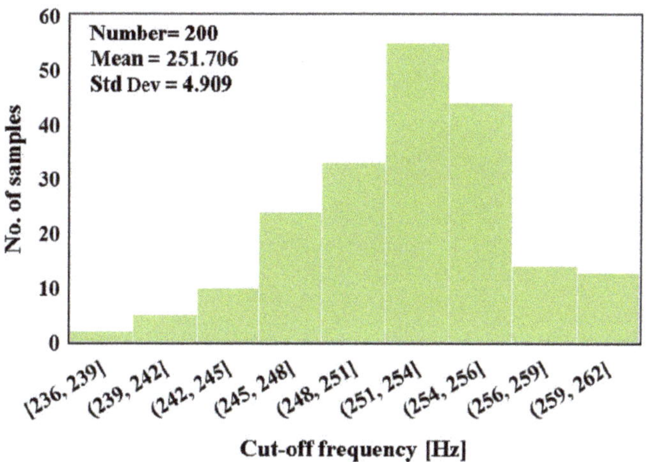

Figure 16. Monte Carlo simulation of the cut-off frequency.

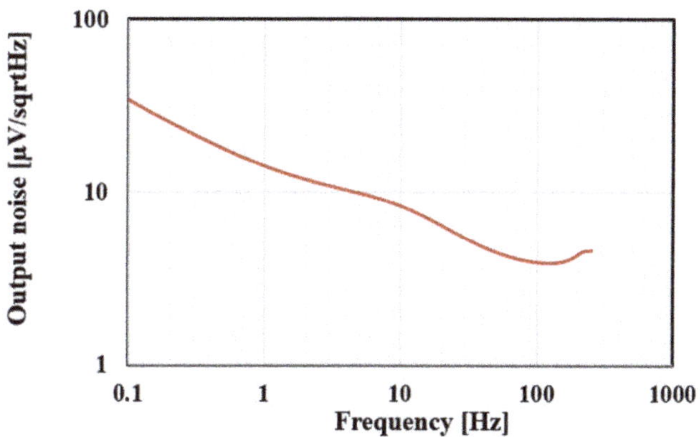

Figure 17. The output referred noise density of the proposed filter.

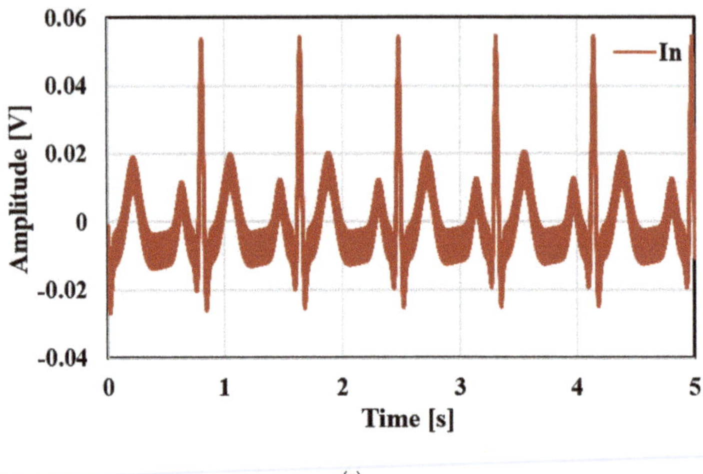

(a)

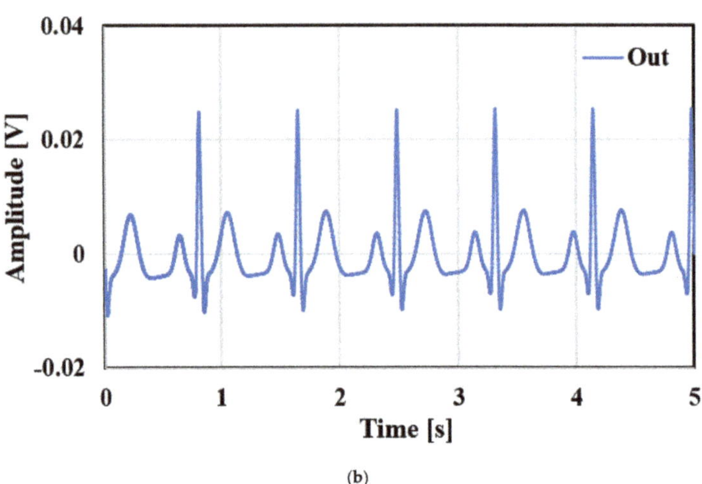

(b)

Figure 18. Transient response of the filter for ECG signal: (**a**) input; (**b**) output.

The summary and comparison between the proposed filter and some previous works are shown in Table 1. Only the fifth-order Butterworth low-pass filters simulated by the LC-ladder type filter and suitable for ECG signal acquisition [7–10] have been selected for comparison. From Table 1, it is clear that the proposed filter has a lower number of active devices, power consumption, and figure-of-merit (FOM). Finally, the FOM versus V_{DD} of fifth-order low-pass filters are shown in Figure 19. Compared with the works in [7,8,10], the proposed filter offers clearly better FOM. The FOM is even slightly lower than the one in [9] with half the value of V_{DD}. It is worth noting that the estimated chip area of 2-inputs and 3-inputs OTA based on the MIGD technique is increased by approximately 5% and 8%, respectively, compared to that of a single-input conventional OTA with the same transistor dimensions. This confirms the advantage of this technique of saving chip area. Note, a similar conclusion of this advantage based on experimental results is stated in [11]. The small chip area of the proposed filter is evident in Table 1 compared with that of [10] that used off-chip capacitors for filter realization.

Table 1. Performance comparison between the proposed filter and other fifth-order low-pass filters for ECG signal acquisition.

Symbol	This Work	MEJ (2019) [10]	IEEE TBioCAS (2019) [9]	IEEE TCAS-II (2018) [8]	IEEE TBioCAS (2009) [7]
V_{DD} [V]	0.5	0.25	1	1	1
Tech [um]	0.18	0.13	0.18	0.18	0.18
V_{TH} [V]	0.5	0.44	0.5	0.5	0.5
Order (N)	5	5	5	5	5
No. of active device	5 MIGD-OTAs	6 FDDTAs	6 OTAs	6 OTAs	11 OTAs
Structure	G_m-C fully-diff.	G_m-C fully-diff.	G_m-C fully-diff.	G_m-C fully-diff.	G_m-C fully-diff.
BW [Hz]	250	100	250	250	250
IRN [μV_{rms}]	167	4.7	134	100	300
DR [dB]	63.24	57.00	61.2	49.9	50
Power (P) [nW]	34.65	603	41	350	453
FOM = P/(N * BW * DR) [pJ]	0.0191	1.7	0.0286	0.896	1.15
LV capability = V_{TH}/V_{DD} * 100 [%]	100	176	50	50	50
Area [mm^2]	0.08 (estim.) (off-chip cap.)	0.67 (off-chip cap.)	0.24	0.12	0.13
Obtained results	Simulation	Measured	Measured	Measured	Measured

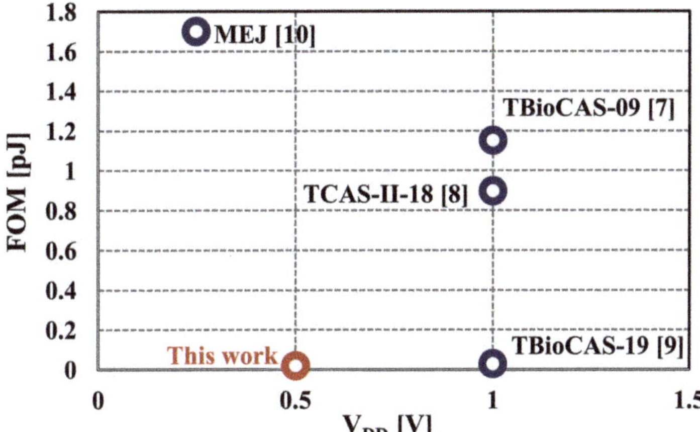

Figure 19. Figure-of-merit (FOM) against V_{DD} of the fifth-order low-pass filters.

4. Conclusions

In this paper, a fifth-order Butterworth low-pass filter using multiple-input OTA was proposed. The design proves that the number of OTAs for realizing the fifth-order low-pass filter architecture can be reduced using multiple-input OTAs. This entails the reduction of both the power consumption and the active area. Comparison with other designs in the literature shows that the proposed structure is the most beneficial, regarding the number of active devices and power consumption. The proposed filter was simulated with a 0.18 μm CMOS process and supplied with 0.5 V, which entailed operation in a subthreshold region. Simulation results including PVT corner and Monte Carlo (MC) analyses confirmed the robustness of the design.

Author Contributions: Conceptualization, F.K. and M.K.; methodology, M.K. and T.K.; software, F.K.; validation, F.K., N.A., and M.K.; formal analysis, M.K. and T.K.; investigation, F.K., M.K., and T.K.; writing—original draft preparation, M.K. and F.K.; writing—review and editing, M.K., F.K., and T.K. All authors have read and agreed to the published version of the manuscript.

Funding: This work was supported by King Mongkut's Institute of Technology Ladkrabang under grant KREF026201. For the research, the infrastructure of the SIX Center was used.

Conflicts of Interest: The authors declare no conflict of interest.

Appendix A

Let us compare noise properties of a fully-differential OTA in Figure A1a, biased with current I_B, and a multiple-input OTA composed of n identical OTAs of the same structure, but biased with currents of I_B/n (Figure A1b). For simplicity, let us consider only the thermal noise.

The mean-square value of the output noise current of the OTA in Figure A1a, operating in a weak-inversion region can be expressed as:

$$\overline{i_n^2} = 2qI_B A \tag{A1}$$

where q is the electron charge and A is a constant depending on the particular structure of the OTA. Consequently, the input referred noise is given by:

$$\overline{v_n^2} = \frac{2qI_B A}{g_m^2} \tag{A2}$$

The output noise current of each OTA in Figure A1b is:

$$\overline{i_n^2} = 2q\frac{I_B}{N}A \tag{A3}$$

However, the total output noise current, equal to the sum of N output currents, is the same as that for the reference OTA in Figure A1a. If the total output noise is referred to one input, we obtain:

$$\overline{v_{ni}^2} = \frac{\overline{i_n^2}}{\left(\frac{g_m}{N}\right)^2} = \frac{2qI_B A}{g_m^2}N^2 \tag{A4}$$

Thus, the rms value of the input noise is given by:

$$\sqrt{\overline{v_{ni}^2}} = N\sqrt{\frac{2qI_B A}{g_m^2}} \tag{A5}$$

If a noiseless passive voltage divider, with N inputs, and a voltage gain of 1/N from each input is added at the input of the OTA in Figure A1a, then the rms value of the i-th input referred noise voltage is given by the same equation, namely, the i-th input-referred noise is the same as that for the OTA in Figure A1b.

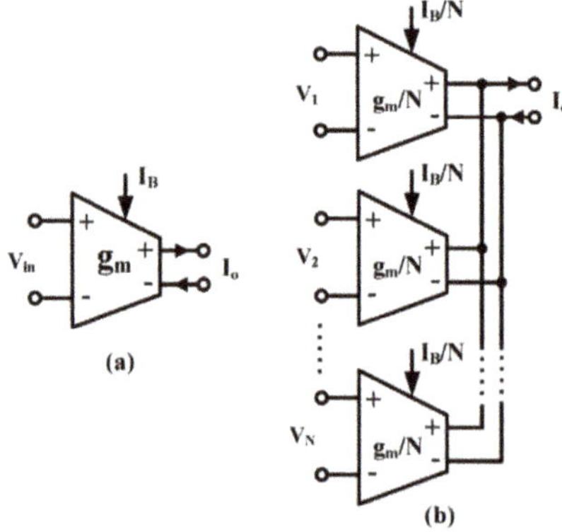

Figure A1. Single-input fully-differential OTA (**a**) and multiple-input fully-differential OTA (**b**).

If we define the dynamic range as the ratio of the maximum input rms voltage, limited by an assumed level of nonlinear distortion (V_{inmax}) to the *i*-th input referred noise, then for the multiple-input OTA in Figure A1b we have:

$$DR_{mi} = \frac{V_{inmax}}{\sqrt{\overline{v_{ni}^2}}} \tag{A6}$$

Since in the subthreshold region the linear range of a differential pair does not depend on the biasing current, then for the OTA in Figure A1a, with an additional passive voltage divider, the linear range will be extended N times, and the DR will be:

$$DR_{pd} = N \cdot \frac{V_{inmax}}{\sqrt{\overline{v_{ni}^2}}} = N \cdot DR_{mi} \tag{A7}$$

Hence, the dynamic range of the OTA with a passive, noiseless voltage divider at the input is N times as large as that for the OTA composed of N identical OTAs, biased with N-times lower current. Similar proof could be concluded for flicker noise, however, in such a case not only the total biasing current, but also the total areas of transistor channels should be equal for the two compared circuits, i.e., the transistor channel areas of each OTA in Figure A1b should be N times smaller than for that of the OTA in Figure A1a.

References

1. Webster, J.G. *Medical Instrumentation: Application and Design*, 4th ed.; John Wiley & Sons: Hoboken, NJ, USA, 2010; pp. 11–12.
2. Zhang, T.-T.; Mak, P.-I.; Vai, M.I.; Mak, P.-U.; Law, M.-K.; Pun, S.-H.; Wan, F.; Martins, R.P. 15-nW biopotential LPFs in 0.35-μm CMOS using subthreshold-source-follower biquads with and without gain compensation. *IEEE Trans. Biomed. Circuits Syst.* **2013**, *7*, 690–702. [CrossRef] [PubMed]
3. Sawigun, C.; Thanapitak, S. A 0.9-nW, 101-Hz, and 46.3-μV$_{rms}$ IRN low-pass filter fir ECG acquisition using FVF biquad. *IEEE Trans. Very Large Scale Integr. Syst.* **2018**, *26*, 2290–2298. [CrossRef]
4. Sawigun, C.; Thanapitak, S. A subthreshold buffer-based biquadratic cell and its application to biopotential filter design. *IEEE Trans. Circuits Syst. I Regul. Pap.* **2018**, *65*, 2774–2783.
5. Sawigun, C.; Thanapitak, S. A nanopower biopotential lowpass filter using subthreshold current-reuse biquads with bulk effect self-neutralization. *IEEE Trans. Circuits Syst. I Regul. Pap.* **2019**, *66*, 1746–1752. [CrossRef]
6. Sreenivasulu, P.; Hanumantha Rao, G.; Rekha, S.; Bhat, M.S. A 0.3 V, 56 dB DR, 100 Hz fourth order low-pass filter for ECG acquisition. *Microelectron. J.* **2019**, *94*. [CrossRef]
7. Lee, S.-Y.; Cheng, C.-J. Systematic design and modelling of a OTA-C filter for portable ECG detection. *IEEE Trans. Biomed. Circuits Syst.* **2009**, *3*, 53–64. [CrossRef] [PubMed]
8. Sun, C.-Y.; Lee, S.-Y. A fifth-order Butterworth OTA-C LPF with multiple-output differential-input OTA for ECG applications. *IEEE Trans. Biomed. Circuits Syst. II Express Briefs* **2018**, *65*, 421–425. [CrossRef]
9. Lee, S.-Y.; Wang, C.-P.; Chu, Y.-S. Low-voltage OTA–C filter with an area and power-efficient OTA for biosignal sensor applications. *IEEE Trans. Biomed. Circuits Syst.* **2019**, *13*, 56–67. [CrossRef]
10. Pinto, P.M.; Ferreira, L.H.C.; Colletta, G.D.; Braga, R.A.S. A 0.25-V fifth-order Butterworth low-pass filter based on fully differential difference transconductance amplifier architecture. *Microelectron. J.* **2019**, *92*. [CrossRef]
11. Khateb, F.; Kulej, T.; Kumngern, M.; Psychalinos, C. Multiple-input bulk-driven MOS transistor for low-voltage low-frequency applications. *Circuits Syst. Signal Process.* **2019**, *38*, 2829–2845. [CrossRef]
12. Khateb, F.; Kulej, T.; Veldandi, H.; Jaikla, W. Multiple-input bulk-driven quasi-floating-gate MOS transistor for low-voltage low-power integrated circuits. *AEU Int. J. Electron. Commun.* **2019**, *100*, 32–38. [CrossRef]
13. Khateb, F.; Kulej, T.; Kumngern, M.; Jaikla, W.; Ranjan, R.K. Comparative performance study of multiple-input Bulk-driven and multiple-input Bulk-driven Quasi-floating-gate DDCCs. *Int. J. Electron. Commun.* **2019**, *108*, 19–28. [CrossRef]

14. Kumngern, M.; Kulej, T.; Stopjakova, V.; Khateb, F. 0.5V sixth-order Chebyshev band-pass filter based on multiple-input bulk-driven OTA. *AEU Int. J. Electron. Commun.* **2019**, *111*. [CrossRef]
15. Salthouse, C.D.; Sarpeshkar, R. A practical micropower programmable bandpass filter for use in bionic ears. *IEEE J. Solid State Circuits* **2003**, *38*, 63–70. [CrossRef]

Publisher's Note: MDPI stays neutral with regard to jurisdictional claims in published maps and institutional affiliations.

© 2020 by the authors. Licensee MDPI, Basel, Switzerland. This article is an open access article distributed under the terms and conditions of the Creative Commons Attribution (CC BY) license (http://creativecommons.org/licenses/by/4.0/).

Article

Design of a High Precision Ultrasonic Gas Flowmeter

Jianfeng Chen, Kai Zhang *, Leiyang Wang and Mingyue Yang

College of Metrology & Measurement Engineering, China Jiliang University, Hangzhou 310018, China;
p1802085205@cjlu.edu.cn (J.C.); p1702085249@cjlu.edu.cn (L.W.); s1902080445@cjlu.edu.cn (M.Y.)
* Correspondence: zkzb3026@cjlu.edu.cn

Received: 23 June 2020; Accepted: 19 August 2020; Published: 26 August 2020

Abstract: Aiming at the problems of substantial pressure loss, small range ratio and contact measurement in traditional gas flowmeters, this paper designs a new type of data-filtering ultrasonic gas flowmeter. The flowmeter is composed of hardware circuits such as STM32F407 (ARM Cortex 32-bit microcontroller) main control chip and high-precision timing chip TDC-GP22 (time to digital converter). The software uses a new data-filtering algorithm combining Kalman filtering algorithm and arithmetic average algorithm to improve the measurement accuracy of ultrasonic gas flowmeter. Through experimental comparison, we find that the filtering algorithm effectively reduces the measurement error of the system. Within the flow range of 0.025–4 m^3/h, the maximum relative error of the system measurement is 2.7404%, which meets the national standard for the measurement error of the 1.5-level instruments. Moreover, it reduces the zero-drift to about one half of the original, which significantly improves the stability of the system. The gas flowmeter has the characteristics of high accuracy, good stability, low power consumption, and the overall performance is significantly improved.

Keywords: ultrasonic gas flowmeter; the principle of time-difference method; data filtering; low-power measurement

---

1. Introduction

Nowadays, flow detection technology has been widely used in various fields of industrial production such as petroleum, chemical industry, energy, etc. [1,2]. As a flowmeter for detecting fluid, flowmeters can be divided into gas flowmeters and liquid flowmeters from the measurement medium. Traditional turbine flowmeters, vortex flowmeters, and orifice flowmeters play an important role in the measurement of liquid flow [3]. However, due to the particularity of the gas medium flow and the complexity of the problems encountered in the signal transmission in the gas, the severe issues in gas flow measurement still plague people [4]. In the field of gas measurement, most domestic gas meters currently used in the country are membrane gas meters. The metering principle of the traditional membrane gas meter is to introduce the gas into a metering chamber with a constant volume and discharge it when it is full. Through a specific transmission mechanism, the number of charges and exhaust cycles converts into a mass, which reflects in the counter of the gas meter on [5]. The measurement technology of membrane gas meter is mature, reliable measurement and stable quality. However, its interior is contact measurement. Due to long-term wear, the measurement accuracy, sensitivity and stability of the measuring element is reduced, and there are problems such as substantial pressure loss, small range ratio and difficult meter reading. It has been challenging to meet the needs of modern society and people's daily life [6].

With the development of electronic information and the Internet of things technology, ultrasonic flowmeters have become a new type of intelligent instrument. Ultrasonic flowmeters achieve remote data transmission in combination with the Internet of things technology, solving the problem of field meter reading [7]. Compared with the traditional membrane gas meter, the ultrasonic gas flowmeter has the outstanding advantages of noncontact type, small pressure loss, wide range ratio, high accuracy, etc. [8].

In recent years, a large number of researchers have made significant achievements in the study of fluid pipeline structure and ultrasonic transducers [9,10]. In 2013, Zhao Xuesong conducted a flow field simulation on the pipeline of a mono ultrasonic flowmeter. The flow field characteristics of different shapes and sizes in the pipeline are studied, and the best tapered pipeline model is designed [11]. In 2015, Li Yuxi studied the influence of different installation methods of ultrasonic transducers on the measurement accuracy of flow meters and determined the optimal installation method of ultrasonic flow meters [12]. Moreover, the data filtering of the ultrasonic gas flowmeter mainly adopts the median filtering algorithm and wavelet algorithm to improve the measurement performance [13]. In 2018, Yao Ping used algorithms such as wavelet analysis and recursive average filtering to study how to improve the measurement accuracy of gas ultrasonic flowmeters in complex flow fields [14]. These research results have promoted the widespread application of ultrasonic gas flowmeters [15].

However, most of the existing ultrasonic gas flowmeters meet the measurement accuracy requirements in the high zone (From the transition flow Q_t (including Q_t) to the maximum flow Q_{max}: $Q_t \leq Q_i \leq Q_{max}$) and the accuracy requirements of small flow in low zone (From minimum flow Q_{min} to transition flow Q_t: $Q_{min} \leq Q_i < Q_t$) cannot be guaranteed [16]. To further improve the overall performance of the ultrasonic gas flowmeter, this paper designs a new data-filtering algorithm combining Kalman filtering and arithmetic average. The small flow in the low zone has especially high measurement accuracy, which significantly improves its measurement stability and can better meet the needs of practical applications.

2. Method of Time-Difference Measurement

Principle of Time Difference Measurement

In this paper, the ultrasonic gas flowmeter adopts the principle of time-difference method. The time-difference method is to indirectly obtain the average flow velocity of the fluid medium by measuring the transmission time interval of the ultrasonic signal in the fluid medium in the forward and reverse directions [17].

Figure 1 shows the measurement principle of the time-difference method [18]. Transducer P_{up} is a forward-flow transducer, and P_{dn} is a reverse-flow transducer. The fluid flows from left to right in Figure 1 as a positive flow direction. The installation angle of the transducers and the pipe is θ, the pipe diameter is D. The linear distance between the two transducers is L. The transmission speed of ultrasonic waves in a gas medium is c. The forward velocity of the gas is v. The upstream time transmitted by the P_{up} transducer and received by the P_{dn} transducer is [19]:

$$t_{up} = \frac{L}{c + v \cos \theta} \tag{1}$$

The downstream time transmitted by the P_{dn} transducer and received by the P_{up} transducer is:

$$t_{down} = \frac{L}{c - v \cos \theta} \tag{2}$$

From Equations (1) and (2), through the downstream time transmitted t_{down} minus the upstream time transmitted t_{up}, the forward and reverse flow time difference Δt of the ultrasonic signal transmission can be obtained.

$$\Delta t = t_{down} - t_{up} = \frac{2Lv \cos \theta}{c^2 - v^2 \cos^2 \theta} \tag{3}$$

Because the sound velocity c of the ultrasonic wave is much greater than the flow velocity v of the gas, so c^2 much larger than $v^2 \cos^2 \theta$, the Equation (3) can be approximated as:

$$v = \frac{\Delta t c^2}{2L \cos \theta} \tag{4}$$

The linear velocity v of the gas can be obtained by Equation (4). When measuring the instantaneous flow rate of gas, it is necessary to use the average surface velocity to calculate. According to the relevant knowledge of fluid mechanics, the average surface velocity has different correction coefficients according to different states of the fluid. Let the correction coefficient be K, then the instantaneous flow Q of the gas is:

$$Q = K \times S \times v = \frac{K \pi D^2}{4} v \tag{5}$$

where S is the cross-sectional area of the pipeline; K is the instrument factor, which is related to the Reynolds number of the fluid state.

It can be seen from Equations (4) and (5), under the condition of ensuring the accuracy of other fixed quantities, accurate measurement of the time difference Δt of the forward and reverse flow is the key to ensure the system's metering accuracy [20].

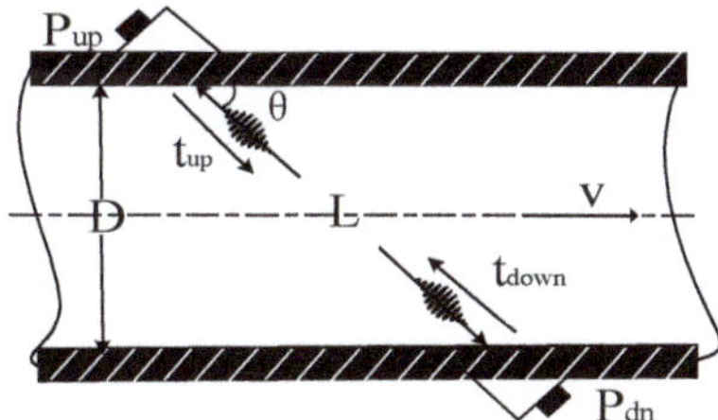

Figure 1. Schematic diagram of the principle of time-difference method.

3. Hardware Design and Signal-Processing Process

3.1. Overall Hardware Design

Figure 2 shows the system hardware block diagram of the ultrasonic gas flowmeter. The system is mainly composed of a control chip, a timing chip, an excitation signal amplifying circuit, a receiving conditioning circuit, an LCD (liquid crystal display) module and a power module.

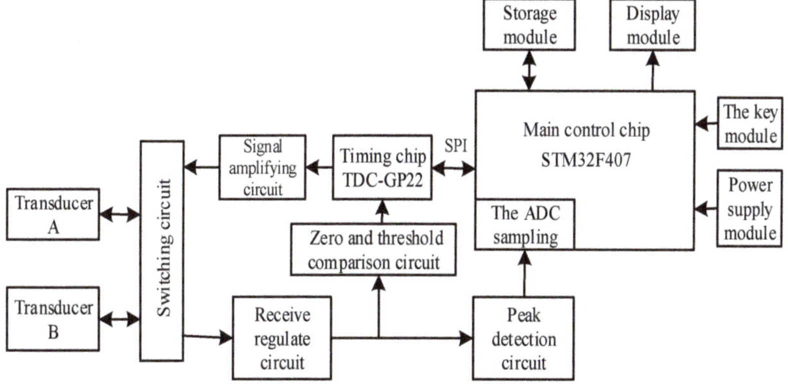

Figure 2. System hardware block diagram.

This system uses a 32-bit single-chip STM32F407 as the calculation and control module, which meets the requirements of high operation accuracy and low-power-consumption measurement. This system selects the TDC-GP22 high-precision timing chip as the time measurement module. Through the SPI (serial peripheral interface) communication mode, the single-chip microcomputer controls the pulse generator inside the TDC-GP22 chip to generate an excitation signal and amplified by the amplifier circuit, which acts on the transmitting transducer to send out the transmit waveform and start the measuring time. Moreover, transmitting in the pipeline for some time, the transmit waveform is received by the receiving transducer. After being amplified and filtered, the ultrasonic echo signal is sent to the threshold comparison circuit, and the zero-crossing generates a signal to stop timing. Finally, it sends to the TDC-GP22 timing chip to complete the time measurement. The TDC-GP22 sends an interrupt signal to the microcontroller after the measurement is completed. The microcontroller reads the time-difference measurement result and brings into the flow calculation formula to display and store the flow data.

3.2. TDC-GP22 Circuit Generates an Excitation Signal

The transit time of the ultrasonic gas flowmeter is at the nanosecond level; the TDC-GP22 can measure at 90-picosecond resolution in the single-precision mode and at 45-picosecond resolution in double precision mode. All meet our requirements for time-difference measurement accuracy. To reduce power consumption, we choose a single-precision measurement mode. The TDC-GP22 high-precision timing chip used in this paper integrates a pulse-generating circuit, which greatly simplifies the configuration of peripheral circuits. The internal logic gate delay is used to measure the time interval to ensure the accuracy of the measurement. Figure 3 shows the excitation signal generated by the TDC-GP22 pulse-generating circuit.

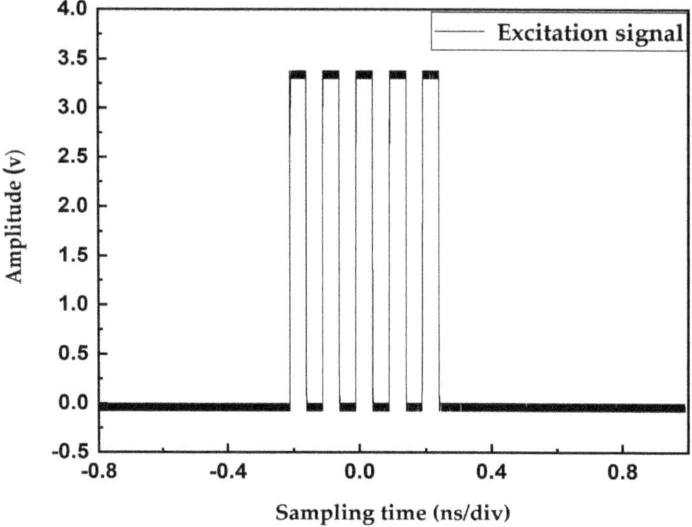

Figure 3. High-precision timing chip (TDC-GP22) circuit generates an excitation signal.

3.3. The Amplified Excitation Signal by the Amplifier Circuit

Figure 3 shows that the amplitude of the excitation signal is 3.3 V. In actual design—because the ultrasonic signal generated by low-amplitude excitation is attenuated severely in the gas medium—an excitation source with an amplitude of more than 15 V is generally required to excite the gas ultrasonic transducer. Therefore, we used the TPS61085 (Booster converter) boost chip to amplify the original signal. Figure 4 shows that the amplifier circuit boosts the 3.3 V voltage of the initial excitation signal to 18 V, which acts on the transmitting transducer to send out the transmit waveform.

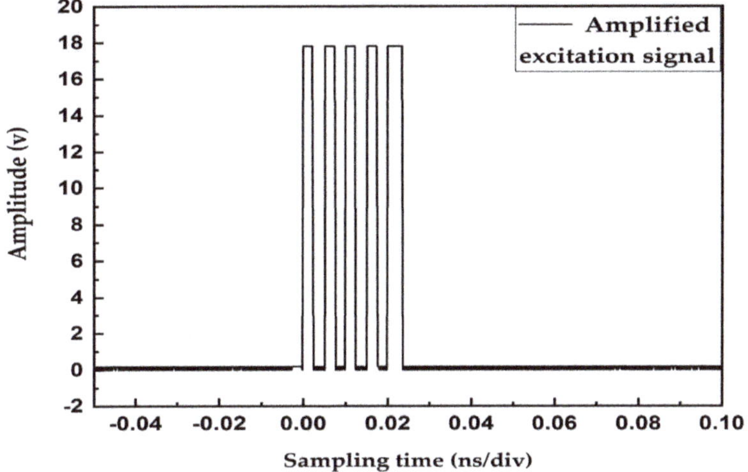

Figure 4. Amplified excitation signal by the amplifier circuit.

3.4. The Receiving Circuit Processes the Echo Signal

After propagating through the fluid medium, the transmit waveform signal is received by the receiving transducer. Figure 5a shows the original echo signal generated by the receiving transducer has an amplitude of several tens of millivolts, and it contains some noise interference signals. To obtain a stable and pure echo signal, the received original signal needs to be conditioned [21]. Therefore, we use the low-noise and high-precision op amp OPA320 (precision operational amplifier) to amplify the weak echo signal containing noise and then pass the active band-pass filter circuit to filter the amplified high and low-frequency interference signals. Finally, a stable and pure echo signal is obtained. Figure 5b shows the echo signal obtained after amplification and filtering by the receiving circuit.

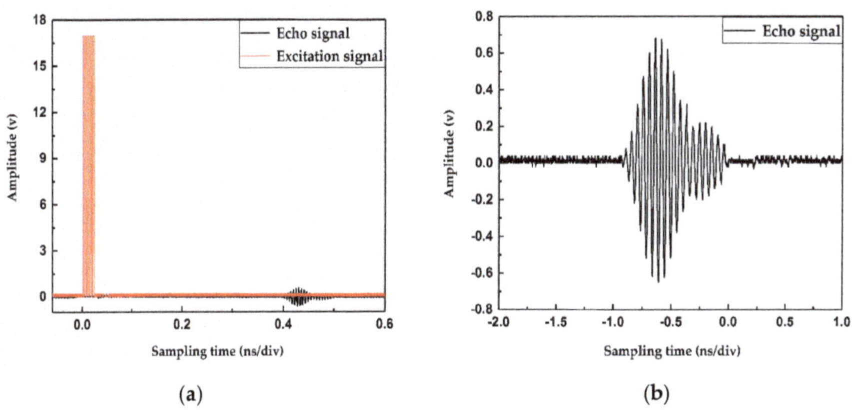

Figure 5. (a) Signal processed before the receiving circuit; (b) signal obtained by the receiving circuit.

3.5. Threshold and Zero-Crossing Comparison Circuit Generates Stop Timing Signal

To avoid some interference signals will also trigger the TDC-GP22 chip timing. We first use a threshold comparison circuit to set a threshold voltage. When the amplitude of the echo signal is higher than the threshold voltage, the enable pin of the zero-crossing comparison circuit is triggered and generates a stop-timing signal and sends it to the stop pin of TDC-GP22 to complete timing [22]. In this way, the interference before the useful signal is shielded by a combination of threshold and a zero-crossing comparison circuit. Figure 6 shows the stop-timing signal generated by the threshold and zero-crossing comparison circuit.

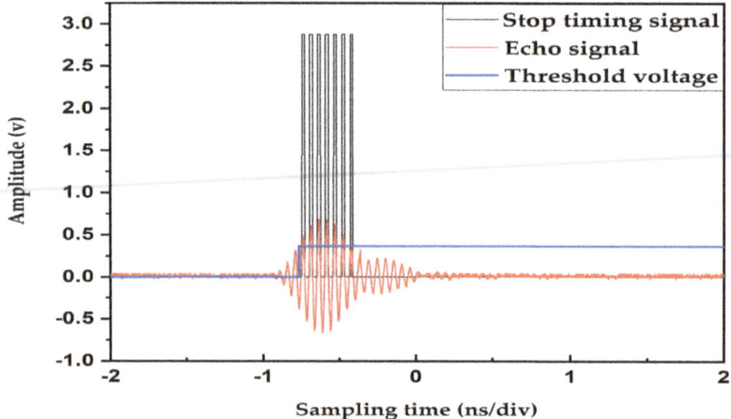

Figure 6. Threshold and zero-crossing comparison circuit generate the stop-timing signal.

4. Software Design

4.1. Software System Design

Figure 7 shows the flow chart of ultrasonic gas flowmeter measurement. When the measurement starts, the single-chip microcomputer needs to be initialized, including initializing the corresponding peripheral configuration and system clock and then configuring the registers of the TDC-GP22. The single-chip microcomputer sends an instruction to start measurement to the TDC-GP22 through the SPI communication method and simultaneously enters the time measurement program. After completing the upstream and downstream time measurement, it enters the time-difference processing module and then runs a filtering algorithm to process the time-measurement data. Finally, the single-chip microcomputer calculates the gas flow and send the data to the LCD. At the same time, the flow rate data are stored in the main control chip's internal flash to prevent data loss after power failure. After completing measurement, the single-chip microcomputer enters the low power STANDBY mode and waits for the timer interrupt to wake up for the next measuring.

The ultrasonic gas flowmeter in this study is battery-powered, so its system must use low-power measurement. The power consumption of the STM32F407 microcontroller in STANDBY mode is less than 2.5 μA and the static power consumption of the TDC-GP22 timing chip is the only 2.2 μA. To reduce the measurement power consumption of the system, the sampling frequency of the system is related to the change of the gas flow rate. When the gas flow rate is considered to be stable, the sampling period sets to 1 s and the rest of the time is in a dormant state. When the gas flow rate changes, the system will increase the sampling frequency to reflect the amount of gas change.

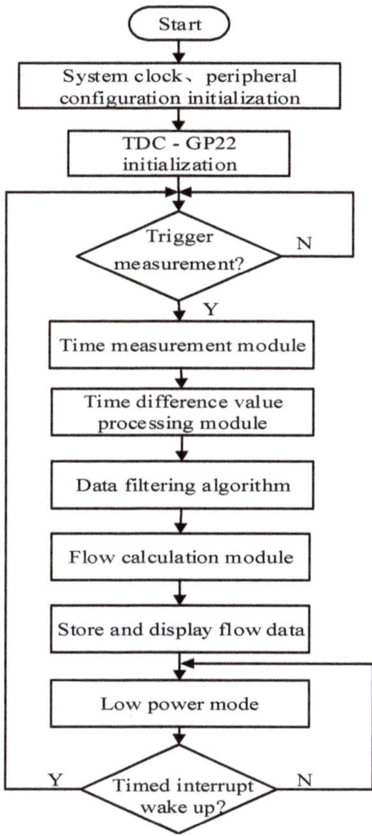

Figure 7. System software flow chart.

4.2. Data Filtering

The purpose of data filtering is to eliminate random errors in the original data. A reasonable data-filtering algorithm must be based on a thorough study of the original data, taking into account data volatility, data distribution characteristics and sudden changes [23,24]. In the measurement of ultrasonic gas flowmeters, measurement deviations are caused by disturbing factors such as random noise and piping structure design errors. For the arithmetic average filtering algorithm, the smoother the filtering effect, the higher the lag of the algorithm and the poor ability to suppress random errors that occur in the system. For the application of the Kalman filter algorithm, it can achieve a strong ability to suppress the arbitrary interference appearing in the order and the filter effect will be worse when the response to sudden changes is realized. Therefore, in the case of frequent changes of gas flow rate, how to make the ultrasonic gas flowmeter respond quickly to the amount of gas change and achieve accurate measurement is another test of the filtering algorithm.

To balance the smoothing effect of filtering and the timeliness of data processing. In the design of the algorithm, it is necessary to discriminate whether the measurement deviations are caused by the interference of external random factors or the design error of the internal structure of the measuring pipe, then adopts different algorithms for different situations. Therefore, this study proposes a new data-filtering algorithm combining a Kalman filtering algorithm and an arithmetic averaging algorithm.

4.2.1. Kalman Filtering Algorithm

Kalman filtering is a time–domain filtering method that is suitable for recursive solving. It can process the data obtained at each sampling instant immediately and based on the state estimates before that moment. The recursive equation gives the new state at any time estimate. Inflow measurement systems, the entire process can be represented by the following discrete models [25]:

$$x(k) = Ax(k-1) + Bu(k-1) + w(k-1) \tag{6}$$

$$z(k) = Cx(k) + v(k) \tag{7}$$

where the $x(k)$ is state variable and the gain matrix A is the state of $k-1$ is linearly mapped to the state of k at the current time, matrix B represents the gain of the optional control input $u(k-1)$ and the random signal $w(k-1)$ is the process excitation noise, because the state of the process has no control inputs, $u = 0$, the matrix C represents the gain of the state variable $x(k)$ against the measured variable $z(k)$ and $v(k)$ represents the observed noise [26].

The Kalman filtering algorithm is as follows:

1. Predict the current state:

$$\hat{x}(k \mid k-1) = A(k-1) \cdot \hat{x}(k-\mid k-1) \tag{8}$$

2. The covariance of prior estimation errors:

$$P(k|k-1) = A(k-1)P(k-1|k-1)A^{T}(k-1) + Q(k-1) \tag{9}$$

3. Gain calculation:

$$K(k) = P(k|k-1)C^{T}(k)\left[C^{T}(k)P(k|k-1)C(k) + R(k)\right] \tag{10}$$

4. Status estimate update:

$$\hat{x}(k|k) = \hat{x}(k|k-1) + K(k)[z(k) - C(k)\hat{x}(k|k-1)] \tag{11}$$

5. The covariance of posterior estimation error:

$$P(k|k) = [1 - K(k)C(k)]P(k|k-1) \tag{12}$$

where $\hat{x}(k|k-1)$ is the prior estimate, $\hat{x}(k|k)$ is the posterior estimate and $P(k|k-1)$ is the covariance of the previous estimate error, $P(k|k)$ is the covariance of the following estimation error, $K(k)$ is the filter gain, $R(k)$ is the observation noise covariance matrix, $Q(k-1)$ is the process excitation noise covariance matrix [27].

4.2.2. Filter Algorithm Combining Kalman and Arithmetic Average

Figure 8 shows a data-filtering algorithm combining the Kalman filtering algorithm and the arithmetic averaging algorithm. First, the system quickly measures eight sets of time-difference data and determine the difference value between the maximum and minimum values. When the ultrasonic gas flowmeter detects that the difference value higher than or equal to a particular threshold δ, the data have a wide range at this time, the gas flow rate is considered to be in a rapidly changing stage due to external interference. Hence, the system increases the sampling frequency Fs to track the change of the gas flow rate. At this time, the system selects a Kalman filter algorithm to process the time-difference data. When the difference value is less than a particular threshold range δ, the range of data changes is small. The gas flow rate is considered to be relatively stable at this time. The sampling frequency Fs is reduced, and we select the arithmetic averaging algorithm to solve the problem that the internal structure design error of the actual pipeline affects the gas flow rate.

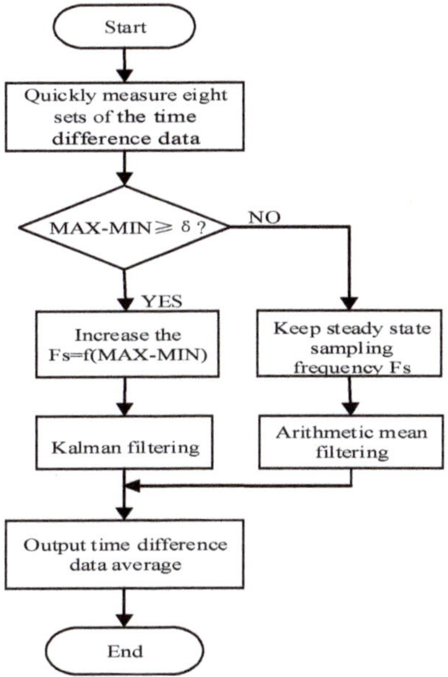

Figure 8. Data-filtering algorithm.

The data-filtering algorithm can effectively improve the measurement accuracy of the system. It can also significantly reduce the zero-drift of the system. Especially in the measurement of small flow, the stability and measurement accuracy of the system was improved.

5. Results and Discussion

5.1. Experimental Device

To test the effect of the comparison data-filtering algorithm, we use a gas flow measurement system for experiments. The schematic diagram of the experimental measurement system is shown in Figure 9, which is mainly composed of the standard flowmeter, the ultrasonic gas flowmeter, the pressure measurement sensor, the temperature measurement sensor, a flow regulating valve, a fan and an outlet valve.

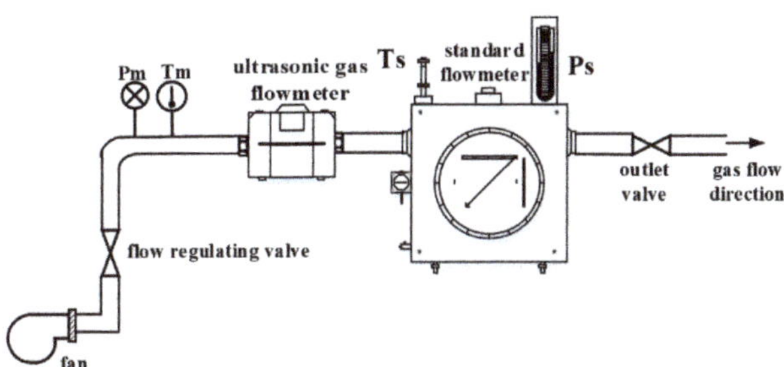

Figure 9. Schematic diagram of the experimental measurement system.

In the experimental measurement, the ultrasonic gas flowmeter was installed on the pipe section of the verified flowmeter and the fan started to blow in air, and the air outlet valve was closed. After the pressure between the standard flowmeter and the verified flowmeter was consistent and the measurement system was stable. We opened the air outlet valve [28]. In the experiment, we adjusted the flow rate by adjusting the opening of the flow regulating valve. After running for some time, we read the flow data of the standard flowmeter and the verified flowmeter.

To verify the measurement performance of ultrasonic gas flow, it is necessary to test the instantaneous flow at different flow points. The measuring range designed is 0.025–4 m^3/h in this study. According to "JJG-1030-2007 Verification regulation of ultrasonic flowmeters", the detection flow points are Q_{min}, Q_t, $0.4\,Q_{max}$, Q_{max}, for each flow point test is no less than 3 times; the experimental results are shown in Table 1.

Table 1. Flow experiment data at different flow points.

Average Value of the Standard Flowmeter (m^3/h)	The Value of the Verified Flowmeter (m^3/h)	Average Value of the Verified Flowmeter (m^3/h)	Relative Error (%)
0.0250	0.0253 0.0259 0.0258	0.0257	2.7404
0.4000	0.3975 0.3922 0.3902	0.3933	−1.6731
1.6000	1.6477 1.6233 1.5813	1.6174	1.0901
4.0000	3.9530 3.9348 4.0238	3.9705	−0.7364

5.2. Zero-Drift Discussion

Zero-drift means that the flow rate is zero under the test conditions: the pipe section is filled with air, and the inlet and outlet of the pipe are sealed. Therefore, the time-difference data of the zero-drift is continuously collected when the flow rate is zero.

Theoretically, when the flow rate of the fluid is zero, the transmission time of the ultrasonic signal in the forward and reverse directions is equal. Actually, it is disturbed by factors such as pipe structure, the performance difference of upstream and downstream transducers, which causes the time difference to not be zero, so the calculated flow rate is not zero, and it will cause errors in the actual flow measurement. Especially for the measurement of small flow, because the time difference of the small flow is close to the time difference of the zero-drift, the error of the measured flow result is greater. Therefore, the zero-drift phenomenon will affect stability and accuracy of the ultrasonic flowmeter measurement.

To compare the effect before and after the data-filtering algorithm, we sampled the original zero-drift time-difference data and the zero-drift time-difference data after data-filtering algorithm. We labeled them as Sample 1, Sample 2, respectively. As shown in Figure 10, the black line represents the original zero-drift data. It can be seen from figure that the zero-drift range of the system remains within ±10 ns. The thick red line represents the zero-drift data obtained after data filtering. The zero-drift data range of the system is within ±5 ns. By comparison, it can be seen that the scope of the zero-drift data becomes significantly smaller.

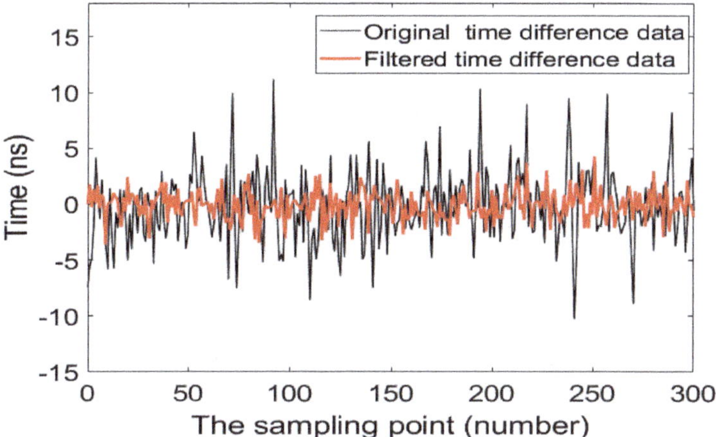

Figure 10. Comparison of zero-drift before and after data filtering.

To describe the degree of dispersion of the zero-drift data after the data filtering, the definition of sample standard deviation is introduced:

$$S = \sqrt{\frac{1}{n-1}\sum_{i=1}^{n}\left(X_i - \overline{X}\right)^2} \tag{13}$$

where S is the samples standard deviation, X_i is the time-difference value of the zero-drift, $\overline{X}$ is the average value of samples and n is the number of samples.

Table 2 analyses the difference value between the maximum and minimum value and standard deviation of the two groups of samples. The difference value indicates the range of the sample data [29]. The smaller the scale of the sample data, the more stable the data; and the smaller the standard deviation, the lower the degree of sample dispersion [30]. It can be seen from Table 1 that the difference value and standard deviation of Sample 2 are the smallest, and the difference value of Sample 1 is the largest. The results in Table 1 show that the data of Sample 2 are the most stable and have the least dispersion.

Table 2. Zero-drift data analysis.

Number	Difference between Maximum and Minimum (ns)	Standard Deviation (ns)
Sample 1	17.258	2.8873
Sample 2	7.893	1.3232

From the above analysis, it can be concluded that the data-filtering algorithm can effectively reduce the zero-drift of the system. The smaller the change range of the zero-point drift data, the more stable the measurement result of the ultrasonic flowmeter, and the less affected by the interference of the actual measurement environment.

5.3. Experimental Data

The range of the ultrasonic gas flowmeter is 0.025–4 m^3/h, the transition flow Q_t is 0.4 m^3/h, the range ratio R-value is 160, the accuracy level is 1.5 and the measuring pipe diameter is nominally diameter 20-mm in diameter. To verify and compare the effectiveness of data-filtering algorithms, we select two flow points distributed in the low zone (0.025 m^3/h $\leq Q_i <$ 0.4 m^3/h) and the high zone

$(0.4 \text{ m}^3/\text{h} \leq Q_i \leq 4 \text{ m}^3/\text{h})$ and test multiple groups time-difference data. Three sets of data were selected for each flow point, as shown in Tables 3 and 4, respectively. The time-difference data of a particular flow point is chosen randomly and the effect of the data-filtering algorithm is tested. The comparison of the results before and after the data filtering is shown in Figure 11.

Table 3. Time-difference signal samples without data filtering.

Particular Flow Point in the Low Zone Time Difference Data (μs)	Relative Error (%)	Particular Flow Point in the High Zone Time Difference Data (μs)	Relative Error (%)
0.2978	−9.4558	1.6293	0.1332
0.3561	8.2700	1.6136	−0.8317
0.3328	1.1858	1.6385	0.6986

Table 4. Time-difference signal samples with data filtering.

Particular Flow Point in the Low Zone Time Difference Data (μs)	Relative Error (%)	Particular Flow Point in the High Zone Time Difference Data (μs)	Relative Error (%)
0.2625	−0.2407	1.5698	−0.1950
0.2612	−0.7347	1.5723	−0.0360
0.2657	0.9754	1.5765	0.2310

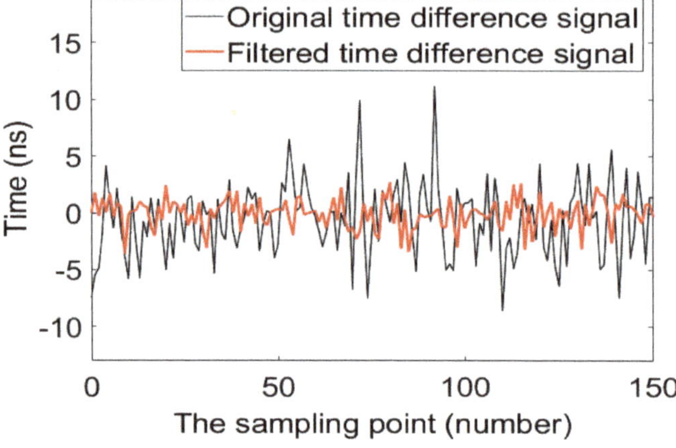

Figure 11. Comparison of time-difference signals before and after data filtering.

This study uses the time-difference method to measure the flow rate of an ultrasonic gas flow meter. Combining Equations (4) and (5), it can be seen that the ultrasonic velocity c in a static fluid is taken as a constant. The installation angle θ of the transducer and the pipe and the linear distance L between the two transducers are directly measurable physical quantities, so the accurate measurement of the time difference between the forward and reverse flow can improve the measurement accuracy of the system. To verify the effect of the filtering algorithm and eliminate the influence of other physical constants on the verification results, this study compares the relative error data of the three sets of time difference before and after filtering and improves the measurement accuracy of the system by accurately measuring the time difference between the forward and reverse flow.

Figure 11 is a comparison of the effect of time-difference signals before and after data-filtering algorithm at a particular flow point. The black line represents the signal before data filtering, and the thick red line represents the signal after data filtering. By comparison, we can find that the time-difference signal after data filtering is smoother than the time-difference signal before data filtering. It shows that the filtered signal is less affected by the random interference.

5.4. Data Discussion

Tables 2 and 3 show three sets of time-difference data before and after data filtering. The formula for calculating relative errors in table is:

$$E = \frac{\Delta T}{T} \times 100\% \tag{14}$$

where E is the actual relative error, generally given as a percentage, ΔT is the absolute error, that is the difference value between the measured value and the real value, and T is the real value. In the actual calculation, the average value is used instead of the real value.

It can be seen from Table 3 that when the flow point is in the high zone, the relative error of time-difference data is small; when the flow point is in the low zone, the relative error of time-difference data is large, and it is difficult to achieve accurate measurement. It can be seen from Table 4 that when the flow point is in the high zone, the relative error is small. Especially the relative error of the low zone flow is significantly reduced, which proves that the relative error of time-difference data after data filtering is smaller, and the measurement accuracy is higher.

To better compare the time-difference data in Tables 3 and 4, the sample averages of the time-difference data of the low zone and high zone flow and the difference ΔE between the maximum and minimum of the relative error were calculated. The results are shown in Table 5.

Table 5. Performance comparison with and without data filtering.

	Before Data Filtering	After Data Filtering
Sample average of a particular flow point in the low zone (μs)	0.3289	0.2631
Sample average of a particular flow point in the high zone (μs)	1.6271	1.5729
Relative error ΔE in the low zone	17.7258%	1.7101%
Relative error ΔE in the high zone	1.5303%	0.4260%

By comparing the data in Table 5, it can be found that the relative error ΔE of a particular flow point in the low zone has decreased from 17.7258% before data filtering to 1.7101% after data filtering. The relative error ΔE of a particular flow point in the high zone decreased from 1.5303% before data filtering to 0.4260% after data filtering. The comparison shows that the data using the filtering algorithm is more stable and more accurate.

6. Conclusions

This paper designs a high-precision ultrasonic gas flow meter and analyses the signal-processing process of the ultrasonic gas flow meter. A data-filtering algorithm combining Kalman filtering and arithmetic averaging is proposed to improve the measurement accuracy and stability of the system.

The zero-drift of the ultrasonic flowmeter is an essential manifestation of the measurement performance of the entire system. The smaller the zero-drift is, the higher the measurement stability of the system is. Therefore, by comparing the effects of the data-filtering algorithm, it is evident that the zero-drift scope of the system is reduced from within ±10 ns to within ±5 ns. The zero-drift is about 1/2 of the original zero-drift of the system, which effectively improves the stability of the system measurement.

To verify the effectiveness of the data-filtering algorithm, it can be seen from the experimental comparison results that the relative error of the time-difference data after the data-filtering algorithm is reduced, especially the relative error of the small flow is reduced more obviously. Within the flow range of 0.025–4 m^3/h, the maximum relative error of the system measurement is 2.7404%, which meets the national standard for the measurement error of the 1.5-level instruments. It shows that the measurement accuracy of the system is higher after data filtering, and its comprehensive measurement performance was significantly improved.

Author Contributions: J.C. did the research, wrote and edited the paper. L.W. provided article writing ideas and data collection. M.Y. completed signal processing. K.Z. provided the experiment conditions and gave many important suggestions on article. All authors have read and agreed to the published version of the manuscript.

Funding: This research was funded by National Natural Science Foundation of China, Grant Number 11472260.

Conflicts of Interest: The authors declare no conflict of interest.

References

1. Hamouda, A.; Manck, O.; Hafiane, M.L.; Bouguechal, N.E. An Enhanced Technique for Ultrasonic Flow Metering Featuring Very Low Jitter and Offset. *Sensors* **2016**, *16*, 1008. [CrossRef] [PubMed]
2. Shen, Z. Study on Key Technology of Multi-Channel Ultrasonic Gas Flowmeter Signal Processing Based on Zero-Crossing Detection. Master's Thesis, Hefei University of Technology, Hefei, China, 2017. (In Chinese).
3. Yan, J.; Bian, K. Design and realization of ultrasonic gas meter. *Ind. Metrol.* **2013**, *23*, 46–48. (In Chinese)
4. Luca, A.; Fodil, K.; Zerarka, A. Full-wave numerical simulation of ultrasonic transit-time gas flowmeters. In Proceedings of the IEEE Ultrasonics Symposium, Tours, France, 18–21 September 2016; pp. 1–4.
5. Arpino, F.; Dell'Isola, M.; Ficco, G.; Vigo, P. Unaccounted for gas in natural gas transmission networks: Prediction model and analysis of the solutions. *J. Nat. Gas Sci. Eng.* **2014**, *17*, 58–70. [CrossRef]
6. Zhang, H.; Guo, C.W.; Lin, J. Effects of Velocity Profiles on Measuring Accuracy of Transit-Time Ultrasonic Flowmeter. *Appl. Sci.* **2019**, *9*, 1648. [CrossRef]
7. Rajita, G.; Mandal, N. Review on transit time ultrasonic flowmeter. In Proceedings of the International Conference on Control, Instrumentation, Energy & Communication, Kolkata, India, 28–30 January 2016; pp. 88–92.
8. Zheng, D.D.; Mei, J.Q.; Wang, M. Improvement of gas ultrasonic flowmeter measurement non-linearity based on ray tracing method. *IET Sci. Meas. Technol.* **2016**, *10*, 602–606. [CrossRef]
9. Zheng, D.D.; Zhang, P.Y.; Xu, T.S. Study of acoustic transducer protrusion and recess effects on ultrasonic flowmeter measurement by numerical simulation. *Flow Meas. Instrum.* **2011**, *22*, 488–493. [CrossRef]
10. Guo, L.; Sun, Y.; Liu, L.; Shen, Z.; Gao, R.; Zhao, K. The flow field analysis and flow calculation of ultrasonic flowmeter based on the fluent software. *Abstr. Appl. Anal.* **2014**, *2014*, 528–602. [CrossRef]
11. Zhao, X. Development of Single-Path Ultrasonic Flowmeter. Master's Thesis, China Jiliang University, Hangzhou, China, 2013. (In Chinese).
12. Li, Y. Ultrasonic Flowmeter Structure Design Based on Structure and Flow Field Analysis. Master's Thesis, Zhejiang University, Hangzhou, China, 2015. (In Chinese).
13. Chen, H.X.; Zuo, M.J.; Wang, X.D.; Hoseini, M.R. An adaptive Morlet wavelet filter for time-of-flight estimation in ultrasonic damage assessment. *Measurement* **2010**, *43*, 570–585. [CrossRef]
14. Yao, P. Measurement Accuracy Improvement Method of Gas Ultrasonic Flowmeter in Complex Flow Field. Master's Thesis, Zhejiang University, Hangzhou, China, 2018. (In Chinese).
15. Willatzen, M.; Kamath, H. Nonlinearities in ultrasonic flow measurement. *Flow Meas. Instrum.* **2008**, *19*, 79–84. [CrossRef]
16. Chen, Y.; Huang, Y.; Chen, X. Acoustic propagation in viscous fluid with uniform flow and a novel design methodology for ultrasonic flowmeter. *Ultrasonics* **2013**, *53*, 595–606. [CrossRef]
17. Chen, Q.; Li, W.; Wu., J. Realization of a multipath ultrasonic gas flowmeter based on transit-time technique. *Ultrasonics* **2014**, *54*, 285–290. [CrossRef] [PubMed]
18. Han, D.; Kim, S.; Park, S. Two-dimensional ultrasonic anemometer using the directivity angle of an ultrasonic sensor. *Microelectron. J.* **2008**, *39*, 1195–1199. [CrossRef]

Sensors **2020**, *20*, 4804

19. Iooss, B.; Lhuillier, C.; Jeanneau, H. Numerical simulation of transit-time ultrasonic flowmeters: Uncertainties due to flow profile and fluid turbulence. *Ultrasonics* **2002**, *40*, 1009–1015. [CrossRef]
20. Li, W.H.; Chen, Q.; Wu, J.T. Double threshold ultrasonic distance measurement technique and its application. *Rev. Sci. Instrum.* **2014**, *85*, 044905. [CrossRef] [PubMed]
21. Tiwari, K.A.; Raisutis, R.; Samaitis, V. Hybrid signal processing technique to improve the defect estimation in ultrasonic non-destructive testing of composite structures. *Sensors* **2017**, *17*, 2858. [CrossRef]
22. Zahiri-Azar, R.; Salcudean, S.E. Time-Delay Estimation in Ultrasound Echo Signals Using Individual Sample Tracking. *IEEE Trans. Ultrason. Ferroelectr. Freq. Control* **2008**, *55*, 2640–2650. [CrossRef]
23. Ding, F.; Zhang, X.; Xu, L. The innovation algorithms for multivariable state-space models. *Int. J. Adapt. Control Signal Process.* **2019**, *33*, 1601–1608. [CrossRef]
24. Zhang, X.; Xu, L.; Ding, F.; Hayat, T. Combined state and parameter estimation for a bilinear state space system with moving average noise. *J. Frankl. Inst.* **2018**, *355*, 3079–3103. [CrossRef]
25. Buchner, H.; Kellermann, W. Improved Kalman gain computation for multichannel frequency-domain adaptive filtering and application to acoustic echo cancellation. In Proceedings of the 2002 IEEE International Conference on Acoustics, Speech, and Signal Processing, Orlando, FL, USA, 13–17 May 2002.
26. Kumar, G.; Prasad, D.; Singh, R.P. Target tracking using adaptive Kalman Filter. In Proceedings of the 2017 International Conference on Smart grids, Power and Advanced Control Engineering, Bangalore, India, 17–19 August 2017; pp. 376–380.
27. Li, L.; Hua, C.; Yang, H. A new adaptive unscented Kalman filter based on covariance matching technique. In Proceedings of the 2014 International Conference on Mechatronics and Control, Jinzhou, China, 3–5 July 2014; pp. 1308–1313.
28. Peng, L.H.; Zhang, B.W.; Zhao, H.C.; Stephane, S.A.; Ishikawa, H.; Shimizu, K. Data integration method for multipath ultrasonic flowmeter. *IEEE Sens. J.* **2012**, *12*, 2866–2874. [CrossRef]
29. Raine, A.B.; Aslam, N.; Underwood, C.P.; Danaher, S. Development of an ultrasonic airflow measurement device for ducted air. *Sensors* **2015**, *15*, 10705–10722. [CrossRef]
30. Wang, Y.J.; Ding, F.; Wu, M.H. Recursive parameter estimation algorithm for multivariate output-error systems. *J. Frankl. Inst.* **2018**, *355*, 5163–5181. [CrossRef]

© 2020 by the authors. Licensee MDPI, Basel, Switzerland. This article is an open access article distributed under the terms and conditions of the Creative Commons Attribution (CC BY) license (http://creativecommons.org/licenses/by/4.0/).

Article

An 8.8 ps RMS Resolution Time-To-Digital Converter Implemented in a 60 nm FPGA with Real-Time Temperature Correction

Zhipeng Song [1], Zhixiang Zhao [2], Hongsen Yu [1], Jingwu Yang [1], Xi Zhang [1], Tengjie Sui [1], Jianfeng Xu [1,*], Siwei Xie [3,*], Qiu Huang [2,*] and Qiyu Peng [4,*]

[1] State Key Laboratory of Digital Manufacturing Equipment and Technology, School of Mechanical Science and Engineering, Huazhong University of Science and Technology, Wuhan 430074, China; m201970747@hust.edu.cn (Z.S.); m201970550@hust.edu.cn (H.Y.); jingwuyang@hust.edu.cn (J.Y.); xizhang@hust.edu.cn (X.Z.); stj@hust.edu.cn (T.S.)

[2] School of Biomedical Engineering, Shanghai Jiao Tong University, Shanghai 200000, China; jasonorzzx@sjtu.edu.cn

[3] Pitech Company, Shenzhen 518000, China

[4] Department of Molecular Biophysics and Integrated Bioimaging, Lawrence Berkeley National Laboratory, Berkeley, CA 94720, USA

* Correspondence: jfxu@hust.edu.cn (J.X.); siwei.xie@tech-pi.com (S.X.); qiuhuang@sjtu.edu.cn (Q.H.); qpeng@lbl.gov (Q.P.)

Received: 21 March 2020; Accepted: 9 April 2020; Published: 11 April 2020

Abstract: This paper presented a non-uniform multiphase (NUMP) time-to-digital converter (TDC) implemented in a field-programmable gate array (FPGA) with real-time automatic temperature compensation. NUMP-TDC is a novel, low-cost, high-performance TDC that has achieved an excellent performance in Altera Cyclone V FPGA. The root mean square (RMS) for the intrinsic timing resolution was 2.3 ps. However, the propagation delays in the delay chain of some FPGAs (for example, the Altera Cyclone 10 LP) vary significantly as the temperature changes. Thus, the timing performances of NUMP-TDCs implemented in those FPGAs are significantly impacted by temperature fluctuations. In this study, a simple method was developed to monitor variations in propagation delays using two registers deployed at both ends of the delay chain and compensate for changes in propagation delay using a look-up table (LUT). When the variations exceeded a certain threshold, the LUT for the delay correction was updated, and a bin-by-bin correction was launched. Using this correction approach, a resolution of 8.8 ps RMS over a wide temperature range (5 °C to 80 °C) had been achieved in a NUMP-TDC implemented in a Cyclone 10 LP FPGA.

Keywords: time-to-digital converter (TDC); field-programmable gate arrays (FPGA); non-uniform multiphase (NUMP) method; temperature correction

---

1. Introduction

High-resolution time-to-digital converters (TDCs) are widely used in medical time-of-flight (TOF) positron emission tomography (PET) cameras [1–3], light detection and ranging (LiDAR) [4–6], large high-altitude air shower observatory systems (LHAASO) [7–9], and high-energy physics (HEP) experiments [10]. Many applications have a requirement for temperature-insensitive high-performance TDCs that will provide stable and reliable operation over a wide range of temperatures.

Field programmable gate array (FPGA)-based TDCs, constructed with off-the-shelf low-cost components, offer a promising practical alternative to conventional application specific integrated circuit (ASIC)-based TDCs. Most high-resolution FPGA-based TDCs are constructed using carry chains (delay chains) in the FPGA. The performance of those FPGA-based TDCs is largely determined by the accuracy and stability of the propagation delays in the delay chain.

The propagation delay in the delay chain is vulnerable to the influence of manufacturing processes and FPGA operating voltages and temperatures (PVT) [11,12]. In practice, variations in delay characteristics can be calibrated through code density tests [13]. Variations in propagation delay due to the operating voltage can be reduced by using high-performance power supplies [14]. However, it is not feasible or practical to control the ambient temperatures in many applications. In wave union TDC [15], a look-up table (LUT) for calculating timestamp has been built in the initial calibration mode. The LUT has been updated using the random event data in normal mode to compensate for the effects of the temperature fluctuations on propagation delays. This correction approach has achieved continuous calibration based on the technique of the ping-pang memories. However, this approach may not be suitable in applications with non-random event data inputs. Variations in propagation delays due to temperature fluctuations can be corrected using on-board LUT containing the variations of width for the time bins [16,17] or the temperature-LUT coefficient [18]. However, these methods require a temperature sensor and a regulating device and require intensive experiments to be performed to determine the relationship between the temperature and the correction parameters. The technology of dual delay lines (DDLs) is used to enable real-time calibrations in [19], which sets up an alternate result of calibrations to offset environmental effects. However, calibrating continuously results in the large power consumption, and the use of DDLs and the double backup of calibration results will consume lots of FPGA resources. A hardware unit consisting of an inverter and buffers is used to detect the ambient temperature and generate the correct values [20] but does not achieve an excellent performance.

In this paper, a novel scheme was introduced for real-time monitoring and compensation of changes in propagation delays. The event that triggered the temperature correction process was the variation of the propagation delay. The detection of the event was performed according to periodic discrete-time measurements, which could be classified as the periodic event-triggered control (PETC) [21]. When propagation delay changes exceeded a certain threshold, the bin-by-bin correction was launched, and LUT was updated on the threshold. Thus, the temperature correction scheme could be considered as an event-based control approach [22–27]. The proposed scheme was validated in a two-channel non-uniform multiphase (NUMP) TDC [28] using an Altera 60 nm Cyclone 10 LP FPGA and achieved a resolution of 8.8 ps root mean square (RMS) over a wide temperature range from 5 °C to 80 °C.

This paper is organized as follows. In Section 2, the principle of the NUMP TDC is described briefly, and the temperature characteristics of the delay chain and the principle of the proposed temperature correction method are described. Section 3 introduces the implementation details. The experimental results are presented in Section 4, and several design key points are discussed in Section 5. Finally, Section 6 summarizes and concludes the paper.

2. Operating Principles

2.1. Top-Level Diagram of NUMP TDC

The structure of the NUMP TDC is shown in Figure 1 [28]. As shown in Figure 1a, a phase-shifted clock generator (PSCG) is used to produce many copies of the system clock (400 MHz) with different phase shifts. A special structure is used to reduce the accumulation of delay chain jitter containing four equal-length sub-delay chains with a fixed phase difference (0.5π) clock feed equivalent to one delay chain. Each sub-delay chain consists of 95 delay units, and each delay unit corresponds to the carry chain of an adder. Every adder in PSCG is paired up with a dedicated register in the register bank, so there are 380 registers in the register bank. When the clock is fed into the first adder, multiple clocks with random phases are generated from the sum-out pins of the adders.

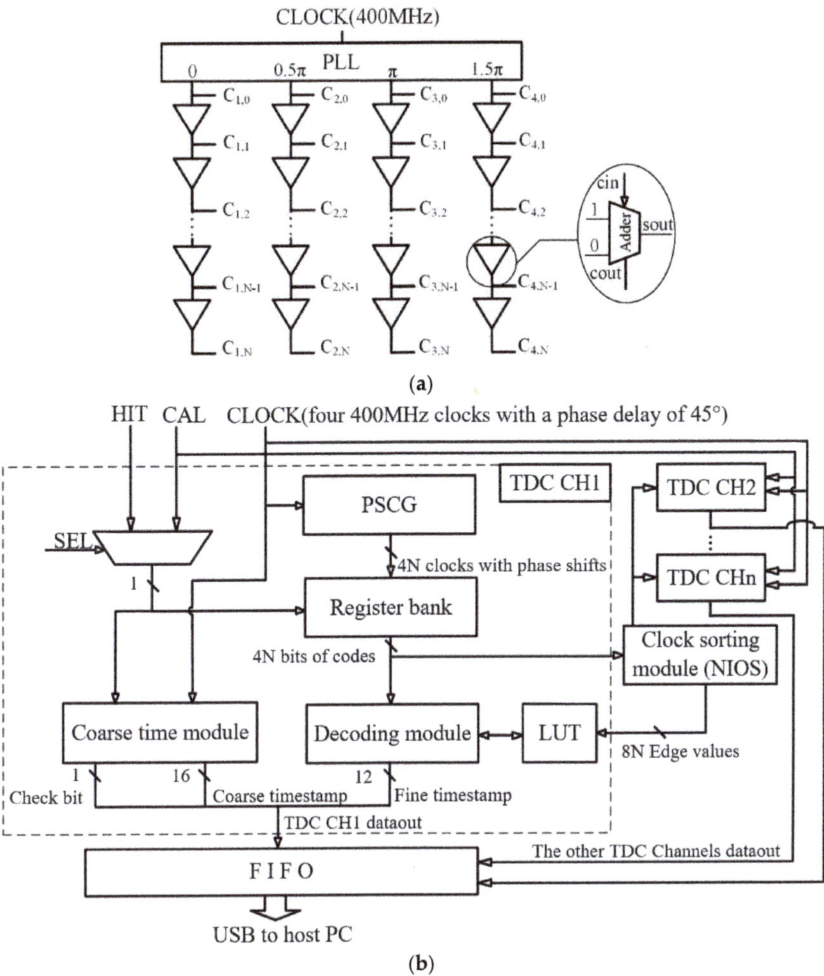

Figure 1. The architecture of NUMP TDC. (**a**) PSCG constructed with multiple delay chains (carry chains). (**b**) Diagram of the NUMP TDC. NUMP, non-uniform multiphase; TDC, time-to-digital converter; PSCG, phase-shifted clock generator.

A key feature of the NUMP TDC is that it is "non-uniform", which means that the phase relationships between the multiple clocks that are output from the PSCG are neither uniformly distributed nor sorted in time order. Thus, as shown in Figure 1b, a clock sorting module is designed to sort the clocks. The rising edges and the falling edges will be captured both, and their values will be used to calculate the fine timestamp in the decode module.

The other modules of the NUMP TDC will be introduced with the temperature correction module later.

2.2. Temperature Characteristics of Delay Chain

In previous work, Pan et al. [18] used a complementary metal oxide semiconductor (CMOS) inverter as an example to qualitatively analyze the effect of temperature on the delay chain of Cyclone II FPGA devices and suggested that carrier mobility and threshold voltage affected the propagation delay. A series of experiments were performed, and results indicated that the temperature fluctuations had similar effects on all bins in a delay chain with a limited length (less than 105 delay units). Although Cyclone 10 LP FPGA is a newer product in the Cyclone series FPGA, its positioning and underlying logical architecture are similar to the Cyclone II FPGA. The logic array block (LAB) consists of logic elements (LEs) and has the same carry chain structure [29,30]. Therefore, their delay chains have similar temperature characteristics. The detail about the delay chain has been discussed further in Section 5.

The principle of the NUMP TDC is quite different from a wave-union TDC [15]. Therefore, the conclusion reached in the previous study could not be directly used to support the temperature correction method for this paper. Based on Pan's research, a further conjecture could be proposed: since the variation in propagation delay of each delay unit (hereinafter called a 'cell') is roughly the same when the temperature changes, if the input of a delay chain is a clock signal instead of a hit signal, the delay at the delay chain's end cell is always longer than at the delay chain's front end cell, since the delay effect gradually accumulates in each cell. As the number of cells increases, it is obvious that the propagation delay of the clock signals will increase linearly.

Experimental studies were performed to validate this hypothesis. The clock sorting operation was performed repeatedly, with the temperature changed from 10 °C to 70 °C at 10 °C per step. The results of the clock sorting operation measured at 10 °C were selected as the references. The measurements at other temperatures were subtracted from the references to obtain the changes of the propagation delays (hereinafter called the 'variation of edge values') for each cell. Figure 2 shows the variation of the edge values measured at temperatures ranged from 20 °C to 70 °C at 10 °C per step.

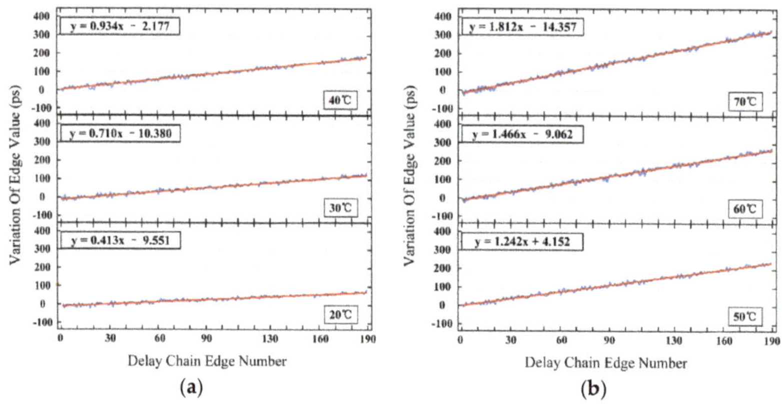

Figure 2. Variation of edge values at different temperatures compared to the clock sorting operation results at 10 °C. The horizontal axis shows the edge number, and the vertical axis is the variation of edge value. The blue lines and red lines represent the true curves and fitted curves, respectively. There are 95 cells in each sub-delay chain, and the first sub-delay chain's variation of edge values are shown here. (**a**) The changes of the propagation delays at 20 °C , 30 °C and 40 °C; (**b**) The changes of the propagation delays at 50 °C , 60 °C and 70 °C.

As shown in Figure 2, the edge values of each cell for FPGA at different temperatures were different from the edge values obtained by the clock sorting operation at 10 °C. Therefore, the temperature of the FPGA (more precisely, the delay chain) could be measured using the variation in edge values.

Figure 3 shows the variation in delay increment with FPGA temperature, which was the slope of the change in Figure 2. This graph had excellent linearity with an R^2 value that reached 0.998. The slope of the approximated line was 0.027, meaning that for each increase (or decrease) temperature by 1 °C, the propagation delay variation would increase (or decrease) by 0.054 ps (0.027 × 2). Although this linear relationship existed between the propagation delay and the temperature, this relationship was not a prerequisite for the temperature correction method proposed in this paper.

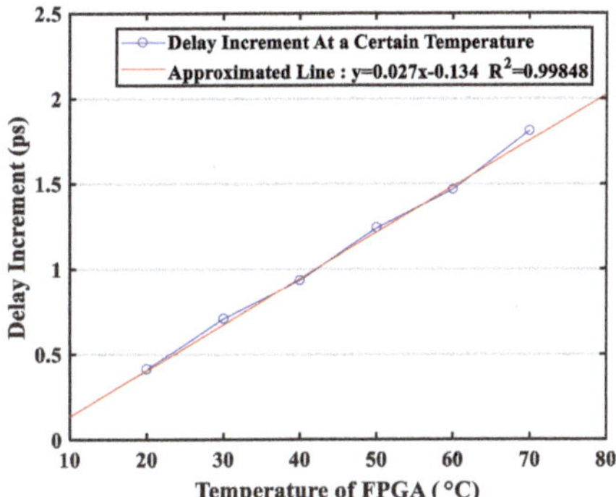

Figure 3. Variation in delay increment with a field-programmable gate array (FPGA) temperature.

2.3. Temperature Measurement Structure

The basic concept of the proposed temperature correction method for the NUMP TDC is as follows. A code density test is used to test the edge values of one or more cell(s) in the delay chain, and the results are compared with the edge values stored in the LUT. If the edge value variation exceeds a specified temperature correction threshold, the NUMP TDC switches to temperature correction mode, and the original LUT is updated to consider the current temperature state.

The FPGA delay chain temperatures could be measured using two methods, as shown in Figure 4.

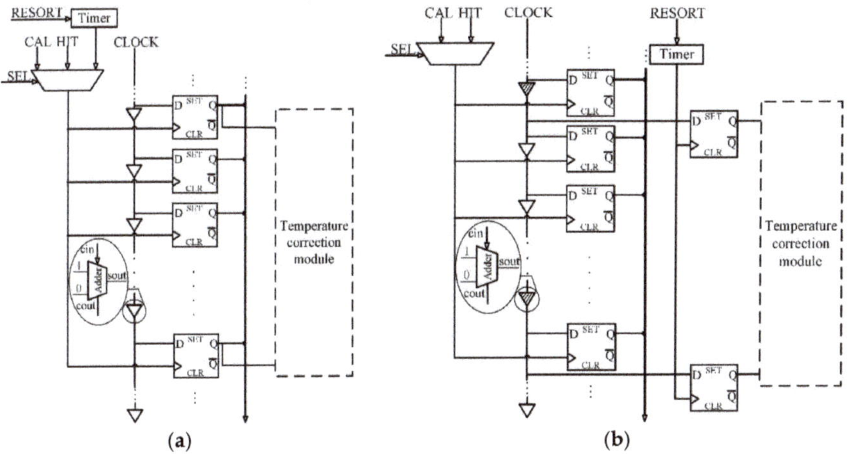

Figure 4. Two methods for temperature measurement. (**a**) Monitoring through dedicated registers. (**b**) Monitoring through the thermometric registers. The shadowed cells in Figure 4b are the thermometric cells; the only difference between these cells and normal cells is that the thermometric cells are extra paired up with thermometric registers.

For the first method, the temperature states are monitored using dedicated registers, as shown in Figure 4a. The advantage of this method is that the results accurately reflect the temperature states of the delay chain. However, this method has an obvious shortcoming: the delay chain states cannot sample simultaneously by the hit signal and the resort signal, so a timer is required to periodically test the temperature. This inevitably results in a larger dead time.

The second method is shown in Figure 4b and uses thermometric registers besides the delay chain to monitor variations in the edge values of the thermometric cells. This allows the temperature state of the delay chain to be monitored at any time. The biggest advantage of this method is that it is independent of time measurement and temperature measurement. The thermometric registers are not as precise as the dedicated registers and may not be as accurate at capturing the delay chain temperature. However, this difference can be minimized by using a logic lock tool in the design platform. By deploying the thermometric registers around the delay chain, the clock states in the delay chain can be almost simultaneously fed into two types of registers. This makes it possible to use additional thermometric registers rather than dedicated delay chain registers to monitor the delay chain temperature. Thus, for our proposed temperature correction module, the second method was utilized.

2.4. Top-Level Diagram of NUMP TDC with Temperature Correction Module

Figure 5 provides a diagram of the NUMP TDC with the temperature correction module. This diagram has two major differences from a conventional NUMP TDC. Firstly, two additional thermometric registers are deployed at both ends of the delay chain to monitor variations in the edge values. Secondly, the LUT is updated in real-time to compensate for temperature effects on the delay chain.

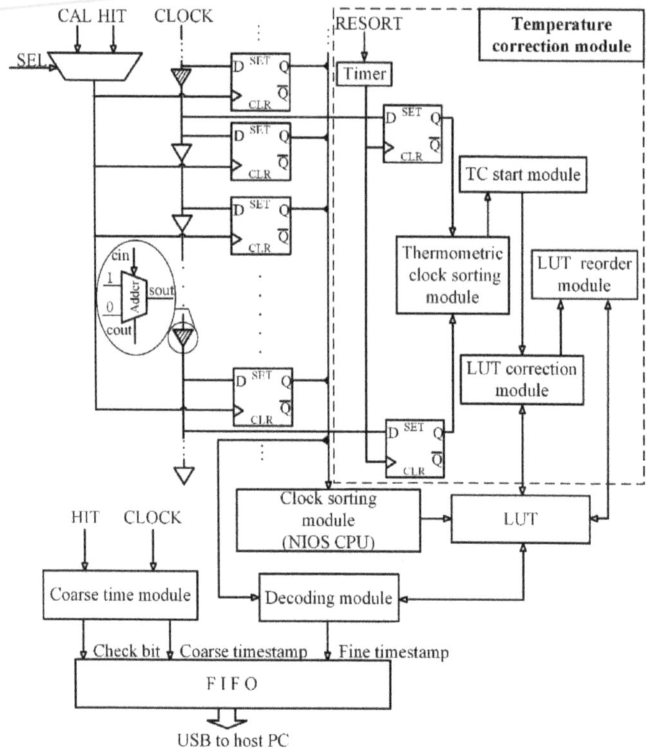

Figure 5. Diagram of NUMP TDC with the temperature correction module. For more clarity, this diagram only describes one channel TDC and omits some of the details already expressed earlier.

The NUMP TDC with the temperature correction module has three operation modes. In the initialization clock sorting mode, a calibration signal CAL is selected to latch the clock states in the registers, and the sampling results are calibrated by the clock sorting module, which is based on the internal netware input/output subsystem (NIOS) CPU of the FPGA. The phases (the edge values) of each cell are calibrated, sorted, and stored in the LUT. Once this step is complete, the time intervals can be measured by the NUMP TDC.

In normal operation mode, clock states with random phase shifts are sampled simultaneously by the hit signal. The sampled states of the clocks are sent to the decoding module to calculate the fine timestamp. The hit signal is also fed to the coarse time module, which outputs a 16-bit course timestamp and a check bit to correct the metastable error. The fine timestamp and the coarse timestamp with its check bit are then combined and buffered in the first input first output (FIFO) buffer and transmitted to the host PC by USB cable. Simultaneously, the thermometric clock sorting module periodically measures the edge values of the thermometric cells. The variation in edge values is fed to the temperature correction start module (hereinafter called the 'TC start module') for analysis. The NUMP TDC will automatically switch to temperature correction mode if the variation is larger than a given threshold.

In the temperature correction mode, the edge values are corrected. The edge values of each cell are calibrated and sorted, and the LUT is updated. Once the process is completed, the TDC switches back to the normal operation mode.

3. Implementation

3.1. Top-Level Diagram of Temperature Correction Module

A diagram of the temperature correction module is shown in Figure 6.

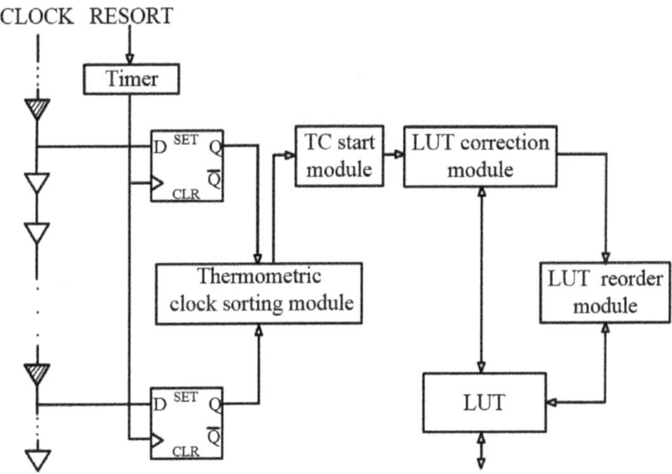

Figure 6. Diagram of the temperature correction module.

Since temperature variations tend to occur very slowly, it is not necessary to continuously monitor the temperature. The timer is used for periodic measurement of the delay chain temperatures using the thermometric clock sorting module. After each measurement, the absolute value of the difference between the edge values of the thermometric cells is fed to the TC start module for analysis of the current temperatures. The analysis process is as follows. If there is no temperature change trend identified, the TDC will remain in normal operation mode. If a temperature change trend is identified, the module will check if the change is larger than the given threshold. If so, the TDC switches to temperature correction mode, i.e., temperature correction will be launched, and the time measurement process will be suspended. Otherwise, the TDC will remain in normal operation mode.

Once the temperature correction starts, the edge values in LUT will be individually corrected by the LUT correction module. This disturbs the original order of the edge values. Therefore, a LUT reorder module is introduced to reorder the edge values in the LUT after each correction. Once the reorder process is completed, and the entire temperature correction process is finished, the TDC switches back to the normal operation mode.

The temperature correction module is performed in real-time and online.

3.2. Clock Sorting Operation for Temperature Measurement

This paper proposed a simple method to monitor the temperature of delay chains. Two thermometric registers are deployed at both ends of the delay chains, and a thermometric clock sorting operation is carried out periodically to obtain the edge values of the thermometric cells. The variation in edge values is calculated and fed into the TC start module to determine whether temperature correction should be performed.

The thermometric clock sorting operation is based on the code density test [13]. Two signals are used during this process—the reference signal C_0 and the signal to be measured C_X—which is the thermometric cell clock. The frequency of the resort signal is 6.7821 MHz, which is not related to the system clock, which is 400 MHz. Thus, the probability for each cell to be sampled is the same. A large number of random samples are performed to evenly distribute the resort signals at various positions across the whole clock cycle. The proportion of measurements is calculated between C_0 and C_X for each of the four types "10", "11", "01", and "00". The results are multiplied by the clock cycle to obtain the edge values of C_X.

There are always two possible relationships between C_0 and C_X. If the ratio between the four cases is 0.2, 0.2, 0.3, and 0.3, the rising and falling edge positions will be restricted to the two cases shown in Figure 7.

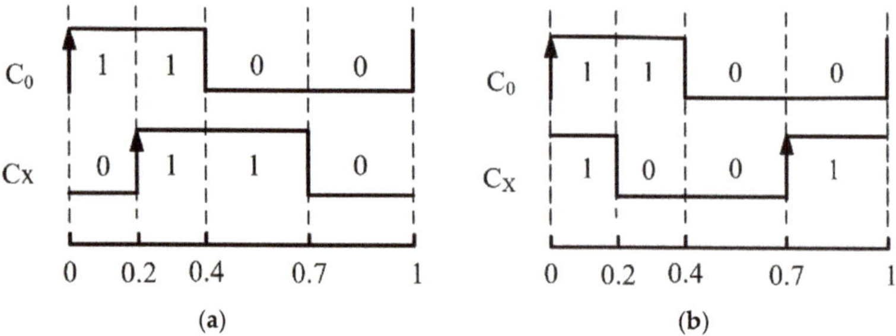

Figure 7. The phase relationship between the reference signal C_0 and the signal being measured C_X: (a) Rising edge of C_X arrives ahead of its falling edge for a clock cycle of C_0; (b) Falling edge of C_X arrives ahead of its falling edge for a clock cycle of C_0.

For the initialization clock sorting operation, CAL_1 is used to obtain two possible relationships between the reference clock C_0 and any other clock C_X. The role of CAL_2 is to distinguish between these two cases and is not required in temperature correction mode. By transmitting the results of clock sorting operation to the host computer for analysis, the phase relationship between each thermometric cell's signal C_X and the reference signal C_0 can be confirmed clearly.

It should be noted that the thermometric cells at both ends of the delay chains referred to in this paper do not represent the first and last cells in the absolute sense, but rather the cells at the start and end positions of the delay chain. In order to save computing resources and time of the whole correction process, the cell with a number that is a power of 2 can be considered as the thermometric cell in priority. For example, the 64th cell can be selected if there are 95 cells in each delay chain. Figure 8 shows the clock signal of the 64th cell at TDC temperatures of 10 °C and 70 °C. As the temperature increases from 10 °C and 70 °C, the value of the rising edge changes from 363 ps to 574 ps, i.e., a 211 ps edge increment occurs, which accounts for a delay of 8.44% of the total clock cycle of 2500 ps.

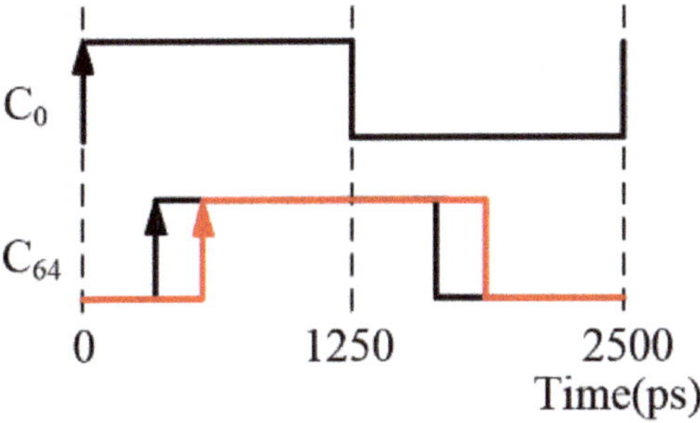

Figure 8. Variation of edge for the 64th cell as the temperature changes. The black line represents the original edge value at 10 °C, and the red line represents the edge value after the temperature increases to 70 °C.

This variation in edge values due to changes in temperature is small relative to the overall clock cycle. Therefore, in contrast with the thermometric clock sorting operation module, it is not necessary to count the proportions of each of the four cases, as counting a single case will be sufficient. This greatly reduces programming complexity and saves FPGA resources. For example, to calculate the variation in the rising edge of the 64th cell, we only need to count the frequency of "10" cases. The frequency of each case is divided by the total number and multiplied by the system clock cycle to get the edge values at the current temperature. After the initialization clock sorting operation is completed, the thermometric clock sorting operation starts after one timer period. For any specific temperature, the variation in edge value can be represented by Equation (1) as:

$$\Delta edge = edge_n^* - edge_n \tag{1}$$

where n is the thermometric cell number, $edge_n^*$ is the rising/falling edge value when the temperature varies, and $edge_n$ is the original rising/falling edge value.

A positive value of $\Delta edge$ indicates that the current temperature of the delay chain is higher than the original temperature and the temperature rise mark bit is asserted. A negative value of $\Delta edge$ indicates that the current temperature of the delay chain is lower than the original temperature and the temperature drop mark bit is asserted. Both $\Delta edge$ and the mark bit are sent to the TC start module to determine the temperature trend.

3.3. Temperature Correction Start Module

The TC start module operates as follows. It firstly assesses whether the current temperature of the delay chain has increased, decreased, or is substantially unchanged compared to the previous temperature. If there is a clear and strong temperature trend, temperature correction is launched. Otherwise, TDC remains in normal operation mode.

For accurate temperature trend measurement, the temperature measurement frequency, which is controlled by a timer, should be set to a suitable frequency. However, a temperature measurement frequency that is too high will increase the randomness of the temperature measurement, which may increase the probability of incorrectly identifying a temperature trend. Therefore, strict trend judgment criteria are required. In this paper, two counters, named *rise_cnt* and *fall_cnt*, were used to determine the temperature trend. Both counters started at zero and were incremented each time its corresponding temperature mark bit was asserted. The difference and sum of *rise_cnt* and *fall_cnt* were both calculated. Before the sum of the two counters reached a pre-defined value, if the difference between the two counters was greater than half this value, a temperature trend was identified with the larger counter, indicating whether the temperature was rising or falling.

Once it is evident that there is a temperature trend, a threshold judgment is performed. If Equation (2) is true, the temperature correction process is launched, and the LUT correction module and the LUT reorder module are sequentially triggered to correct and sort edge values and store the new values in the LUT.

$$\left| \Delta edge_{tail} - \Delta edge_{van} \right| > threshold \tag{2}$$

where $\Delta edge_{tail}$ and $\Delta edge_{van}$ are the variations of the edge values of the thermometric cells matched with the thermometric registers, and *threshold* is a pre-defined temperature correction threshold.

Note that once temperature correction is launched, the LUT used for the decoding process must be updated, so time measurement cannot be performed during the temperature correction process and must wait until it has been completed. The introduction of a *threshold* prevents the TDC from performing unnecessary temperature corrections for temperature fluctuations within a small range. The *threshold* is a parameter that should be determined by executing the clock sorting operation repeatedly while the ambient temperature varies and can be set according to the FPGA temperature adaptability and the working environment.

3.4. LUT Correction Module

Since the delay chain has a cumulative effect on the delay of the input clock signal, the delay effect of temperature on the delay chain also has a cumulative effect, i.e., cells at the back of the delay chain will occur larger edge value variation than the cells at the front of the delay chain. The main task of the LUT correction module is to correct the edge values according to the position of the cells in the delay chain.

For example, for the first sub-delay chain, the correction formula will be as follows:

$$edge_i^* = edge_i \pm \frac{threshold}{num_{tail} - num_{van}} \times num_n \tag{3}$$

where num_n is the number of the nth cell of the delay chain, which is equal to n, num_{van} and num_{tail} are the numbers of the thermometric cells, $edge_i$ is the current edge value, and $edge_i^*$ is the edge value after correction. Each cell has a rising and falling edge value, so there will be twice as many edge values as cells, i.e., i is equal to $2n$ or $2n - 1$. The effect of temperature on the delay chain considers each cell as a minimum element, so the same correction parameters are used to correct both types of edges for the same cell. The other three sub-delay chains have a similar correction process to the first sub-delay chain. Each cell is corrected using the same correction parameters for that cell position, e.g., the edge values of the last cell of the second sub-delay chain are corrected using the same correction parameters as the last cell of the first sub-delay chain.

Because of the cross-clock cycle phenomenon, $edge_i^*$ need to be checked after corrected by Formula (3). If $edge_i^*$ is larger than the system clock cycle, the real $edge_i^*$ is equal to its value less than the system clock cycle. If $edge_i^*$ is smaller than 0, the real $edge_i^*$ is equal to its value plus the system clock cycle. The cross-clock cycle phenomenon has been discussed further in Section 3.5.

In addition, to avoid repeatedly triggering temperature corrections at the same temperature, the parameter $edge_n$ in Formula (1) should be updated by reading the corrected LUT after each LUT correction.

3.5. LUT Reorder Module

After executing the initialization clock sorting operation and obtaining the edge values of the clock signal corresponding to each cell, the edge values are sorted and stored for each cell number in LUT. Once the length of the delay chain reaches a certain value, the delay to the input clock signal will be large enough to result in a cross-clock cycle phenomenon. Therefore, there may be multiple cells with similar phases in the delay chain, as shown in Figure 9.

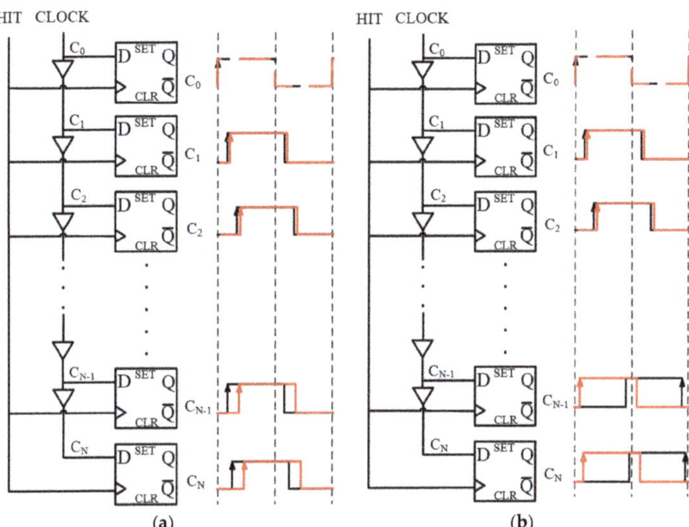

(a)　　　　(b)

Figure 9. Schematic diagram of the edge values of cells when the temperature rises. The black line represents the original edge value, and the red line represents the edge value when the temperature rises. (**a**) Edge values increase but do not cross clock cycles; (**b**) Edge values increase and cross clock cycles.

As shown in Figure 9a, in its original state, the rising edge value of C_N is slightly smaller than C_2. Therefore, the rising edge value of C_N must be placed before the rising edge value of C_2 during the sorting process, before being stored in the LUT. When the temperature rises, the edge increment of the delay chain's back end cells must be larger than the edge increment of the front end cells due to the cumulative effect of the delay. Therefore, after performing edge correction, the rising edge value of C_N will be larger than the rising edge value of C_2. In contrast, as shown in Figure 9b, if the edge value is very close to the clock cycle, then as the temperature rises, the cross-clock cycle phenomenon will occur. (Note the situation when the temperature drop is similar but has not been described here). As a result, the order of the edge values in the original LUT will be disturbed. Therefore, the LUT reorder module should be introduced.

The FPGA sort operation is more complicated than a software sort, as the use of arrays requires a high logic resource consumption. Additionally, the design platform may get trapped in the analysis and synthesis process due to the use of the large array and the loop structure.

Therefore, pairwise comparison is adopted to construct the reorder module. Each time two edge values are read from the LUT, the two adjacent edge values are compared, and the larger value is retained. Once the pairwise comparison is completed, the larger value is saved. The new and the old larger values are further compared, and the larger value is retained as the alternative to the biggest edge value for the current round of comparison. The alternative biggest edge value is constantly updated for each pairwise comparison. When all the edge values in LUT have been compared, the alternative biggest edge value (the biggest edge value in the current round of comparison) is placed in the storage address of the last read data value. In order to prevent data loss, the data originally at this address is stored in the address of the biggest edge value. In summary, each round of comparison will identify the biggest edge value for that round and store that value and its cell number at the end of the LUT. The edge value and cell number, which were previously stored at this position, will be stored in the original address of the biggest edge value.

In order to illustrate this process clearly, information on the first 10 edges was extracted from a real LUT so that the comparison process could be explained in detail. The edge value units were picosecond.

The first round of comparison is shown in Table 1. The cell number 346 and its edge value 92 were retained in the 10th memory address. The edge values stored in the first 9 memory addresses would be sorted by the second round of comparison. The updated LUT for the second round of comparison is shown in Table 2.

Table 1. First round comparison.

Memory Address	Cell Number	Edge Value (Uncorrected)	Edge Value (Corrected)	The Bigger Edge Value	Alternative	Alternative Address
1	1	0	0	46	46	2
2	144	3	46			
3	202	5	22	53	53	4
4	259	7	53			
5	143	8	52	52	53	4
6	26	10	45			
7	201	20	41	41	53	4
8	11	21	33			
9	291	26	36	92	92	10
10	346	34	92			

Table 2. Second round comparison.

Memory Address	Cell Number	Edge Value (Uncorrected)	Edge Value (Corrected)	The Bigger Edge Value	Alternative	Alternative's Address
1	1	0	0	46	46	2
2	144	3	46			
3	202	5	22	53	53	4
4	259	7	53			
5	143	8	52	52	53	4
6	26	10	45			
7	201	20	41	41	53	4
8	11	21	33			
9	291	26	36	36	53	4
10	346	—	—	92	—	—

For the second round of comparison, the even memory address (10th) data was missing for the fifth data comparison. Therefore, the edge value of the last even memory address (8th) was read instead and compared with the odd memory address (9th). After the second round of comparison, the cell number 259 and its edge value 53, which were stored in the 4th memory address, were used to replace the cell number 291 and its edge value 36 stored in the 9th memory address. The third round of sorting is shown in Table 3, which would sort the edge values stored in the first 8 memory addresses.

Table 3. Third round comparison.

Memory Address	Cell Number	Edge Value (Uncorrected)	Edge Value (Corrected)	The Bigger Edge Value	Alternative	Alternative's Address
1	1	0	0	46	46	2
2	144	3	46			
3	202	5	22	36	36	4
4	291	7	53			
5	143	8	52	52	52	5
6	26	10	45			
7	201	20	41	41	52	5
8	11	21	33			
9	259	—	—	53	—	—
10	346	—	—	92	—	—

Based on this process, the sorting process will not be repeated for each memory location. If there are m edge values, then $m - 1$ rounds of comparison are required. Once the reorder process is completed, the overall temperature correction process is finished, and the TDC switches back to the normal operation mode.

4. Experimental Validation

A two-channel NUMP TDC was implemented in a 60 nm Altera Cyclone 10 LP FPGA (10CL120YF780C8G) to validate the proposed temperature correction method. Each channel contained 380 cells to generate 380 delayed 400 MHz clocks.

4.1. Variation of Edge Values with Temperature Changes

The initialization clock sorting operation was executed at 40 °C. As the temperature varied, Figure 10 shows the 760 edge values in the LUT with and without the temperature correction module. The ideal edge value distribution was also shown in these graphs for ease of comparison.

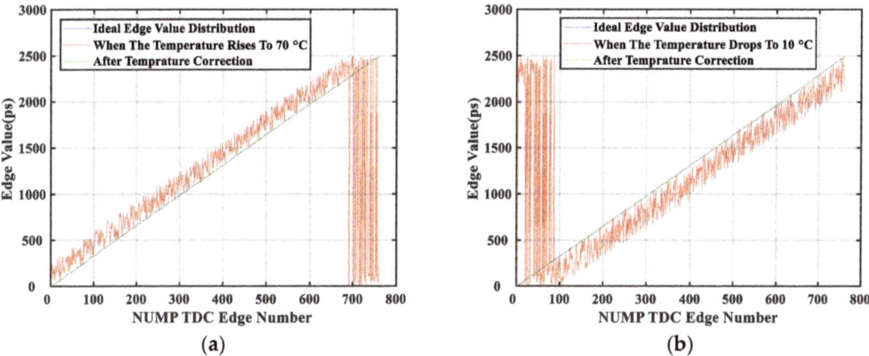

Figure 10. Distribution of corrected and uncorrected edge values when the temperature varies: (a) Temperature increases to 70 °C; (b) Temperature drops to 10 °C.

For discussion purposes, Figure 10a is taken as an example, which showed the effect of temperature rises. The red line in this figure showed that the edge values at the front had larger variations than the edge values at the back. The reason for this was that although the edge values at the front were small, there were a large number of cells match them. Due to the cumulative delay effect, the latter part of the delay chain was more affected by temperature than the front part of the delay chain. This was reflected in the edge values by a higher change in variation. However, the edge values of the cells at the end of the delay chain spanned a whole system clock period due to the rise of temperature, so its edge values were restarted from zero. Therefore, the temperature correction module was required. After LUT correction and reordering was completed, the distribution of the edge values in the LUT was shown by the green line in Figure 10. The offside edges had disappeared, and the LUT could be reused for decoding to calculate the fine timestamp. Figure 10b shows the effect of temperature decreases, which was similar to the effect of temperature rises, and is not described here to avoid repetition.

4.2. Nonlinearities

TDC nonlinearities would directly affect the resolution of the system. The most commonly-used metrics to characterize TDC nonlinearities are differential nonlinearity (DNL) and integral nonlinearity (INL) [31,32]. DNL describes the difference between each bin width and the ideal bin width, which is also known as the least significant bit (LSB) and reflects the degree of nonlinearity of single bin width.

INL represents the cumulative offset, and its value can be obtained by integrating the DNL values for some conventional TDC.

However, the edges of all the time bins of the NUMP TDC were measured independently. Thus, the DNL errors of the NUMP TDC did not accumulate to produce INL errors like conventional TDCs. In the decoding stage, using the binary search algorithm and the sampled delayed clock states latched by hit signal, the fine timestamp region could be narrowed constantly and located finally when searching two adjacent memory addresses. The mean of the corresponding edge values was calculated as the fine timestamp. Note that the edge values used in the decoding process were confirmed accurately by the initialization clock sorting operation. In conclusion, due to the introduction of the clock sorting module, the bin-by-bin calibration was executed to avoid the negative influence of INL that might be caused by uneven bin sizes. Edge values were sorted and stored in the LUT during the initialization clock sorting mode. In normal operation mode, each fine timestamp measurement result only depended on the two adjacent edge values obtained at the final stage of decoding but was not related to the other edge values in the delay chain. Thus, the NUMP TDC had only DNL because the last bin obtained by the decoding module might be not equal to the LSB and had no INL.

The DNL of the delay chain when the temperature changed from 40 °C to 70 °C and 10 °C are shown in Figure 11a,b, respectively.

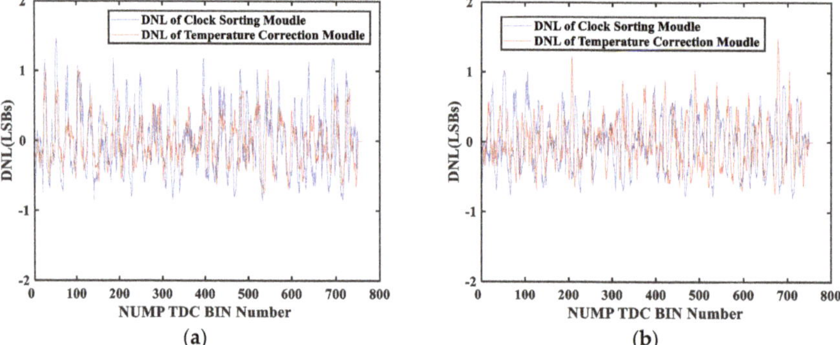

Figure 11. DNL (differential nonlinearity) of the NUMP TDC: (**a**) Temperature rises from 40 °C to 70 °C; (**b**) Temperature drops from 40 °C to 10 °C.

When the temperature rose from 40 °C to 70 °C, and after the temperature correction, the DNL was within [−0.81, 1.20] of LSB. The RMS of the DNL was 0.34 ps. In contrast, the DNL of the initialization clock sorting operation was tested, and the DNL was within [−0.84, 1.48] of LSB. The RMS of the DNL was 0.48 ps. When the temperature dropped from 40 °C to 10 °C, and after the temperature correction, the DNL was within [−0.74, 1.48] of LSB. The RMS of the DNL was 0.29 ps. For the initialization clock sorting operation at 10 °C, the DNL was within [−0.81, 1.01] of LSB. The RMS of the DNL was 0.42 ps. The DNL obtained by the temperature correction module provided a similar level of accuracy as the DNL values obtained by the initialization clock sorting operation, which indicated that the LUT updated by the temperature correction module was practical.

4.3. TDC Measurement

As shown in Figure 12, the external pulses generated by an Analog Devices Inc. clock generation board (AD9548/PCBZ) were sent to the FPGA through a low-voltage differential signaling (LVDS) port and fed into the two TDCs. The differences between the two TDC measurements were calculated in the FPGA and transmitted to the PC through a USB cable.

The performances of the two TDCs were characterized by calculating the RMS of the differences between the two TDC measurements. As shown in Figure 13, the TDC resolution reduced significantly due to temperature changes. However, the proposed temperature correction method successfully eliminated the influence of temperature in the range of 5 °C to 80 °C on the TDC resolution.

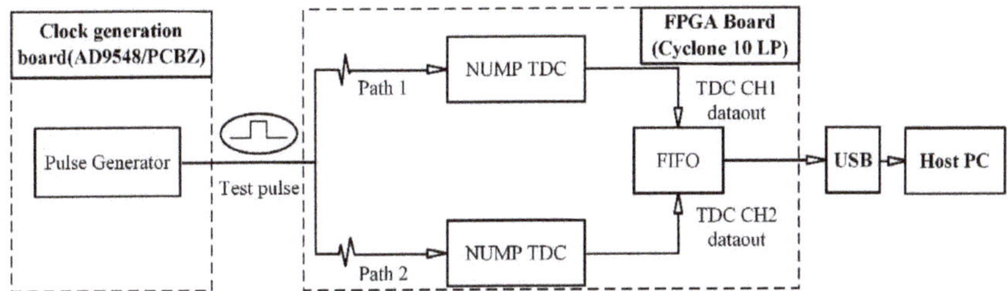

Figure 12. The diagram of TDC measurement. The external test pulses were sent to the FPGA through a low-voltage differential signaling (LVDS) port and fed to the two TDCs.

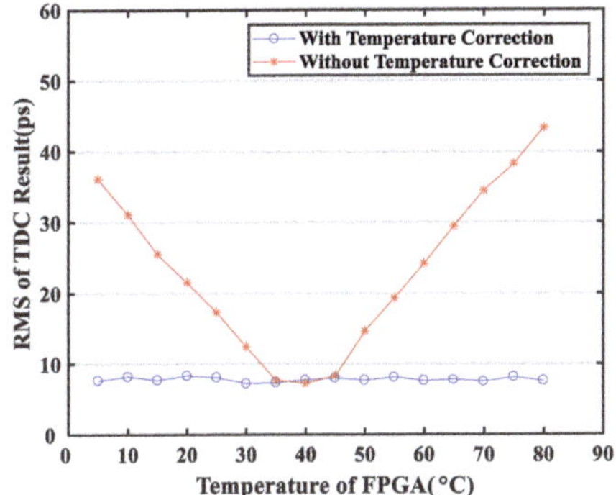

Figure 13. Single-shot resolutions of TDC for temperature variations from 5 °C to 80 °C. The initialization clock sorting operation was executed at 40 °C.

Figure 14 shows a statistic histogram of more than 100,000 measurement results collected during this experiment at different temperatures. This histogram clearly showed that although the temperature changed, the TDC measurement results still showed an excellent Gaussian distribution. The overall TDC resolution was 8.8 ps (RMS), which contained all the error factors, including DNL and jitter. Note that all RMS values given in this paper represented the single-shot resolution, which was $1/\sqrt{2}$ of the dual-channel resolution.

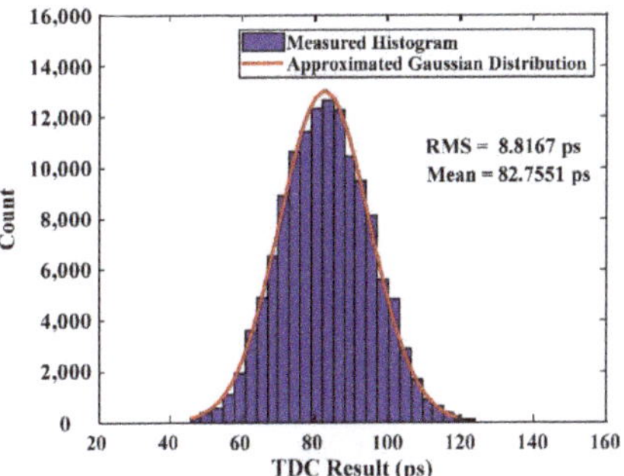

Figure 14. Distribution of the differences between the two TDC measurements in the temperature range from 5 °C to 80 °C. Automatic temperature corrections were applied, and the overall TDC resolution was 8.8 ps.

5. Discussion

5.1. Correction Parameters of Temperature Correction Module

In our design, four sub-delay chains of the same channel were corrected simultaneously. This method had a potential problem: the four sub-delay chains of the same channel might not show similar delay characteristics for temperature changes. In other words, when a phase shift occurred in one sub-delay chain, which was sufficient to trigger a temperature correction, it was not clear whether the other three sub-delay chains would change to the same extent.

To examine this issue further, the linear fitting results of delay chain_1 to delay chain_4 are listed in Table 4, which were similar to the equations shown in Figure 2. Furthermore, based on the equations shown in Table 4, the linear fitting results of slopes are shown in Table 5, which were similar to the equations shown in Figure 3.

Table 4. Fitting results of edge values' variation.

Temperature	Delay Chain_1	Delay Chain_2	Delay Chain_3	Delay Chain_4	ΔK_{MAX}
20 °C–10 °C	$y = 0.412x -$ 9.551	$y = 0.416x -$ 10.838	$y = 0.416x -$ 11.524	$y = 0.398x -$ 11.503	0.014
30 °C–10 °C	$y = 0.710x -$ 10.38	$y = 0.727x -$ 14.268	$y = 0.732x -$ 16.601	$y = 0.709x -$ 16.551	0.022
40 °C–10 °C	$y = 0.934x +$ 2.176	$y = 0.942x -$ 0.444	$y = 0.954x -$ 3.858	$y = 0.923x -$ 3.355	0.020
50 °C–10 °C	$y = 1.242x +$ 4.152	$y = 1.253x +$ 0.278	$y = 1.242x +$ 4.152	$y = 1.234x -$ 3.393	0.011
60 °C–10 °C	$y = 1.465x -$ 9.062	$y = 1.489x -$ 13.876	$y = 1.490x -$ 17.602	$y = 1.463x -$ 18.126	0.025
70 °C–10 °C	$y = 1.811x -$ 14.357	$y = 1.837x -$ 20.103	$y = 1.843x -$ 24.844	$y = 1.794x -$ 24.417	0.032

Table 5. Fitting results of the slope.

	Delay Chain_1	Delay Chain_2	Delay Chain_3	Delay Chain_4	Δ_{MAX}
Fitting results	$Y = 0.027X -$ 0.134	$Y = 0.028X -$ 0.136	$Y = 0.028X -$ 0.125	$Y = 0.027X -$ 0.141	0.001
R^2	0.99848	0.99833	0.99856	0.99862	0.0029

As shown in Table 4, the four sub-delay chains had similar delay characteristics, and the slopes of the fitted functions were only slightly different. ΔK_{MAX} in Table 4 was used to set Delay chain_1 as the reference. Although the intercepts were quite different, the correction process only used the correction parameters related to each cell number and added the original edge values. Thus, the correction effect was not related to the intercepts. As shown in Table 4, the maximum difference in slope was 0.032 when the temperature varied from 10 °C to 70 °C. At 70 °C, the maximum correction error was $0.032 \times 95 = 3.04$ ps, which was smaller than the LSB (3.2895 ps).

As shown in Table 5, the slope of the fitted equations varied between 0.027 to 0.028, which further indicated that the four sub-delay chains had similar delay characteristics across a wide temperature range. Each cell's delay would increase (or decrease) by 0.054 ps to 0.056 ps when the temperature rose (or drops) by one degree Celsius.

5.2. Carry Chain Differences between Cyclone V and Cyclone 10 LP

The NUMP TDC that we proposed was based on the Altera 28 nm Cyclone V FPGA (5CEBA4F23C7N) and implemented with 400 delayed 500 MHz clocks, where each sub-delay chain had 100 cells. NUMP temperature stability was also tested. The NUMP TDC based on the Cyclone V FPGA was not sensitive to temperature fluctuations. The resolution of the TDC without temperature correction was only slightly different and changed from 5.4 ps to 7.3 ps when the temperature was increased from 20 °C to 56 °C [28]. In contrast, the NUMP TDC based on the Cyclone 10 LP FPGA was very sensitive to temperature fluctuations. As could be seen from the experimental results in Figure 12, the resolution of the NUMP TDC without temperature correction was dramatically changed from 7.2 ps to 43.4 ps when the temperature increased from 40 °C to 80 °C.

The 60 nm Cyclone 10 LP FPGAs are the new generation of 28 nm Cyclone V FPGAs due to their unique advantages, including low power consumption and low cost. These FPGAs also differ in terms of their underlying logical resources [29,33]. Each logic array block (LAB) of a Cyclone V FPGA consists of 10 adaptive logic blocks (ALMs), whereas a Cyclone 10 LP FPGA has 16 logical elements (LEs). In the TDC field, the main difference between the two FPGAs is reflected in the carry chain processing, which is a constituent part of the delay chain. For the Cyclone V FPGA, the carry chain consists of "dedicated full adders". However, in Cyclone 10 LP FPGA, the carry chain is formed by the LUT. Figure 15 shows these two distinct carry chain processing methods.

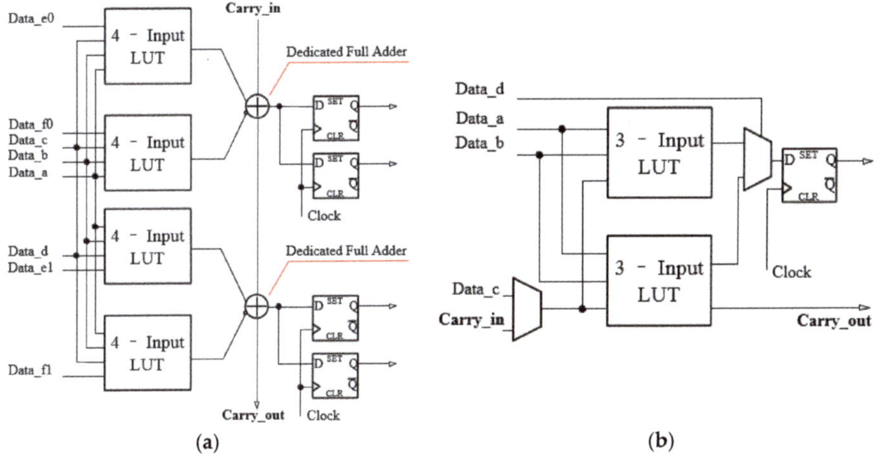

Figure 15. Comparison of the carry chain: (**a**) carry chain in Cyclone V FPGA; (**b**) carry chain in Cyclone 10 LP FPGA.

It was evident that the Cyclone V FPGA and Cyclone 10 LP FPGA had different carry chain structures. Although the difference was not obvious in terms of the propagation delay, there were huge temperature stability differences, which might result in a large difference in temperature characteristics for the same TDC scheme implemented on different devices.

5.3. Resource Usage and Power Consumption

The resources used in NUMP TDC with temperature correction majorly included logic cells, logic registers memory bits, phase-locked loops (PLLs), and routing resources. The logic utilizations of NIOS CPU, NUMP TDC, and temperature correction module are compared in Table 6. Note that multiple TDC channels could share one temperature correction module, and one NIOS CPU could support all the TDC channels in the FPGA. Theoretically, 40 to 70 channels of NUMP TDCs could be implemented with temperature correction in a single Cyclone 10 LP FPGA.

The consumption of a TDC was determined by both the resource usages and the event rate. The power consumption of the NUMP TDC system worked at an event rate of 1 MHz was calculated using the Cyclone 10 LP Early Power Estimator. The results showed the delay chains operated with a clock frequency of 400 MHz dominated the power consumption of the NUMP TDC (39 mW). The power consumptions of the NIOS CPU, a single channel of NUMP TDC, and the temperature correction module were 58 mW, 40 mW, and 1 mW, respectively.

Table 6. Resource utilization of a NUMP [1] TDC [2] implemented in Cyclone 10 LP FPGA [3].

	Resources	Used	Utilization
NIOS CPU	logic cells	8826	7.411%
	logic registers	5730	4.812%
	memory bits	104,016	2.612%
NUMP TDC without temperature correction (single channel)	logic cells	1377	1.156%
	logic registers	708	0.595%
	memory bits	23,552	0.592%
Temperature correction module	logic cells	1230	1.033%
	logic registers	689	0.577%
	memory bits	3072	0.077%

[1] NUMP, non-uniform multiphase; [2] TDC, time-to-digital converter; [3] Cyclone 10 LP FPGA, Cyclone 10 low power field programmable gate array.

5.4. Features of the Temperature Correction Method

The temperature correction method proposed in this paper has some unique features compared to many of the conventional temperature correction methods.

Firstly, it supports asynchronous temperature correction between multiple channels. Although the overall FPGA chip will have the same ambient temperature, the temperature states of the individual channels may not be identical, so the temperature characteristics of the individual delay chains may be different. Therefore, whether the temperature is measured by an external temperature sensor [16,17] or a dedicated delay chain [18], any temperature correction will be triggered for all delay chains, which will inevitably introduce a certain level of error. The temperature correction method proposed in this paper could measure the temperature state of each channel's delay chain and determine whether each channel needs to be corrected, thus providing individualized delay chain correction for each channel. Figure 16 compares the effect of synchronous correction and asynchronous correction through an internal signal test.

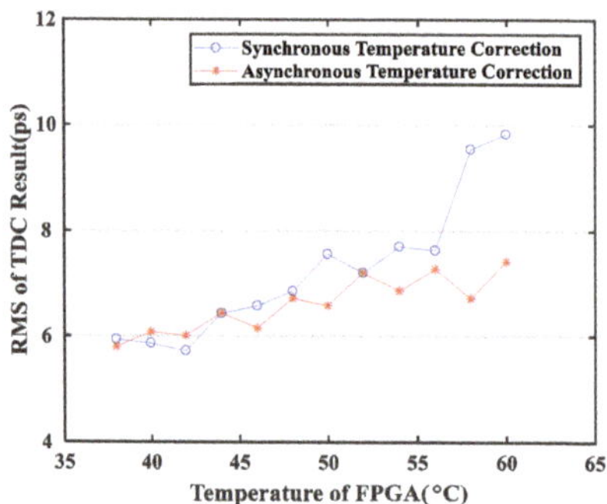

Figure 16. Test results of TDC at the same temperatures for both synchronous and asynchronous temperature correction.

Secondly, there is no requirement for intensive experiments at an early stage to determine the relationship between the temperature and correction parameters, which is convenient for reproducibility with different FPGA boards. For different production batches using the same series of FPGA, the delay characteristics and temperature characteristics of the delay chain may not be identical due to slight variations in the manufacturing process. The difference in delay characteristics can be fully calibrated during the initialization clock sorting operation. For the temperature sensitivity, to explore the corresponding relationship between the correction parameters and temperature, for other temperature correction methods, a special temperature control device is required to calibrate the correction parameters for multiple temperature states. This process is cumbersome, and the correction parameters cannot be generalized, i.e., the exact same experiments need to be repeated for each FPGA board [16–18]. However, these experiments are not necessarily using the temperature correction method proposed in this paper. Although the temperature characteristics of different FPGAs are not linear, as shown in Figure 3, the cumulative characteristics of the delay chain for the clock signal input is still constant, as shown in Figure 2.

Thirdly, there is no requirement for an external component, such as a temperature sensor. The method proposed in this paper accurately measures the temperature measurement state of the delay chain by deploying dedicated thermometric registers at both ends of the delay chain. This greatly reduces the error introduced by temperature measurement.

Finally, the method proposed in this paper could offer excellent flexibility. While deploying more thermometric registers can improve the accuracy of the temperature measurement, this can also be achieved by obtaining the mean of the edge value variation. The strategy of using only two thermometric registers at both ends of a single sub-delay chain is more practical and can save resources while still achieving a good temperature correction effect. Additionally, a larger threshold can be set to reduce the frequency of correction and dead time. For applications that are not sensitive to the dead time, a smaller threshold can also be used to increase the temperature stability of the TDC.

Most FPGA devices have logic cells close to each other with very similar performances. Thus, it is expected that there are no large thermal gradients along the delay line. Therefore, the temperature correction method presented in this paper is probably applicable to other technologies/devices, although it was only validated in Cyclone 10 LP FPGA in this study.

6. Conclusions

In order to achieve high performance within a wide temperature range, this paper proposed a temperature correction scheme that integrated a temperature monitor and automatic correction for NUMP TDC. The effect of temperature on the delay chain had been discussed, and a method to monitor the temperature state of the delay chain by measuring changes in the edge value of the thermometric cells had been proposed. When the variation in the edge value exceeded a given threshold, temperature correction was launched. The temperature correction was performed in the FPGA in real-time.

In conclusion, this paper developed a low-cost and high-performance method that could effectively reduce the effect of temperature fluctuations on an FPGA-based NUMP TDC. Using this method, a resolution of 8.8 ps RMS over a wide temperature range from 5 °C to 80 °C had been achieved in a NUMP-TDC implemented in a Cyclone 10 LP FPGA.

Author Contributions: Z.S., J.X., S.X., Q.H., and Q.P. conceived and designed the technical proposal; Z.S., Z.Z., H.Y., J.Y., T.S., X.Z., and S.X. designed and debugged the VHDL codes and NIOS programs to implement the TDC in the FPGA; Z.S., J.X., S.X., Q.H., and Q.P. designed the experimental schemes and analyzed the data. Z.S., Z.Z., H.Y., J.Y., X.Z., and S.X. performed the experiments and collected the data; Z.S. wrote the manuscript; J.X., S.X., Q.H., and Q.P. reviewed and edited the manuscript; J.X., S.X., Q.H., and Q.P. guided and administrated the research project jointly. All authors have read and agreed to the published version of the manuscript.

Funding: This research was supported by the National Natural Science Foundation of China (51627807), the National Natural Science Foundation-Guangdong Joint Funds of China (U1501256), the 111 Project (B16019), Shenzhen Science and Technology Program (Grant No. KQTD2017033015530192), and Hubei International Cooperation Project (2018AHB001).

Acknowledgments: The authors would like to acknowledge Shichao Yu for the experiment support.

Conflicts of Interest: All the authors declare no conflict of interest. The funding sponsors had no role in the design of the study; in the collection, analyses, or interpretation of data; in the writing of the manuscript, and in the decision to publish the results.

References

1. Venialgo, E.; Lusardi, N.; Garzetti, F.; Geraci, A.; Brunner, S.E.; Schaart, D.R.; Charbon, E. Toward a Full-Flexible and Fast-Prototyping TOF-PET Block Detector Based on TDC-on-FPGA. *IEEE Trans. Radiat. Plasma Med. Sci.* **2019**, *5*, 538–548. [CrossRef]

2. Won, J.Y.; Lee, J.S. Highly Integrated FPGA-Only Signal Digitization Method Using Single-Ended Memory Interface Input Receivers for Time-of-Flight PET Detectors. *IEEE Trans. Biomed. Circuits Syst.* **2018**, *6*, 1401–1409. [CrossRef] [PubMed]

3. Carra, P.; Bertazzoni, M.; Bisogni, M.G.; Cela Ruiz, J.M.; Del Guerra, A.; Gascon, D.; Gomez, S.; Morrocchi, M.; Pazzi, G.; Sanchez, D.; et al. Auto-Calibrating TDC for an SoC-FPGA Data Acquisition System. *IEEE Trans. Radiat. Plasma Med. Sci.* **2019**, *5*, 549–556. [CrossRef]

4. Carrara, L.; Fiergolski, A. An Optical Interference Suppression Scheme for TCSPC Flash LiDAR Imagers. *Appl. Sci.* **2019**, *11*, 2206–2220. [CrossRef]

5. Yoshioka, K.; Kubota, H.; Fukushima, T.; Kondo, S.; Ta, T.T.; Okuni, H.; Watanabe, K.; Hirono, M.; Ojima, Y.; Kimura, K.; et al. A 20-ch TDC/ADC Hybrid Architecture LiDAR SoC for 240 x96 Pixel 200-m Range Imaging with Smart Accumulation Technique and Residue Quantizing SAR ADC. *IEEE J. Solid-St. Circ.* **2018**, *11*, 3026–3038. [CrossRef]

6. Li, X.; Yang, B.; Xie, X.; Li, D.; Xu, L. Influence of Waveform Characteristics on LiDAR Ranging Accuracy and Precision. *Sensors* **2018**, *18*, 1156–1171. [CrossRef]

7. Li, H.; Gong, G.; Li, J. Portable Calibration Node for LHAASO-KM2A Detector Array. *IEEE Trans. Nucl. Sci.* **2017**, *6*, 1363–1366. [CrossRef]

8. Du, Q.; Gong, G.; Pan, W.; on behalf of the LHAASO Collaboration. A packet-based precise timing and synchronous DAQ network for the LHAASO project. *Nucl. Instrum. Methods Phys. Res. A* **2013**, *732*, 488–492. [CrossRef]

9. Liang, Y.; Zhao, L.; Guo, Y.; Qin, J.; Yang, Y.; Cheng, B.; Liu, S.; An, Q. A multichannel front end ASIC for PMT readout in LHAASO WCDA. *J. Instrum.* **2018**, *13*. [CrossRef]

10. Bifulco, P.; Izzo, V.; Mastroianni, S.; Aloisio, A.; Giordano, R.; Ameli, F.; Bocci, V.; Cadeddu, S.; Casu, L.; Lai, A.; et al. A fully-digital and fully synthetizable TDC for high energy physics experiments. In Proceedings of the 2nd International Conference on Event-Based Control, Communication, and Signal Processing (EBCCSP), Krakow, Poland, 13–15 June 2016; pp. 1–4.

11. Zhang, M.; Wang, H.; Liu, Y. A 7.4 ps FPGA-Based TDC with a 1024-Unit Measurement Matrix. *Sensors* **2017**, *17*, 865–883. [CrossRef]

12. Szyduczyński, J.; Kościelnik, D.; Miśkowicz, M. A Successive Approximation Time-to-Digital Converter with Single Set of Delay Lines for Time Interval Measurements. *Sensors* **2019**, *19*, 1109. [CrossRef]

13. Szplet, R.; Kalisz, J.; Szymanowski, R. Interpolating time counter with 100 ps resolution on a single FPGA device. *IEEE Trans. Instrum. Meas.* **2000**, *4*, 879–833. [CrossRef]

14. ANALOG DEVICES Inc. *5A, Low Noise, Programmable Output, 85mV Dropout Linear Regulator*; ANALOG DEVICES, Inc.: Norwood, MA, USA, 2019.

15. Wu, J.; Shi, Z. The 10ps wave union TDC: Improving FPGA TDC resolution beyond its cell delay. In Proceedings of the IEEE International Symposium Circuits and Systems, Como, Italy, 5–9 October 2008; pp. 3440–3446.

16. Qin, X.; Wang, L.; Liu, D.; Zhao, Y.; Rong, X.; Du, J. A 1.15ps Bin Size and 3.5ps Single-Shot Precision Time-to-Digital Converter with On-Board Offset Correction in an FPGA. *IEEE Trans. Nucl. Sci.* **2017**, *12*, 2951–2957. [CrossRef]

17. Qin, X.; Feng, C.; Zhang, D.; Miao, B.; Zhao, L.; Hao, X.; Liu, S.; An, Q. Development of a High Resolution TDC for Implementation in Flash-Based and Anti-Fuse FPGAs for Aerospace Application. *IEEE Trans. Nucl. Sci.* **2013**, *5*, 3550–3556. [CrossRef]

18. Pan, W.; Gong, G.; Li, J. A 20ps Time-to-Digital Converter (TDC) Implemented in Field-Programmable Gate Array (FPGA) with Automatic Temperature Correction. *IEEE Trans. Nucl. Sci.* **2014**, *3*, 1468–1473. [CrossRef]

19. Chen, Y. Time Resolution Improvement Using Dual Delay Lines for Field-Programmable-Gate- Array-Based Time-to-Digital Converters with Real-Time Calibration. *Appl. Sci.* **2018**, *9*, 20. [CrossRef]

20. Van, L.; Xuan, T.; Lee, H. A Novel FPGA Implementation of a Time-to-Digital Converter Supporting Run-Time Estimation and Compensation. *ACM Trans. Reconfig. Technol. Syst.* **2019**, *12*, 10.

21. Miskowicz, M. *Event-Based Control and Signal Processing*; CRC Press: Boca Raton, FL, USA, 2016.

22. Miskowicz, M. Send-On-Delta Concept: An Event-Based Data Reporting Strategy. *Sensors* **2006**, *6*, 49–63. [CrossRef]

23. Pérez-Torres, R.; Torres-Huitzil, C.; Galeana-Zapién, H. A Cognitive-Inspired Event-Based Control for Power-Aware Human Mobility Analysis in IoT Devices. *Sensors* **2019**, *19*, 832. [CrossRef]

24. Barrios-Avilés, J.; Rosado-Muñoz, A.; Medus, L.D.; Bataller-Mompeán, M.; Guerrero-Martínez, J.F. Less Data Same Information for Event-Based Sensors: A Bioinspired Filtering and Data Reduction Algorithm. *Sensors* **2018**, *18*, 4122. [CrossRef]

25. Leon-Garcia, F.; Palomares, J.M.; Olivares, J. D2R-TED: Data—Domain Reduction Model for Threshold-Based Event Detection in Sensor Networks. *Sensors* **2018**, *18*, 3806. [CrossRef]

26. Gao, Y.; Li, Y.; Peng, L.; Liu, J. Design of Event-Triggered Fault-Tolerant Control for Stochastic Systems with Time-Delays. *Sensors* **2018**, *18*, 1929. [CrossRef]

27. Shen, Y.; Kong, Z.; Ding, L. Flocking of Multi-Agent System with Nonlinear Dynamics via Distributed Event-Triggered Control. *Appl. Sci.* **2019**, *9*, 1336. [CrossRef]

28. Sui, T.; Zhao, Z.; Xie, S.; Xie, Y.; Zhao, Y.; Huang, Q.; Xu, J.; Peng, Q. A 2.3ps RMS Resolution Time-to-Digital Converter Implemented in a Low-Cost Cyclone V FPGA. *IEEE Trans. Instrum. Meas.* **2019**, *10*, 3647–3660. [CrossRef]

29. Intel Inc. *Cyclone 10 LP Device Handbook: Logic Array Blocks*; C10LP-51003; Intel Inc.: Santa Clara, CA, USA, 2020.

30. Intel Inc. *Cyclone II Device Handbook: Logic Array Blocks*; CII-51002; Intel Inc.: Santa Clara, CA, USA, 2007.

31. Tancock, S.; Arabul, E.; Dahnoun, N. A Review of New Time-to-Digital Conversion Techniques. *IEEE Trans. Instrum. Meas.* **2019**, *10*, 3406–3417. [CrossRef]

32. Machado, R.; Cabral, J.; Alves, F.S. Recent Developments and Challenges in FPGA-Based Time-to-Digital Converters. *IEEE T. Instrum. Meas.* **2019**, *11*, 4205–4221. [CrossRef]

33. Intel Inc. *Cyclone V Device Handbook: Logic Array Blocks*; CV-52001; Intel Inc.: Santa Clara, CA, USA, 2019.

© 2020 by the authors. Licensee MDPI, Basel, Switzerland. This article is an open access article distributed under the terms and conditions of the Creative Commons Attribution (CC BY) license (http://creativecommons.org/licenses/by/4.0/).

Article

Design of Reconfigurable Time-to-Digital Converter Based on Cascaded Time Interpolators for Electrical Impedance Spectroscopy [†]

Sounghun Shin [1,‡,§], Yoontae Jung [2,‡], Soon-Jae Kweon [2,*], Eunseok Lee [2], Jeong-Ho Park [3,§], Jinuk Kim [2], Hyung-Joun Yoo [2] and Minkyu Je [2]

[1] Foundry Business, Samsung Electronics Co., Ltd., Hwaseong-si, Gyeonggi-do 18448, Korea; sh01.shin@samsung.com

[2] School of Electrical Engineering, Korea Advanced Institute of Science and Technology (KAIST), Daejeon 34141, Korea; yoontae.jung@kaist.ac.kr (Y.J.); eunseoklee@kaist.ac.kr (E.L.); ray1206@kaist.ac.kr (J.K.); hjyoo53@kaist.ac.kr (H.-J.Y.); mkje@kaist.ac.kr (M.J.)

[3] System LSI Business, Samsung Electronics Co., Ltd., Hwaseong-si, Gyeonggi-do 18448, Korea; jeongho.park@samsung.com

[*] Correspondence: hksn0226@kaist.ac.kr

[†] This paper is an extended version of "An efficient, wide range time-to-digital converter using cascaded time-interpolation stages for electrical impedance spectroscopy" published in Proceedings of the IEEE Asia Pacific Conference on Circuits and Systems (APCCAS) on Jeju, South Korea, 25–28 October 2016.

[‡] Both authors contributed equally to this work.

[§] Both authors were with School of Electrical Engineering, KAIST.

Received: 17 February 2020; Accepted: 26 March 2020; Published: 29 March 2020

Abstract: This paper presents a reconfigurable time-to-digital converter (TDC) used to quantize the phase of the impedance in electrical impedance spectroscopy (EIS). The TDC in the EIS system must handle a wide input-time range for analysis in the low-frequency range and have a high resolution for analysis in the high-frequency range. The proposed TDC adopts a coarse counter to support a wide input-time range and cascaded time interpolators to improve the time resolution in the high-frequency analysis without increasing the counting clock speed. When the same large interpolation factor is adopted, the cascaded time interpolators have shorter measurement time and smaller chip area than a single-stage time interpolator. A reconfigurable time interpolation factor is adopted to maintain the phase resolution with reasonable measurement time. The fabricated TDC has a peak-to-peak phase error of less than 0.72° over the input frequency range from 1 kHz to 512 kHz and the phase error of less than 2.70° when the range is extended to 2.048 MHz, which demonstrates a competitive performance when compared with previously reported designs.

Keywords: electrical impedance spectroscopy (EIS); time-to-digital converter (TDC); time interpolator; phase; polar demodulator; quantization; reconfigurability

---

1. Introduction

Electrical impedance spectroscopy (EIS), which measures the impedance over a range of frequencies, has been widely used in today's biomedical applications such as body composition analysis [1–3], cancer diagnosis [4–6], and detection of allergic contact reaction [7], and so on. The frequency range from 1 kHz to several MHz, which is associated with the polarization of macromolecules [8], is required for these applications. With growing demands on portable EIS systems, today's research is focused on designing fully integrated EIS systems that support a wide frequency range.

The impedance spectrum can be obtained by measuring the impedance for a particular frequency and then repeating the measurement with sweeping the frequency, which is called the frequency response analyzing (FRA) method [9]. When a sinusoidal current signal is injected into the material under the experiment, the resulting sinusoidal voltage signal is generated with the magnitude and time delay that depend on the magnitude and phase of the material's impedance, respectively [10,11]. The EIS system is generally composed of two parts: a sinusoidal signal generator (SSG) and a demodulator. The SSG injects the sinusoidal current signal into the target material, and the demodulator measures the resulting voltage signal. Considering that the EIS system that supports sensor array includes several demodulators with one SSG [9,12,13], it is important to design a demodulator to be hardware-efficient and to have high precision.

Conventional quadrature demodulators extract real and imaginary parts of impedance using quadrature mixers and low-pass filters (LPFs) [12–14]. When the resulting signal depending on the target impedance is multiplied by a signal that is in phase with the injected signal and then low-pass filtered, the output value represents the real part of the impedance. When multiplied by a signal that is 90° out of phase with the injected signal and then low-pass filtered, the imaginary part of the impedance is provided. In such a process, incomplete synchronization among the aforementioned signals becomes a dominant source of error, and the need for phase-correction circuitry increases the implementation complexity [15]. To resolve this synchronization issue, polar demodulators, which asynchronously measure the magnitude variation and time delay of the resulting signal, have been proposed [10,11,15–17]. In particular, the polar demodulators in [10,11,17] mitigate the complexity of analog front-end significantly because the magnitude variation is measured without large-time-constant LPFs and the time delay is measured with one of simple logics such as XOR, XNOR, and latch. In quadrature demodulators, the large-time-constant LPFs limit measurement speed [17] and occupy large area. In polar demodulators, contrarily, the simple logic produces a pulse having the width that corresponds to the time delay, and then a time-to-digital converter (TDC) is used to quantize the pulse width. Note that the pulse width is proportional not only to the phase of impedance but also to the period of the injected signal. Therefore, the TDC should be able to handle wide input-time range to support the analysis at low frequencies, and also achieve a high resolution to support the analysis at high frequencies. It is, therefore, important to design the TDC to meet these requirements in a simple and efficient way for implementing a low-complexity polar demodulator.

Two types of TDCs have been proposed for the EIS system: TDCs based on a time-to-voltage converter (TVC) [15] and a counter [17–19]. The TVC converts the pulse width to the voltage by charging a capacitor with a current source, and the following analog-to-digital converter (ADC) quantizes the output of the TVC [15]. However, the phase error increases with the frequency of the injected signal because the full-scale output voltage of the TVC decreases as the frequency of the injected signal increases. Since counter-based TDCs have an input-time range that is theoretically unlimited [20], this type of TDCs appear to be attractive phase quantizers in the EIS system. However, the TDC in [17] requires high-speed counting clock for high-frequency analysis because the counting speed solely decides the time resolution. This TDC adopts a 3.3-GHz counting clock to support the maximum frequency of up to 10 MHz.

We have proposed two reconfigurable TDCs combining counters with time interpolators to improve resolution without increasing the counting clock speed and verified them by simulation [18,19]. Both TDCs employ a coarse counter to secure wide input-time range. The TDC in [18] uses cascaded time interpolators with a reconfigurable time interpolation factor. When the same large interpolation factor is used, the cascaded time interpolators have shorter measurement time and smaller chip area than a single-stage time interpolator. Since both the required input-time range and time resolution vary with the frequency of the injected signal, the reconfigurable time interpolation factor is employed to maintain phase error within an appropriate level while providing reasonable measurement time. Instead of the cascaded time interpolators, the TDC in [19] utilizes a time interpolator with a low interpolation factor and a chain delay line to further reduce the phase error. Although the TDC in [19] shows improved precision, the TDC in [18] allows more efficient implementation in terms of chip size.

This paper presents the reconfigurable TDC based on cascaded time interpolators [18], which has been fabricated in 0.25-μm CMOS process, with providing comprehensive explanation, detailed analysis, and measurement results. The fabricated TDC quantizes the pulse width corresponding to the phase of 0° to 90° with the peak-to-peak phase error of under 0.72° up to 512-kHz frequency and the phase error of under 2.70° up to 2.048-MHz frequency, demonstrating competitive performances compared to previously reported designs.

2. Background and Design Specifications

This section describes the impedance measurement principle and phase measurement scheme in EIS systems. From this background, design specifications are derived at the end of the section.

2.1. Impedance Measurement Principle

When a sinusoidal current signal $i_{in}(t)$ is injected into the target material, a resulting sinusoidal voltage signal $v_b(t)$ is generated, and its magnitude and time delay with respect to $i_{in}(t)$ depend on the impedance of the material, as shown in Figure 1. Z_b is the impedance of the target material, and $|Z_b|$ and θ are the magnitude and phase of Z_b, respectively.

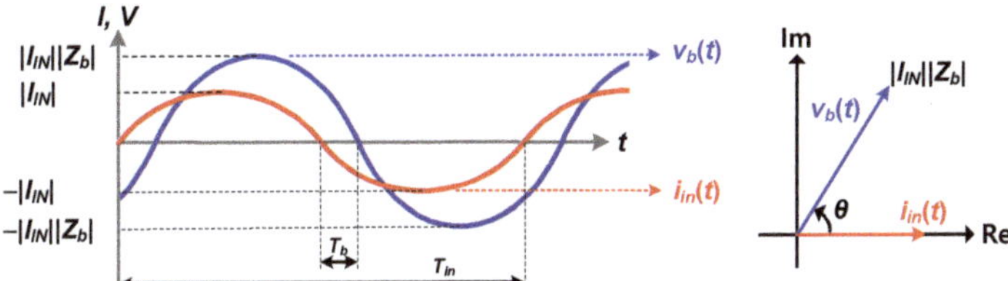

Figure 1. Magnitude and phase of the resulting voltage determined by the impedance under measurement.

Polar demodulators measure the magnitude and time delay of $v_b(t)$ for a particular frequency. $|Z_b|$ and θ can be calculated as follows:

$$|Z_b| = \frac{|v_b(t)|}{|I_{IN}|}, \tag{1}$$

$$\theta = \frac{T_b}{T_{in}}, \tag{2}$$

where $|v_b(t)|$ and T_b are the magnitude and time delay of $v_b(t)$, respectively. $|I_{IN}|$ is the magnitude of $i_{in}(t)$, and T_{in} is the period of $i_{in}(t)$ and $v_b(t)$. Impedance spectrum is obtained by using the FRA method, which analyzes one frequency at a time, repeatedly with sweeping the frequency [9].

2.2. Phase Measurement Scheme in Polar Demodulators

Figure 2 shows the phase measurement scheme adopted in polar demodulators [10,11,15–17]. When $i_{in}(t)$ is injected into the target material and reference resistor, $v_r(t)$ with the same phase as $i_{in}(t)$ is generated from the resistor in addition to $v_b(t)$. Comparators convert $v_r(t)$ and $v_b(t)$ to clock signals, ϕ_r and ϕ_b, from which an XOR gate and an SR latch create clock signals, ϕ_{XOR} and ϕ_{SR}, respectively. These clock signals have the pulse width of T_b, which corresponds to θ, as shown in Figure 2. From Equation (2), when the frequency of the injected current, $f_{in} = 1/T_{in}$ is known, θ can be determined by measuring T_b.

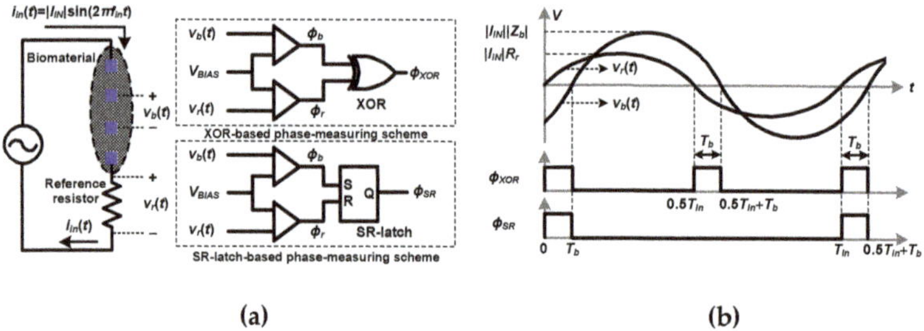

Figure 2. (a) Block diagram and (b) waveforms of the phase measurement scheme (reproduced from [19] with permission from the IEEE).

2.3. Design Specifications

Design specifications of TDC proposed in this paper are presented in Table 1. f_{in} is set from 1 kHz to 2.048 MHz because the proposed TDC is implemented in the form of a fully integrated chip for EIS systems in biomedical applications. Considering that the impedance measured in such applications [15] usually has capacitive reactance, θ ranges from 0° to 90°, and the corresponding input-time range is 0 to 122 ns at the shortest when f_{in} is 2.048 MHz and 0 to 250 µs at the widest when f_{in} is 1 kHz. Referring to performances of the previous works reported in [15,17], this TDC aims to have maximum phase error under 1° and phase resolution over 10 bits for the suggested range of f_{in}. This amount of phase error corresponds to 1.35 ns in the worst case when f_{in} is 2.048 MHz. Lastly, the frequency of the reference clock (f_{clk}) is set to 32.768 MHz, which is only 16 times the maximum f_{in}. Compared to the TDC in [17], where f_{clk} is 330 times higher than the maximum f_{in}, the proposed TDC aims to achieve competitive phase error performance with much lower f_{clk}.

Table 1. Design specifications of the proposed time-to-digital converter (TDC).

Parameter	Target Level
Application	EIS for biomedical applications
Range of the injected signal frequency (f_{in})	1 kHz to 2.048 MHz
Corresponding input-time range for 0° to 90°	0–122 ns (f_{in} = 2.048 MHz) to 0–250 µs (f_{in} = 1 kHz)
Phase resolution	>10 bit
Maximum phase error	1°
Corresponding time error	<~1.35 ns (f_{in} = 2.048 MHz), <~2.8 µs (f_{in} = 1 kHz)
Reference clock frequency (f_{clk})	32.768 MHz

3. Architecture of the Proposed TDC

As shown in Table 1, the requirements of both the input-time range and time resolution vary with f_{in}. The proposed TDC operates across three different modes to meet the design specifications with maintaining reasonable measurement time. This section describes overall architecture and operation in each mode.

3.1. Overall Architecture and Operation

The block diagram of the proposed TDC is presented in Figure 3. The input pulse signal whose pulse width carries θ is denoted as ϕ_{in}. ϕ_{clk} is the reference clock, and the outputs of the system are the digital bits, D_c, D_{f1}, and D_{f2}. This TDC is composed of three stages, namely a coarse stage, the first fine stage, and the second fine stage. The coarse stage consists of a 12-bit coarse counter with digital logics. Each fine stage consists of a time splitter, a reconfigurable time interpolator, and a 4-bit counter. The coarse counter is used to implement a wide input-time range, and two fine stages are employed to improve resolution without increasing f_{clk}. In each fine stage, the time splitter extracts quantization error of the preceding stage and the time interpolator stretches the quantization error. The resolution can be improved by quantizing the stretched quantization error through the fine counters and ϕ_{clk}.

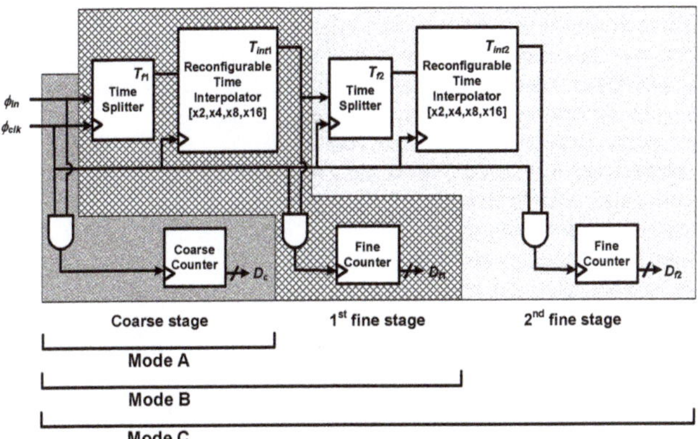

Figure 3. Block diagram of the proposed TDC (reproduced from [18] with permission from the IEEE).

The proposed TDC operates in one of the three modes to achieve the phase resolution over 10 bits. In mode A, only the coarse stage is used, and the time resolution is T_{clk}. For low f_{in} of 1 kHz and 2 kHz, f_{clk} is high enough to achieve target phase resolution. For 4-kHz f_{in} upwards, the fine stages are used with the coarse stage to further improve the time resolution without increasing f_{clk}. In mode B, the coarse stage and the first fine stage are used for f_{in} from 4 kHz to 32 kHz. The first fine stage further quantizes the quantization error of the coarse stage with a time interpolation factor of A_{T1}. The time resolution in mode B is T_{clk}/A_{T1}, which is A_{T1} times higher than the highest resolution in mode A while keeping f_{clk} = 32.768 MHz. In mode C, for 32-kHz f_{in} upwards, the coarse stage and the first fine stage operate in the same manner. In addition, the second fine stage further quantizes the quantization error of the first fine stage. The time resolution in mode C is $T_{clk}/(A_{T1}A_{T2})$, which is $A_{T1}A_{T2}$ times higher than the highest one in mode A, still keeping f_{clk} = 32.768 MHz. When f_{in} = 2.048 MHz, the coarse stage only achieves a 2-bit resolution for θ range from 0° to 90° with f_{clk} = 32.768 MHz. Therefore, the total interpolation factor (A_T) of up to 256 is required to achieve a 10-bit resolution for the whole f_{in} range. A_{T1} and A_{T2} are set, as shown in Table 2, across three different modes, A, B, and C, for varying values of f_{in}.

Table 2. Assignment of the interpolation factor for different operation modes and values of f_{in}.

Mode	f_{in} (Hz)	0° to 90°	f_{clk} (Hz)	Coarse	A_{T1}	A_{T2}
	2.048M	0~0.1221 μs	32.768M	2 bit	16	16
	1.024M	0~0.2441 μs	32.768M	3 bit	16	16
C	512k	0~0.4883 μs	32.768M	4 bit	16	16
	256k	0~0.9766 μs	32.768M	5 bit	16	8
	128k	0~1.9531 μs	32.768M	6 bit	16	4
	64k	0~3.9063 μs	32.768M	7 bit	16	2
	32k	0~7.8125 μs	32.768M	8 bit	16	N/A
B	16k	0~15.625 μs	32.768M	9 bit	8	N/A
	8k	0~31.25 μs	32.768M	10 bit	4	N/A
	4k	0~62.5 μs	32.768M	11 bit	2	N/A
A	2k	0~125 μs	32.768M	12 bit	N/A	N/A
	1k	0~250 μs	16.384M	12 bit	N/A	N/A

When utilizing the time interpolator, A_T is determined by the ratio of discharging capacitance and discharging current [20]. If the current increases for implementing a large A_T, the power consumption of TDC increases accordingly. Moreover, since the pulse width of the interpolated signal increases, the conversion time increases significantly, which leads to the degraded conversion rate. Thus, the large A_T is realized in two steps by dividing A_T into A_{T1} and A_{T2} to offer much more relaxed design conditions. A_{T1} and A_{T2} can be adjusted between 2, 4, 8, and 16 to provide 1, 2, 3, and 4 additional bits, respectively. Reconfigurable A_{T1} and A_{T2} maintain the phase resolution over 10 bits with reasonably short measurement time.

3.2. Operation in Mode A

This mode offers only counter-based time quantization, and the timing diagram of its operation is depicted in Figure 4.

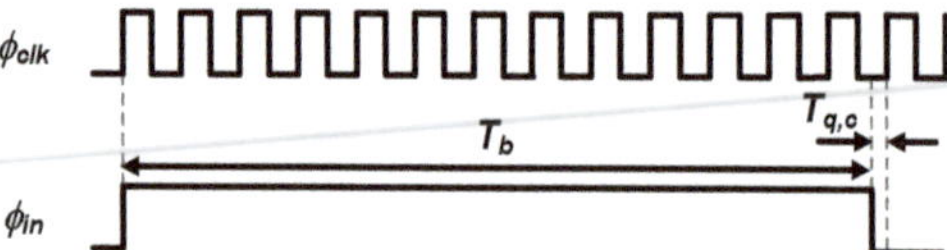

Figure 4. Timing diagram of the proposed TDC in mode A.

The proposed TDC in mode A outputs digital bits D_c with the relation as follows:

$$T_b = D_c \times T_{clk} - T_{q,c},\tag{3}$$

where $T_{q,c}$ is the quantization error of the coarse stage and smaller than T_{clk}. The time resolution in mode A becomes one period of the clock signal, T_{clk}. To achieve a 12-bit phase resolution for θ range from 0° to 90° at low f_{in}, this mode is used for f_{in} of up to 2 kHz.

3.3. Operation in Mode B

Mode B uses the coarse stage and the first fine stage, each generating output digital bits, D_c and D_{f1}, respectively. The timing diagram of its operation is presented in Figure 5.

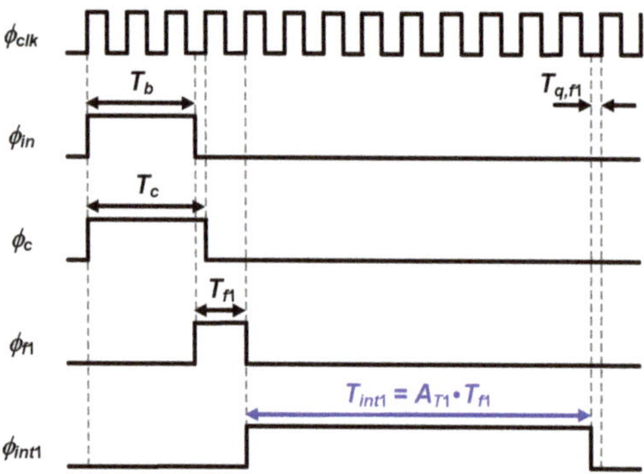

Figure 5. Timing diagram of the proposed TDC in Mode B.

The pulse width of T_{f1} is generated by the time splitter in the first fine stage and expressed as follows:

$$T_{f1} = T_{q,c} + T_{clk}.\tag{4}$$

The time interpolator in the first stage stretches T_{f1} to T_{int1}, which is described by:

$$T_{int1} = A_{T1} \cdot T_{f1}. \tag{5}$$

The fine counter quantizes T_{int1} with the reference clock, to output up to four fine digital bits, D_{f1}. The relation among T_b, D_c, and D_{f1} is given by:

$$T_b = T_c - \left(T_{f1} - T_{clk}\right) = (D_c \times T_{clk}) - \left(\frac{T_{int1}}{A_{T1}} - T_{clk}\right) = (D_c \times T_{clk}) - \left(\frac{D_{f1} \times T_{clk} - T_{q,f1}}{A_{T1}} - T_{clk}\right), \tag{6}$$

where $T_{q,f1}/A_{T1}$ is always smaller than T_{clk}. The time resolution in mode B is improved from T_{clk} to T_{clk}/A_{T1}, which is A_{T1} times higher than the highest resolution in mode A while keeping $f_{clk} = 32.768\,\text{MHz}$. To achieve a 12-bit phase resolution for θ range from 0° to 90°, f_{in} must satisfy the following condition:

$$\frac{1}{f_{in}} \cdot \frac{90}{360} \cdot \frac{1}{2^{12}} < \frac{T_{clk}}{A_{T1}}. \tag{7}$$

With the maximum A_{T1} of 16, this mode is used until f_{in} increases up to 32 kHz.

3.4. Operation in Mode C

The mode C uses the coarse stage, the first fine stage, and the second fine stage, each generating output digital bits, D_c, D_{f1}, and D_{f2}, respectively. The timing diagram of its operation is presented in Figure 6.

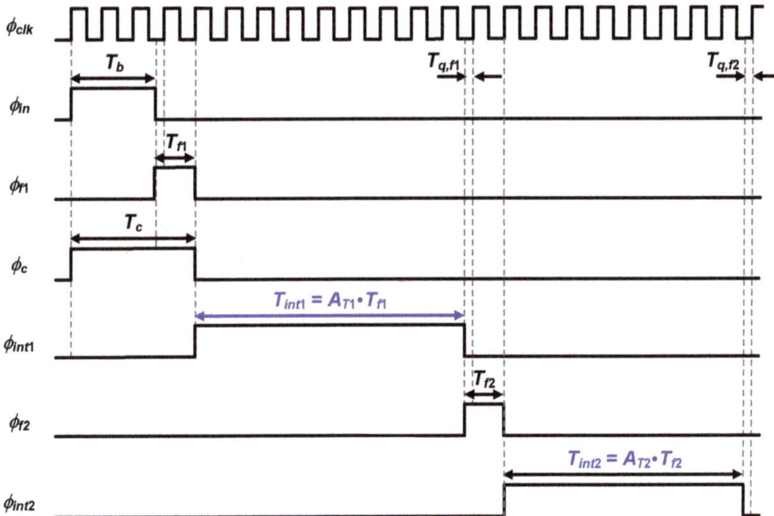

Figure 6. Timing diagram of the proposed TDC in mode C.

The coarse stage and the first fine stage operate in the same manner as in mode B. The time splitter in the second stage generates T_{f2}, which is expressed as:

$$T_{f2} = T_{q,f1} + T_{clk}. \tag{8}$$

The time interpolator in the second stage stretches T_{f2} to T_{int2}, which is described by:

$$T_{int2} = A_{T2} \cdot T_{f2}. \tag{9}$$

The fine counter in the second fine stage quantizes T_{int2} with the reference clock, to output up to four fine digital bits, D_2. The relation among T_b, D_c, D_{f1}, and D_{f2} is given by:

$$
\begin{aligned}
T_b &= (D_c \times T_{clk}) - \left(\frac{D_{f1} \times T_{clk} - T_{q,f1}}{A_{T1}} - T_{clk} \right) = (D_c \times T_{clk}) - \left\{ \frac{D_{f1} \times T_{clk} - (T_{f2} - T_{clk})}{A_{T1}} - T_{clk} \right\} \\
&= (D_c \times T_{clk}) - \left\{ \frac{D_{f1} \times T_{clk} - \left(\frac{D_{f2} \times T_{clk} - T_{q,f2}}{A_{T2}} - T_{clk} \right)}{A_{T1}} - T_{clk} \right\} \\
&= \{(D_c + 1) \times T_{clk}\} - \left\{ \frac{(D_{f1} - 1) \times T_{clk}}{A_{T1}} \right\} + \left(\frac{D_{f2} \times T_{clk} - T_{q,f2}}{A_{T1} A_{T2}} \right),
\end{aligned}
\tag{10}
$$

where $T_{q,f2}/(A_{T1} A_{T2})$ is always smaller than T_{clk}. The time resolution in mode C is improved from T_{clk} to $T_{clk}/(A_{T1} A_{T2})$, which is $A_{T1} A_{T2}$ times higher than the highest resolution in mode A, even though f_{clk} = 32.768 MHz is kept. To achieve a 10-bit phase resolution for θ range from 0° to 90°, f_{in} must satisfy the following condition:

$$
\frac{1}{f_{in}} \cdot \frac{90}{360} \cdot \frac{1}{2^{10}} < \frac{T_{clk}}{A_{T1} A_{T2}}.
\tag{11}
$$

with the maximum A_{T1} and A_{T2} of 16, this mode is used for f_{in} of up to 2.048 MHz.

4. Circuit Design

This section describes how the first fine stage is designed to realize time interpolation with reconfigurable A_{T1}. The second fine stage is designed to be identical to the first one.

4.1. Time Splitter

Figure 7 shows the structure of the time splitter, which consists of three D flip-flops and one NOR gate. As shown in Figures 5 and 6, the time splitter extracts the quantization error of the coarse stage with an offset of T_{clk}. This offset is employed to avoid the metastability issue of the D flip-flops [20]. Therefore, T_{f1} takes the value from T_{clk} to $2T_{clk}$, corresponding to 30.52 ns to 61.04 ns. As shown in Equations (6) and (10), in mode B and C, the offset is compensated when T_b is calculated from the digital outputs.

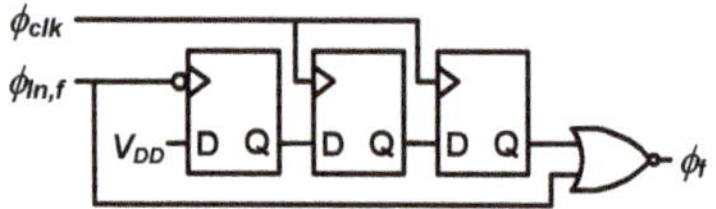

Figure 7. Structure of the time splitter (reproduced from [18] with permission from the IEEE).

4.2. Reconfigurable Time Interpolator

Figure 8 shows the block diagram and timing diagram of the reconfigurable time interpolator. The reconfigurable time interpolator is similar to the time interpolator in [20]. A variable discharging capacitor, which has the capacitance of C_{INT}, is added to obtain reconfigurable A_{T1}.

While ϕ_{f1}, which has the pulse width of T_{f1}, is high, a capacitor which has the capacitance of C_F is discharged by a constant current, I_F, such that $v_1(t)$ drops by ΔV. ΔV is expressed as follows:

$$
\Delta V = \frac{I_F \cdot T_{f1}}{C_F}.
\tag{12}
$$

From the falling edge of ϕ_{f1}, a capacitor of C_{INT} is discharged by a constant current, I_{INT}, during $\phi_{f1,q}$ is high. In the same manner with Equation (12), the pulse width of $\phi_{f1,q}$ is expressed as follows:

$$
A_{T1} \cdot T_{f1} = \frac{C_{INT}}{I_{INT}} \cdot \Delta V = \frac{C_{INT}}{I_{INT}} \cdot \frac{I_F}{C_F} \cdot T_{f1} = \frac{C_{INT}}{C_F} \cdot \frac{I_F}{I_{INT}} \cdot T_{f1} = (M \cdot N) \cdot T_{f1},
\tag{13}
$$

where M is the capacitance ratio, C_{INT}/C_F, N is the current ratio, I_F/I_{INT}, and A_{T1} is $M \cdot N$.

In the proposed TDC, N is kept constant while M is controlled to change A_{T1} between 2, 4, 8, and 16. Adjustment of the capacitance value is selected over the current value because controlling the capacitance ratio is more accurate than controlling the current ratio in IC implementation. In each fine stage, since N is fixed to 2, C_F is kept constant as 3.6 pF, and C_{INT} is changed across 3.6 pF, 7.2 pF, 14.4 pF, and 28.8 pF. I_F and I_{INT} are 80 µA and 40 µA, respectively.

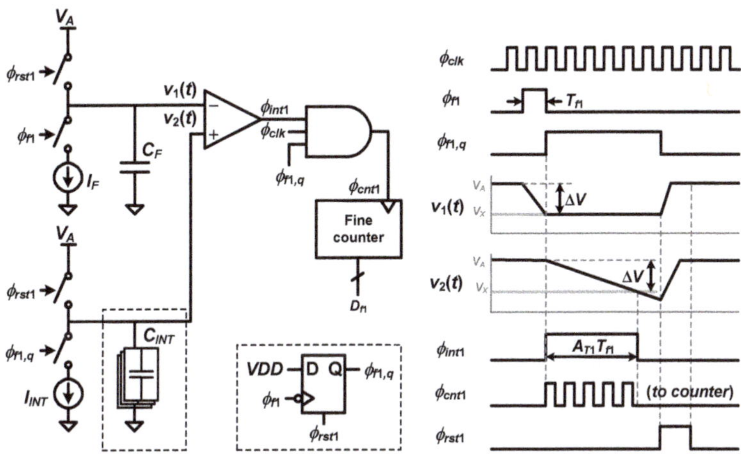

Figure 8. Block diagram and timing diagram of the reconfigurable time interpolator (reproduced from [18] with permission from the IEEE).

4.3. Novel Features of the Proposed TDC

The proposed TDC employs time interpolation technique, which can improve time resolution without increasing f_{clk} in the counter-based TDC. However, two inherent issues are associated with large A_T. Although the resolution becomes much higher when the front-stage quantization error is interpolated with A_T, the chip size or power consumption increases by a substantial amount because A_T is determined by capacitance ratio or current ratio. Also, the conversion time increases significantly because the interpolated pulse width increases as A_T increases. A novel structure of cascading two separate fine stages resolves these two issues at the same time.

When A_T of 256 is obtained by using a single-stage time interpolator with C_F of 1 pF and C_{INT} of 256 pF, this results in excessively large chip size and poor area efficiency. In the proposed TDC, two interpolation stages are cascaded. As a result, the interpolation factor of only 16 is required in each stage instead of 256. In other words, this system requires two capacitors with the size of C_{INT}, which is equal to 16 C_F. Compared to the single stage with A_T of 256, this approach reduces the area used for implementing the discharging capacitors by a factor of 257/34 = ~7.6 times considering that one time interpolator has two discharging capacitors with the sizes of C_{INT} and C_F. As the time interpolators in fine stages occupy a significant portion of the chip size, the area efficiency of the system is greatly improved. On the other hand, when A_T of 256 is set by the capacitance ratio, the current consumption of the cascaded time interpolation stages is two times higher than that of the single-stage time interpolator. For the single interpolator with A_T of 256, from the falling edge of ϕ_c, the conversion time of $A_T T_{f1}$ is required for fine conversion, and the maximum conversion time is $2A_T T_{clk}$ considering the offset of the time splitter. For two cascaded interpolation stages with A_{T1} and A_{T2} of 16, the conversion time of each fine stage is $A_{T1} T_{f1}$ or $A_{T2} T_{f1}$, and the maximum conversion time of each stage is $2A_{T1} T_{clk}$ or $2A_{T2} T_{clk}$. Compared to the single interpolator with A_T of 256, when $A_{T1} = A_{T2} = 16$, the conversion time for fine conversion is reduced by approximately 8 times.

When N_f of time interpolation stages are cascaded and A_T is set by the capacitance ratio, the area efficiency (E_A), power efficiency (E_P), and conversion-time efficiency for fine stages (E_C) can be defined and expressed as follows:

$$E_A = \frac{\text{Total capacitance in an interpolation stage when } N_f = 1}{\text{Total capacitance in interpolation stages when } N_f > 1} = \frac{(1 + A_T)C_F}{N_f\left(1 + \sqrt[N]{A_T}\right)C_F} = \frac{1 + A_T}{N_f\left(1 + \sqrt[N]{A_T}\right)}, \quad (14)$$

$$E_P = \frac{\text{The current consumption of an interpolation stage when } N_f = 1}{\text{The current consumption of interpolation stages when } N_f > 1} = \frac{I_{fine}}{N_f \cdot I_{fine}} = \frac{1}{N_f}, \quad (15)$$

$$E_C = \frac{\text{The max. conversion time through a fine stage when } N_f = 1}{\text{The max. conversion time through fine stages when } N_f > 1} \approx \frac{A_T \cdot 2T_{clk}}{N_f \cdot \sqrt[N]{A_T} \cdot 2T_{clk}} = \frac{A_T}{N_f \cdot \sqrt[N]{A_T}}, \quad (16)$$

where I_{fine} is the current consumption of a single time interpolation stage.

Table 3 summarizes E_A, E_P, and E_C calculated using Equations (14)–(16). Although the maximum $E_A \cdot E_P \cdot E_C$ value is obtained when $N_f = 4$, we chose $N_f = 2$ to minimize the current consumption while taking advantages of the cascaded time interpolators in terms of area and conversion time. The largest capacitor with size of 28.8 pF is small enough to integrate on chip and $2A_{T1}T_{clk} = \sim 1$ μs of conversion time for the fine stage is short enough.

Table 3. E_A, E_P, and E_C versus N_f when A_T is set by the capacitance ratio.

A_T	N_f	E_A	E_P	E_C	$E_A \cdot E_P \cdot E_C$
256	2	7.6	8	0.5	30.4
256	4	12.9	16	0.25	51.6
256	8	10.7	16	0.125	21.4

It is also possible to obtain A_T through the current ratio of I_F/I_{INT} in Figure 8. In this case, setting the current ratio to 256 directly affects the static power, severely degrading the power efficiency of the system. Total discharging current of cascaded time interpolation stages when $N_f = 2$ could be 7.6-times smaller than that of a single time interpolation stage when $N_f = 1$, and the total capacitance of cascaded time interpolation stages when $N_f = 2$ could be two times larger than that of a single time interpolation stage when $N_f = 1$. Compared to the single interpolator with A_T of 256 and $N_f = 1$, when $N_f = 2$, the time for fine conversion is reduced by approximately eight times. The optimization process when A_T is set by the current ratio would be similar to that when A_T is set by the capacitance ratio.

5. Measurement Results

The proposed TDC has been fabricated with a 0.25-μm CMOS process. The size of the circuit is 787 μm × 524 μm (0.412 mm^2). The chip photograph and layout of the fabricated IC are presented in Figure 9.

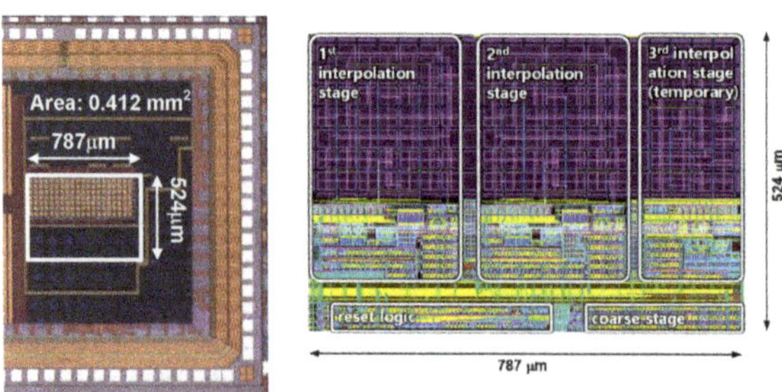

Figure 9. Chip photograph and layout.

Figure 10 shows the measurement setup. Two Agilent 33250A function generators are used, one for generating ϕ_{clk} and another for generating ϕ_{in} whose pulse width varies from 0 to 0.25 of the period which corresponds to the phase of 0° to 90°. These two function generators are synchronized with each other. In the case of generating ϕ_{in} in the form of pulse train, the period can be adjusted from 20.00 ns to 2000.0 s, and its pulse width can be controlled from 8.0 ns to 1999.9 s. Therefore, when f_{in} = 2.048 MHz, the phase above 5.9° could be measured. Arduino DUE was selected for a microcontroller unit because it has 54 digital I/O pins which are enough for receiving digital output bits and transmitting assignment codes through SPI for the reconfiguration of TDC. Moreover, its clock speed of 84 MHz allows generating required signals to initialize the SPI communication and select between the read and write modes. Monitoring pads are placed on the main propagation path of signal to examine whether each block and each stage operate as expected. Keysight Technologies DSO7104A oscilloscope was used throughout the measurement. The oscilloscope has enough sample rate of 4 Gsps, bandwidth of 1 GHz, and four scope channels.

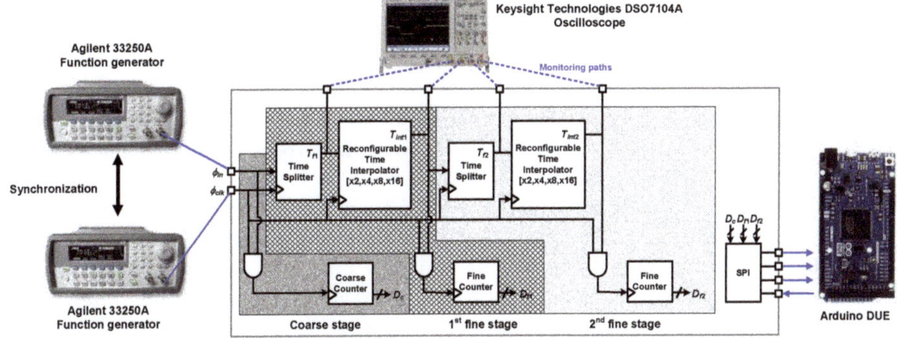

Figure 10. Measurement setup.

In this section, measurement results for one frequency in mode A, one in mode B, and one in mode C are shown. The input-output characteristics of the fabricated TDC are described in Figures 11a, 12a and 13a. Figures 11b, 12b and 13b present the phase errors in each frequency.

5.1. Mode A

Measurement results for f_{in} = 1 kHz are shown in Figure 11. Only the coarse stage is used, and the TDC operates as a counter-based TDC with f_{clk} = 16.384 MHz, halved from the 32.768-MHz clock signal provided by the function generator. The input-output characteristic of the TDC is shown in Figure 11a, where the horizontal axis indicates the pulse width of the input signal in seconds, and the vertical axis shows the TDC output code in units of LSB. The solid line represents theoretical output values for given pulse widths of the input. Figure 11b presents the phase error calculated from the digital output code as a function of the actual phase injected at the input. The input phase is varied with a step size of 0.0005°, which corresponds to about 1.4 ns. The difference between the maximum and minimum phase errors is about 0.022°, implying that the error of the measured output lies within 12-bit-resolution quantization error for θ range from 0° to 90° and that the TDC operates as expected.

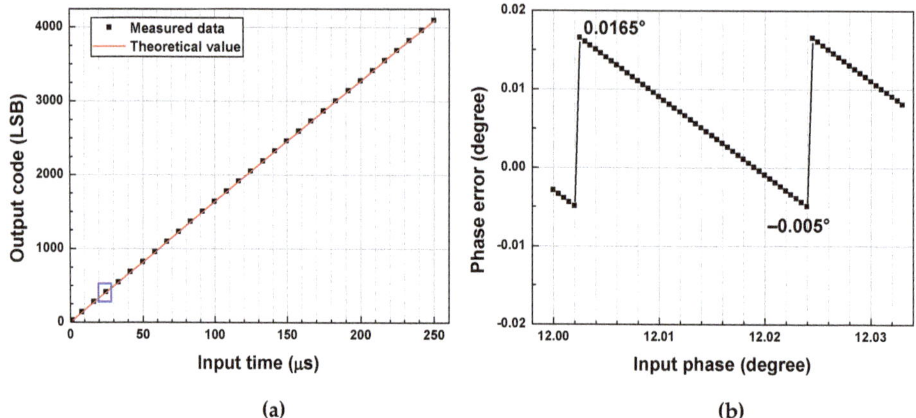

(a)   (b)

Figure 11. (**a**) Measured input-output characteristic of TDC, and (**b**) phase errors for f_{in} = 1 kHz.

5.2. Mode B

Measurement results for f_{in} = 8 kHz are shown in Figure 12. The coarse stage and the first fine stage are used with f_{clk} = 32.768 MHz and A_{T1} = 4. However, A_{T1} of 4.05 is obtained from the measured T_{int1} and T_{f1}, and this value is substituted to Equation (6) to derive T_b using output code. The input-output characteristic of the TDC is shown in Figure 12a, where the horizontal axis, vertical axis, and solid line are as described in Figure 11a. Figure 12b presents the phase error calculated from the digital output code as a function of the actual phase injected at the input. The input phase is varied with a step size of 0.0003°, which corresponds to 0.1 ns. The peak-to-peak phase error is about 0.022°, implying that the error of the measured output lies within 12-bit-resolution quantization error for θ range from 0° to 90°. This result shows a good agreement with the theoretical prediction.

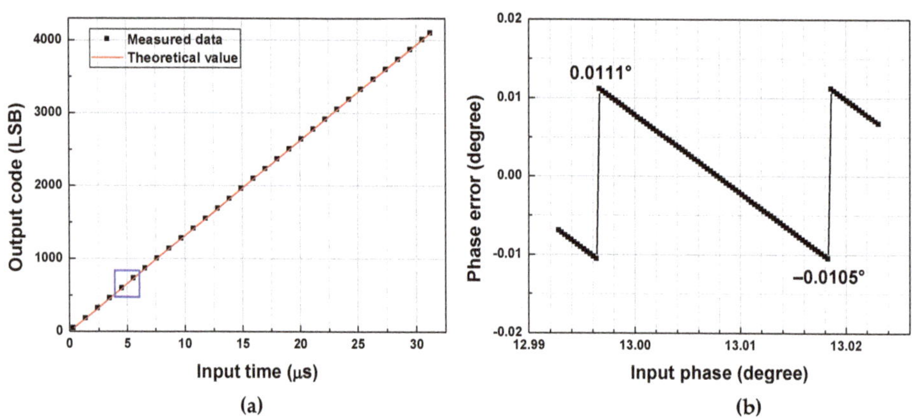

(a)   (b)

Figure 12. (**a**) Measured input-output characteristic of TDC, and (**b**) phase errors for f_{in} = 8 kHz.

5.3. Mode C

Measurement results for f_{in} = 2.048 MHz are shown in Figure 13. For f_{in} = 2.048 MHz, A_{T1} and A_{T2} are both set to 16 to achieve the largest A_T of 256. However, A_{T1} of 15.9 and A_{T2} of 15.9 are obtained, and these values are substituted to Equation (10) to derive T_b using output code. The input-output characteristic of the TDC is shown in Figure 13a, where the horizontal axis, vertical axis, and the solid line are as described in Figure 11a. Figure 13b presents the phase error calculated from the digital output code as a function of the actual phase injected at the input. The input phase is varied with a step size of 0.07°, which corresponds to 0.1 ns. The peak positive phase error is 1.123°, and the peak negative phase error is −1.846°, resulting in the peak-to-peak phase error of 2.969°. This phase error is beyond the target quantization error of 0.088° and target phase error of 1°. The error analysis for the mode-C operation is presented in the next sub-section with a summary of the measurement results.

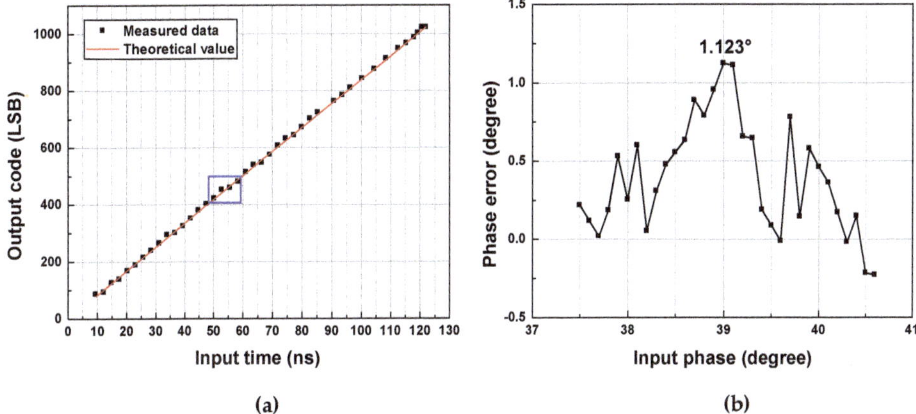

Figure 13. (**a**) Measured input-output characteristic of TDC, and (**b**) the maximum positive phase error for f_{in} = 2.048 MHz.

5.4. Error Analysis

The peak-to-peak phase error and corresponding time error for f_{in} from 1 kHz to 2.048 MHz are described in Figure 14. The peak-to-peak phase error is proportional to f_{in} and exceeds 1° for only when f_{in} = 1.024 MHz and f_{in} = 2.048 MHz.

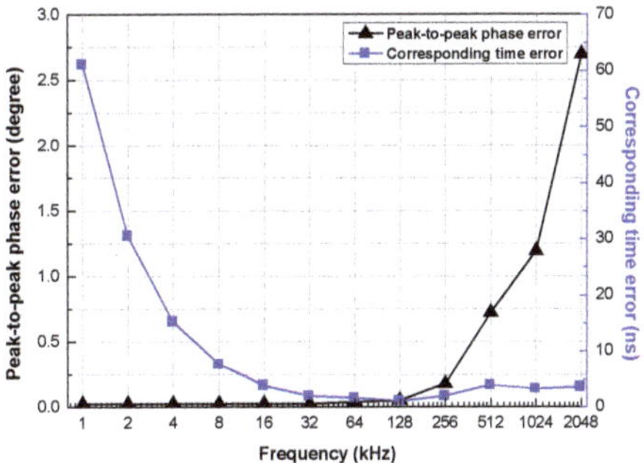

Figure 14. Peak-to-peak phase error and corresponding time error for f_{in} from 1 kHz to 2.048 MHz.

Table 4 summarizes the exact values of the peak-to-peak phase error and corresponding time error as a function of f_{in}. It also shows the values of A_{T1} and A_{T2} set for the measurement. Although A_{T2} increases, the corresponding time error does not decrease below 1.06 ns.

Table 4. Peak-to-peak phase error and corresponding time error in mode C.

f_{in} (Hz)	A_{T1}	A_{T2}	Target Phase Error (°)	Peak-to-Peak Phase Error (°)	Corresponding Time Error (ns)
64k	16	2	0.022	0.037	1.61
128k	16	4	0.022	0.049	1.06
256k	16	8	0.022	0.178	1.93
512k	16	16	0.022	0.721	3.91
1.024M	16	16	0.088	1.194	3.29
2.048M	16	16	0.088	2.696	3.66

These extra errors would come from the uncertainty of comparator operation in time interpolators. The schematic of comparators and the uncertainty caused by the comparator operation during time interpolation are shown in Figure 15a,b, respectively. The comparators are designed by cascading two self-biased inverters as in [21,22].

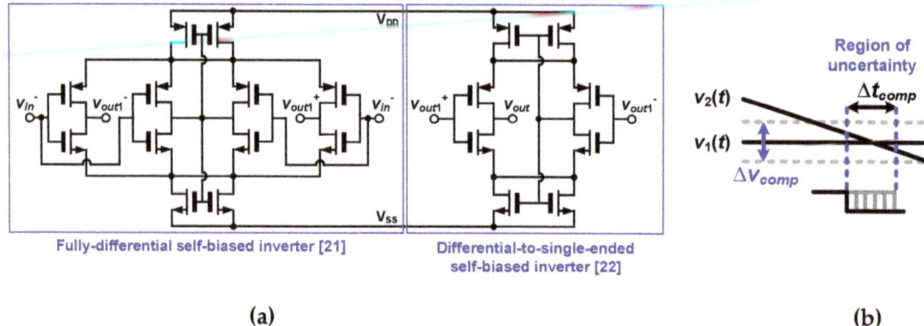

Figure 15. (**a**) Schematic of comparators and (**b**) the uncertainty caused by comparators during time interpolation.

As shown in Figure 15b, if there is a voltage-domain uncertainty of Δv_{comp} when $v_1(t)$ and $v_2(t)$ cross each other, Δv_{comp} causes a time-domain uncertainty of Δt_{comp}, which is given by:

$$\Delta t_{comp} = \Delta v_{comp} \frac{C_{INT}}{I_{INT}}. \tag{17}$$

Therefore, Δv_{comp} should be minimized in order to reduce Δt_{comp} and hence extra time errors. The finite resolution and noise of the comparator would cause the uncertainty in the voltage domain, which, in turn, is translated into the uncertainty in the time domain.

The resolution of the comparator ($\Delta v_{min,comp}$), which means the minimum input difference that saturates the output, is expressed as follows [23]:

$$\Delta v_{min,comp} = \frac{V_{OH} - V_{OL}}{A_{v0}}, \tag{18}$$

where A_{v0} is the open-loop gain of the comparator, V_{OH} is the output voltage when the output is high, and V_{OL} is the output voltage when the output is low. V_{OH} and V_{OL} should be large and small enough, respectively, so that the following digital logic can distinguish binary states. That is, when the transient behavior of $v_1(t)$ and $v_2(t)$ is not fast enough or A_{v0} is not sufficiently large, the comparator suffers from a substantial uncertainty in time domain (Δt_{comp}) because it will take relatively longer time for the following digital logic to determine binary states.

Considering that T_{f1} and T_{f2} vary within the range from T_{clk} to $2T_{clk}$, the crossing point of $v_1(t)$ and $v_2(t)$ in Figure 15b falls within the voltage range from 0.9 V to 1.4 V. In this voltage range, A_{v0} varies between 41.8 dB and 58.0 dB depending on the voltage level where $v_1(t)$ and $v_2(t)$ intersect.

From A_{v0} = 41.8 dB and supply voltage of 2.5 V, $\Delta v_{min,comp}$ is about 20.3 mV. For f_{in} = 128 kHz with A_{T1} = 15.9 and A_{T2} = 4.05, Δt_{comp} in the first stage caused by $\Delta v_{min,comp}$ is calculated as about 14.6 ns by substituting C_{INT} = 28.8 pF and I_{INT} = 40 µA into Equation (17). This Δt_{comp} in the first fine stage is divided by A_{T1} after being quantized by the fine counter. Therefore, the fabricated TDC could not achieve the time error below 0.92 ns.

The noise of comparator would become another source of uncertainty. The input-referred noise voltage of the differential-to-single-ended self-biased inverter in Figure 15a is expressed as follows [24]:

$$\overline{V_{n,inv}^{2}(f)} = \left[\frac{1}{W_P L_P} + \frac{1}{W_N L_N}\right] \cdot \frac{2K}{f \cdot C_{OX}} + \frac{8kT\gamma}{g_{m,N} + g_{m,P}}, \tag{19}$$

where W_P and L_P are the width and length of the input PMOS transistor, respectively, while W_N and L_N are the width and length of the input NMOS transistor, respectively. K is the process-dependent flicker noise constant. $g_{m,N}$ and $g_{m,P}$ are the transconductances of the NMOS and PMOS input transistors, respectively. The input-referred noise of the fully differential self-biased inverter is also given by Equation (19).

Since the noise of the first stage is dominant compared to that of the second stage and the 1/f noise can be ignored because of the wide bandwidth of the comparator, the input-referred noise voltage of the comparator can be approximated as follows:

$$\overline{V_{n,comp}^{2}(f)} \approx \frac{8kT\gamma}{g_{m,N} + g_{m,P}}. \tag{20}$$

When T = 300 K, γ = 1, and noise bandwidth = 100 MHz, the input-referred noise of the comparator is about 1.8 mV$_{rms}$, 180 µV$_{rms}$, and 18 µV$_{rms}$ for $g_{m,N} + g_{m,P}$ = 1 µS, $g_{m,N} + g_{m,P}$ = 10 µS, $g_{m,N} + g_{m,P}$ = 100 µS, respectively. Since our TDC consumes enough current, the uncertainty due to the comparator noise is not significant in our design. However, if the bandwidth of the comparator is very large and $g_{m,N} + g_{m,P}$ is small, considerable uncertainties may occur. For f_{in} = 128 kHz with A_{T1} = 15.9 and A_{T2} = 4.05, the voltage-domain uncertainty of 1.8 mV$_{rms}$ causes Δt_{comp} = 1.3 ns$_{rms}$, which corresponds to the time error of 0.08 ns$_{rms}$. In particular, it is important to ensure that the value of $g_{m,N} + g_{m,P}$ is sufficiently large when the bandwidth of the comparator is wide.

Since the extra time errors are mainly due to the resolution of comparators, the errors would be mitigated if the comparators based on multi-stage amplifiers are used as in [15]. In such comparators, the optimum number of amplifier stages, N_{OPT}, is expressed as follows [15]:

$$N_{OPT} \approx 1.1 \times \ln\left(\frac{V_{OH}}{\Delta v_{min,comp}}\right) + 0.79. \tag{21}$$

By replacing the two-stage high-gain amplifiers with multiple stages of low-gain amplifiers, the improved Δt_{comp} can be obtained as $\Delta v_{min,comp}$ is reduced.

Moreover, when A_{T2} increases from 8 to 16, the time error increases rather than decreases. Therefore, another way of improving the extra time error is measuring the quantization error of the first stage without performing time interpolation of the second fine stage. In [19], we employed a chain delay line in the second fine stage instead of the time interpolator, and the results were verified by simulation. Since the mismatches between unit delay cells would occur, the proposed architecture needs to be verified by measurement.

5.5. Performance Summary and Comparison

Table 5 summarizes performances of the presented TDC, together with those of two polar demodulators and two TDCs, reported previously. In comparison with the TDC in [20] that consists of a coarse counter and a single-stage time interpolator with large A_T of 250, our TDC offers lower complexity and shorter conversion time for achieving the same phase resolution, as analyzed in

Section 4.3. Compared to the TDC presented in this manuscript, another kind of our TDC in [19] seems to achieve smaller phase error. However, the TDC in [19] has not yet been verified by measurement. Through our future work, the TDC in [19] will be fabricated and measured. The phase error performance of our TDC presented in this paper is competitive when compared with previously reported designs in [15,17].

Table 5. Performance summary and comparison.

	This Work	*IEEE Sensors J.'* **2013** [15]	**IEEE MWSCAS' 2013** [17]	*IEEE TNS.'* **2006** [20]	**IEEE MWSCAS' 2017** [19]
Tech.	0.25 μm	0.35 μm	0.35 μm	0.35 μm	0.18 μm
Application	EIS	EIS	EIS	N/A	EIS
Implementation scope	TDC	Polar demodulator	Polar demodulator	TDC	TDC
Architecture	Counter + Cascaded time interpolator	TVC with ADC	Counter	Counter + time stretchers	Counter+ time stretcher + chain-delay-line
f_{in}	1 kHz–2.048 MHz	0.1 kHz–100 kHz	0.1 kHz–10 MHz	N/A	1 kHz–2.048 MHz
f_{clk}	32.768 MHz	No use	3.33 GHz	80 MHz	32.768 MHz
Power	7.5 mW	21 mW	*28 mW	0.75 mW	**2.4 mW**
Supply	2.5 V	2.5 V	1.8 V	3 V–4 V	1.8 V
Area	0.41 mm^2	*0.40 mm^2	*0.40 mm^2	0.23 mm^2	0.35 mm^2
Time resolution	N/A	N/A	300 ps	50 ps	103 ps–244 ns
Phase error	<0.72° (@512 kHz) <2.70° (@2.048 MHz)	<3.95°	<2.2°	N/A	<0.088°
Remarks	Meas.	Meas.	Meas.	Meas.	Sim.

* Total power consumption or size of entire polar demodulator for EIS system.

6. Conclusions

A reconfigurable time-to-digital converter (TDC) used to quantize the phase of impedance in electrical impedance spectroscopy (EIS) is introduced in this manuscript and verified through the fabricated IC. This TDC adopts a coarse counter to have a wide input-time range and cascaded time interpolators to improve resolution in the high-frequency analysis without increasing counting clock speed. When the same large interpolation factor is assumed, the cascaded time interpolators have shorter measurement time and smaller chip area than a single-stage time interpolator. The reconfigurable time interpolation factor maintains phase resolution within an appropriate level while providing reasonable measurement time. The fabricated TDC achieves the peak-to-peak phase error of under 0.72° for the input frequency range from 1 kHz to 512 kHz and the peak-to-peak phase error of 2.70° when it covers up to 2.048 MHz, demonstrating competitive performances in comparison with previously reported designs. Two precision improvement methods are also proposed: revision of the comparator, and revision of the second fine stage. As future work, we plan to fabricate the TDC that uses chain delay lines in the second fine stage and compare it with the TDC based on cascaded time interpolators.

Sensors **2020**, *20*, 1889

Author Contributions: S.S., S.-J.K., and J.-H.P. implemented and measured the IC. S.S. and Y.J. analyzed the measurement results and wrote the draft of the manuscript through E.L. and J.K.'s editing and discussion. H.-J.Y. and M.J. helped in setting the target specification. S.-J.K. is the corresponding author, reviewed and edited this manuscript through discussion with H.-J.Y. and M.J. All authors discussed the results and commented on the manuscript. All authors have read and agreed to the published version of the manuscript.

Funding: This research was supported in part by Basic Science Research Program under grant number NRF-2018R1A6A3A01013018 and BK21 plus program through the National Research Foundation (NRF) of Korea funded by the Ministry of Education. In addition, this research was supported in part by the Convergence Technology Development Program for Bionic Arm (2017M3C1B2085296) through NRF of Korea funded by the Ministry of Science & ICT.

Acknowledgments: The EDA tool was supported by the IC Design Education Center (IDEC), Korea.

Conflicts of Interest: The authors declare no conflict of interest.

References

1. Choi, A.; Kim, J.Y.; Jo, S.; Jee, J.H.; Heymsfield, S.B.; Bhagat, Y.A.; Kim, I.; Cho, J. Smartphone-based bioelectrical impedance analysis devices for daily obesity management. *Sensors* **2015**, *15*, 22151–22166. [CrossRef] [PubMed]

2. Shin, S.-C.; Lee, J.; Choe, S.; Yang, H.I.; Min, J.; Ahn, K.-Y.; Jeon, J.Y.; Kang, H.-G. Dry electrode-based body fat estimation system with anthropometric data for use in a wearable device. *Sensors* **2019**, *19*, 2177. [CrossRef] [PubMed]

3. Ferreira, J.; Pau, I.; Lindecrantz, K.; Seoane, F. A handheld and textile-enabled bioimpedance system for ubiquitous body composition analysis. An initial functional validation. *IEEE J. Biomed. Health Informat.* **2017**, *21*, 1224–1232. [CrossRef] [PubMed]

4. Halter, R.J.; Hartov, A.; Heaney, J.A.; Paulsen, K.D.; Schned, A.R. Electrical impedance spectroscopy of the human prostate. *IEEE Trans. Biomed. Eng.* **2007**, *54*, 1321–1327. [CrossRef] [PubMed]

5. Åberg, P.; Nicander, I.; Hansson, J.; Geladi, P.; Holmgren, U.; Ollmar, S. Skin cancer identification using multifrequency electrical impedance–a potential screening tool. *IEEE Trans. Biomed. Eng.* **2004**, *51*, 2097–2102. [CrossRef] [PubMed]

6. Kerner, T.E.; Paulsen, K.D.; Hartov, A.; Soho, S.K.; Poplack, S.P. Electrical impedance spectroscopy of the breast: Clinical imaging results in 26 subjects. *IEEE Trans. Med. Imag.* **2002**, *21*, 638–645. [CrossRef] [PubMed]

7. Nyrén, M.; Kuzmina, N.; Emtestam, L. Electrical impedance as a potential tool to distinguish between allergic and irritant contact dermatitis. *J. Am. Acad. Dermatol.* **2003**, *48*, 394–400. [CrossRef] [PubMed]

8. Dean, D.A.; Ramanathan, T.; Machado, D.; Sundararajan, R. Electrical impedance spectroscopy study of biological tissues. *J. Electrostat.* **2008**, *66*, 165–177. [CrossRef] [PubMed]

9. Wei, C.-L.; Wang, Y.-W.; Liu, B.-D. Wide-range filter-based sinusoidal wave synthesizer for electrochemical impedance spectroscopy measurements. *IEEE Trans. Biomed. Circuits Syst.* **2014**, *8*, 442–450.

10. Kweon, S.-J.; Park, J.-H.; Shin, S.-H.; Yoo, H.-J. A low-power polar demodulator for impedance spectroscopy based on a novel sampling scheme. In Proceedings of the 2015 International SoC Design Conference (ISOCC), Gyungju, South Korea, 3–7 November 2015; pp. 331–332.

11. Kweon, S.-J.; Shin, S.; Park, J.-H.; Suh, J.-H.; Yoo, H.-J. A CMOS low-power polar demodulator for electrical bioimpedance spectroscopy using adaptive self-sampling schemes. In Proceedings of the 2016 IEEE Biomedical Circuits and Systems Conference (BioCAS), Shanghai, China, 17–19 October 2013; pp. 284–287.

12. Manickam, A.; Chevalier, A.; McDermott, M.; Ellington, A.D.; Hassibi, A. A CMOS electrochemical impedance spectroscopy (EIS) biosensor array. *IEEE Trans. Biomed. Circuits Syst.* **2010**, *4*, 379–390. [CrossRef] [PubMed]

13. Hong, S.; Lee, K.; Ha, U.; Kim, H.; Lee, Y.; Kim, Y.; Yoo, H. A 4.9 mΩ-sensitivity mobile electrical impedance tomography IC for early breast-cancer detection system. *IEEE J. Solid-State Circuits* **2015**, *50*, 245–257. [CrossRef]

14. Li, H.; Liu, X.; Li, L.; Mu, X.; Genov, R.; Mason, A.J. CMOS electrochemical instrumentation for biosensor microsystems: A review. *Sensors* **2017**, *17*, 74. [CrossRef] [PubMed]

15. Kassanos, P.; Triantis, I.F.; Demosthenous, A. A CMOS magnitude/phase measurement chip for impedance spectroscopy. *IEEE Sens. J.* **2013**, *13*, 2229–2236. [CrossRef]

16. Ramos, J.; Ausín, J.L.; Torelli, G. A 1-MHz analog front-end for a wireless bioelectrical impedance sensor. In Proceedings of the 6th Conference on Ph.D. Research in Microelectronics Electronics, Berlin, Germany, 18–21 July 2010; pp. 1–4.
17. Huang, H.; Palermo, S. A TDC-based front-end for rapid impedance spectroscopy. In Proceedings of the 2013 IEEE 56th International Midwest Symposium on Circuits and Systems (MWSCAS), Columbus, OH, USA, 4–7 August 2013; pp. 169–172.
18. Shin, S.; Kweon, S.-J.; Park, J.-H.; Choi, Y.-C.; Yoo, H.-J. An efficient, wide range time-to-digital converter using cascaded time-interpolation stages for electrical impedance spectroscopy. In Proceedings of the 2016 IEEE Asia Pacific Conference on Circuits and Systems (APCCAS), Jeju, South Korea, 25–28 October 2016; pp. 425–428.
19. Kweon, S.-J.; Park, J.-H.; Shin, S.; Yoo, S.-S.; Yoo, H.-J. A reconfigurable time-to-digital converter based on time stretcher and chain-delay-line for electrical bioimpedance spectroscopy. In Proceedings of the 2017 IEEE 60th International Midwest Symposium on Circuits and Systems (MWSCAS), Boston, MA, USA, 6–9 August 2017; pp. 1037–1040.
20. Chen, P.; Chen, C.-C.; Shen, Y.-S. A low-cost low-power CMOS time-to-digital converter based on pulse stretching. *IEEE Trans. Nucl. Sci.* **2006**, *53*, 2215–2220. [CrossRef]
21. Milovanovic, V.; Zimmermann, H. Analyses of single-stage complementary self-biased CMOS differential amplifiers. In Proceedings of the NORCHIP 2012, Cpenhagen, Denmark, 12–13 November 2012.
22. Tan, M.-T.; Chang, J.S.; Tong, Y.-C. A process-independent threshold voltage inverter-comparator for pulse width modulation applications. In Proceedings of the 6th IEEE International Conference on Electronics, Circuits and Systems, Pafos, Cyprus, 5–8 September 1999; pp. 1201–1204.
23. Allen, P.E.; Holberg, D.R. *CMOS Analog Circuit Design*, 2nd ed.; Oxford University Press: New York, NY, USA, 2002.
24. Kim, J.; Maitra, R.; Pedrotti, K.D.; Dunbar, W.B. A patch-clamp ASIC for nanopore-based DNA analysis. *IEEE Trans. Biomed. Circuits Syst.* **2013**, *7*, 285–295. [CrossRef] [PubMed]

© 2020 by the authors. Licensee MDPI, Basel, Switzerland. This article is an open access article distributed under the terms and conditions of the Creative Commons Attribution (CC BY) license (http://creativecommons.org/licenses/by/4.0/).

Article

Miniaturized 0.13-µm CMOS Front-End Analog for AlN PMUT Arrays

Iván Zamora, Eyglis Ledesma, Arantxa Uranga and Núria Barniol *

Departament d'Enginyeria Electrònica, Universitat Autónoma de Barcelona, 08193 Bellaterra, Spain;
ivan.zamora@uab.es (I.Z.); eyglis.ledesma@uab.es (E.L.); arantxa.uranga@uab.es (A.U.)
* Correspondence: nuria.barniol@uab.cat

Received: 30 January 2020; Accepted: 20 February 2020; Published: 22 February 2020

Abstract: This paper presents an analog front-end transceiver for an ultrasound imaging system based on a high-voltage (HV) transmitter, a low-noise front-end amplifier (RX), and a complementary-metal-oxide-semiconductor, aluminum nitride, piezoelectric micromachined ultrasonic transducer (CMOS-AlN-PMUT). The system was designed using the 0.13-µm Silterra CMOS process and the MEMS-on-CMOS platform, which allowed for the implementation of an AlN PMUT on top of the CMOS-integrated circuit. The HV transmitter drives a column of six 80-µm-square PMUTs excited with 32 V in order to generate enough acoustic pressure at a 2.1-mm axial distance. On the reception side, another six 80-µm-square PMUT columns convert the received echo into an electric charge that is amplified by the receiver front-end amplifier. A comparative analysis between a voltage front-end amplifier (VA) based on capacitive integration and a charge-sensitive front-end amplifier (CSA) is presented. Electrical and acoustic experiments successfully demonstrated the functionality of the designed low-power analog front-end circuitry, which outperformed a state-of-the art front-end application-specific integrated circuit (ASIC) in terms of power consumption, noise performance, and area.

Keywords: ultrasound; PMUT; high-voltage (HV) transmitter; low-voltage receiver (RX) amplifier; ultrasound application-specific integrated circuit (ASIC); monolithical integration; CMOS; MEMS

---

1. Introduction

Ultrasound, since it was discovered, has been a widely used tool for multiple applications such as medical echography and nondestructive testing. However, today the ultrasound sensing market is showing an impressive resurgence: new applications along with improved manufacturing capabilities and advanced technological readiness are driving the growth of micromachined ultrasound transducers (MUTs). Volumetric medical imaging [1], in vivo and in vitro neuromodulation ultrasound [2], fingerprint sensing [3], and gesture recognition [4] are some new applications based on MUTs.

Nowadays, two different micromachined ultrasound transducers can be found in the literature [5]: the first one is based on a capacitive resonant element (CMUT) that consists of a thin metallized suspended membrane over a cavity with a rigid metallized substrate. A DC voltage is applied between two electrodes, which causes the membrane to deflect toward the substrate due to the electrostatic force. Therefore, an ultrasound can be generated in the surrounding medium from the vibration of the membrane when AC voltage is imposed. CMUTs, although they can be integrated over a preprocessed complementary-metal-oxide-semiconductor (CMOS) substrate, have the following drawbacks: they require a large DC bias, different transmission and reception arrays may be required for imaging applications, and they have higher equivalent capacitance. The second transducer is based on piezoelectric materials (PMUTs), in which (in contrast to CMUTs) the deflection of the membrane is produced by the lateral strain generated from a piezoelectric actuation, whereby the membrane must include at least one piezoelectric layer as well as a passive elastic layer. Although PMUTs have low electromechanical coupling coefficients and produce little output acoustic pressure with narrow bandwidths compared to CMUTs, PMUTs do not require DC bias, they have fewer geometric and design constraints, and the same transmission and reception arrays with equal coupling may be used for imaging applications [5]. Additionally, the capability of using a low thermal process for the piezoelectrical material, such as aluminum nitride (AlN), allows for the possibility of monolithically integrating PMUTs on CMOS.

Figure 1 shows a typical block diagram for ultrasound imaging applications using PMUTs. On the transmission side, the high-voltage (HV) pulser is able to drive the ultrasound transducer to emit enough acoustic pressure. In order to focus on a specific point, a transmission beamforming controller is used to generate the respective delays. A high-voltage DC–DC Converter may be required to generate the supply voltage of the HV pulser. On the reception path, isolation switches included in the front-end amplifier functional block are used to isolate the HV present in the transmission of low-voltage circuits used in the reception. Due to the weak electrical signals generated by the PMUTs, a low-noise amplifier (LNA) is employed to amplify these signals. After amplification, the amplitudes of these signals are extracted by an envelope detector and captured by a sample-and-hold circuit, which will later be digitized and processed. In order to obtain an image, most ultrasound measurements are based on the pulse echo method, where a generated short ultrasound pulse is propagated in a specified direction and is partly reflected wherever there is a change in the acoustic properties of the medium. The reflected ultrasound echo is processed in order to obtain information about distance, shape, and the physical properties of the target. An image is formed by mapping the echo strength versus travel time (proportional to the distance).

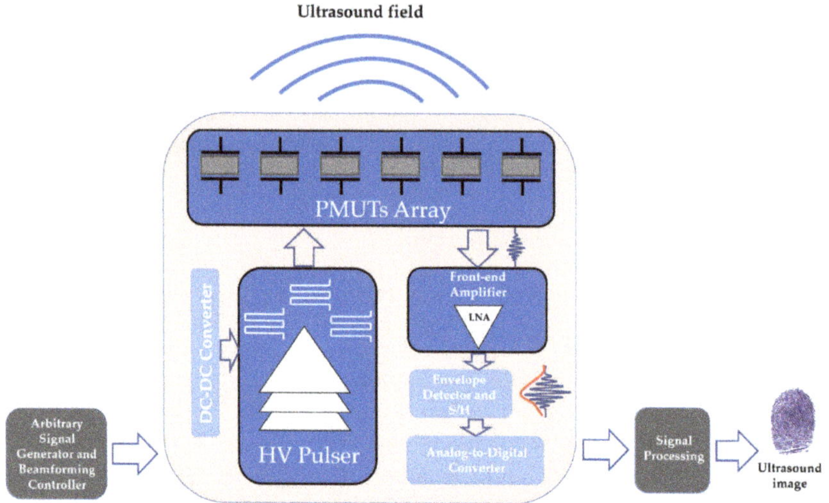

Figure 1. Image of an ultrasonic system block diagram.

The integration of PMUTs and CMOS to form a single chip allows for a reduction in parasitic elements, size, weight, and power consumption of the overall system. Recent works [3,6–8] have presented a single-chip ultrasonic fingerprint sensor based on PMUTs that are directly bonded to a CMOS readout application-specific integrated circuit (ASIC) using Al-Ge eutectic bonding. This integration strategy (presented by Reference [8]) is a considerable improvement over the wire bonding method and reduces the electrical parasitic elements, but it is a very expensive process, and due to the presence of special eutectic bonding, the fill factor is affected. In order to overcome these drawbacks, the Silterra foundry has implemented a MEMS-on-CMOS platform, where an AlN-based PMUT can be fabricated over CMOS preprocessed wafers. This monolithic integration of a PMUT over the CMOS reduces fabrication complexity, minimizes the interconnections between PMUT neighbors and the CMOS electronic circuitry, and reduces the parasitic elements; consequently, a higher fill factor and improved signal integrity are possible. In this work, we present the design of a 0.13-μm CMOS front-end transmitter and receiver using the MEMS-on-CMOS technological platform from Silterra [9]. This paper is organized as follows: Section 2 describes the design of the system, and in particular, Section 2.1 presents the ultrasonic transducer; Section 2.2 describes the high-voltage transmitter circuitry and its simulation results; Section 2.3 describes two of the front-end amplifiers that are most used to convert the electric charge generated by PMUT into voltage; and Section 2.4 compares both reception (RX) approaches, making use of the simulation results. Section 3 presents the electrical characterization of the circuitry, and Section 4 shows the acoustic experiments, including an ultrasound image.

2. Front-End Analog Circuit Design

2.1. Ultrasonic Transducer

As proof of concept for the integrated front-end circuitry, an 80-μm/side AlN PMUT in a 6 × 6 array configuration was used [10]. The electrical capacitance of a single PMUT was estimated at around 200 fF, adding up to a total of 1.2 pF per row. The obtained resonance frequency in fluorinert (FC-70) was in the 3-MHz range, with a bandwidth around 1 MHz [10].

2.2. High-Voltage Transmitter

The transmitter circuit is in charge of the generation of a large voltage signal in order to excite the transducer at its resonance frequency (f_r) and provide enough acoustic pressure for a high signal-to-noise ratio in the echo measurements. We set the needed actuation voltage to a maximum voltage of 32 V to guarantee enough output pressure [10]. Additionally, and taking into account the fact that the square wave increases the effective amplitude 1.27 times compared to a sine wave, we chose a squared-wave transmitter circuit with a 32-V peak voltage and with rising and falling times less than 100 ns to guarantee at least a 50% duty cycle and a 3-MHz driving frequency (according to the expected PMUT resonance frequency in fluorinert).

Previous transmitter circuits [11,12] have used a conventional level shifter that consists of an nMOS differential pair and cross-coupled pMOS transistors. This structure, although it is very simple, has two main problems. On the one hand, the input transistors (nMOS) need to be large in order to overpower the pMOS load when the inputs change, resulting in high input capacitance and therefore limiting the maximum operation speed. On the other hand, a large crowbar current flows when both the nMOS and pMOS transistors are conducting, affecting the efficiency and power consumption of the system. In Reference [13], a charge-recycling HV pulser (CRHV) was presented in order to improve the power consumption, although it needs four control signals with different logic levels per channel, which makes the circuitry more complex.

In order to solve the issues mentioned above, we chose the circuit proposed by Reference [14]. This circuit has the basic structure of a conventional level shifter (see Figure 2), but introduces additional high-voltage devices (Msw+ and Msw-) in a series with latching transistors (M3 and M4) in order to avoid the crowbar current. These new transistors work as switches controlled by the high-voltage signals Vsw+ and Vsw-, which inhibit the current of M3 and M4 when the nMOS devices (M1 and M2) are driving the current. The control signals (Vsw+ and Vsw-) are obtained from the output of a second dynamic level shifter, where its input signal is delayed with respect to Vin+ [15].

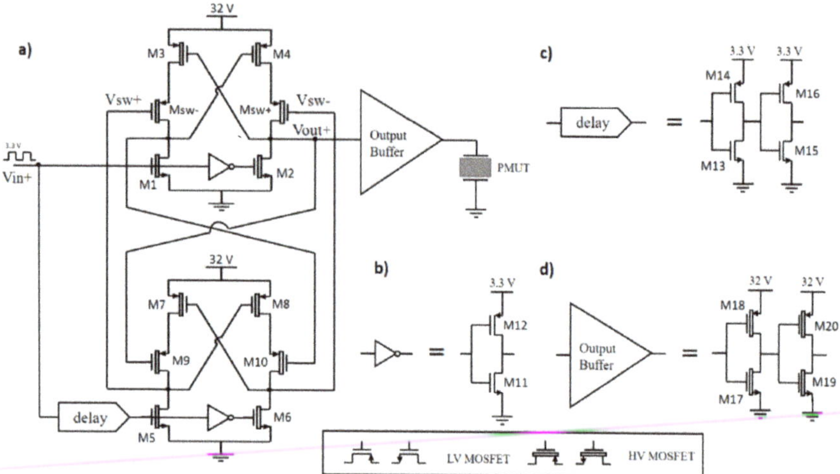

Figure 2. Schematic of the high-voltage transmitter. (**a**) Complete architecture to excite the piezoelectric micromachined ultrasonic transducer (PMUT). (**b**) An inverter circuit. (**c**) Cascaded inverter used as a delay element and (**d**) an output buffer using an inverter string to drive large capacitive loads.

In order to optimize the maximum operation speed and dynamic power consumption, a proper dimensioning of the nMOS and pMOS transistor sizes was done. Figure 3 shows the simulated source and sink currents and the rise and fall times in the output node (Vout+ node). It can be seen how the presence of the Msw+ and Msw- devices allows for the removal of the crowbar current. A high-voltage inverter buffer was designed to drive large capacitive loads. The dimensions of the MOSFETs selected for this circuit are shown in Table 1.

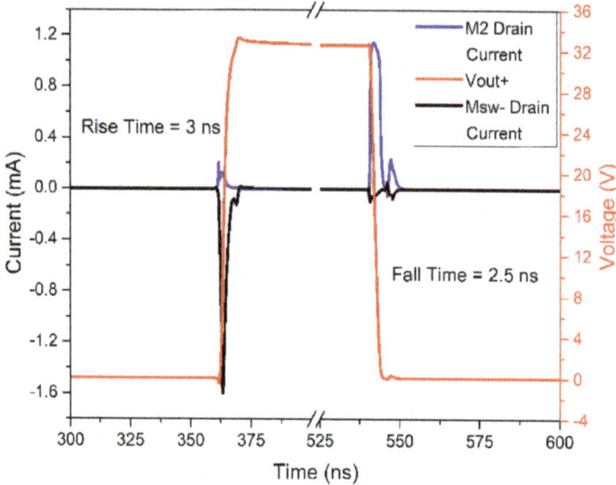

Figure 3. Voltage and source and sink currents in the output node (Vout+) [15].

Table 1. Dimensions of the MOSFET devices used in the transmitter.

MOSFETs	Aspect Ratio (W/L) (µm/µm)
M1, M2	20/5.75
M3, M4, Msw+, Msw−	10/4
M5, M6	10/5.75
M7, M8, M9, M10	4.7/4
M11	10/1
M12	25/1
M13	1/4
M14	2.5/4
M15	1/3
M16	2.5/3
M17	10/4.8
M18	25/4
M19	20/4.8
M20	50/4

In most imaging ultrasound applications, the transducer is an array of PMUTs. To obtain enough acoustic pressure level, several PMUTs of the array emit simultaneously, forming a transmit channel. In order to reduce the size of the die, one HV transmitter is used per channel, and therefore it has to be able to drive a large load capacitance at the resonance frequency. Figure 4 shows the simulation results of the maximum operation frequency of our HV transmitter, taking into account the designed output buffer, as a function of the load capacitance. Taking into account that the capacitance associated with an individual PMUT was in the range of 200 fF, we could make Figure 4, which indicates the maximum number of PMUTs simultaneously actuated at each frequency (e.g., in the case of 3 MHz, the maximum capacitance corresponded to 37 pF, which is equivalent to 185 individual PMUTs, a reasonable number for a linear array configuration).

With this circuit, 2.9 ns of a 10–90% rise/fall time was obtained for 1.2 pF (equivalent to one row of a 6 × 6 PMUT array) of load capacitance, achieving the design requirements.

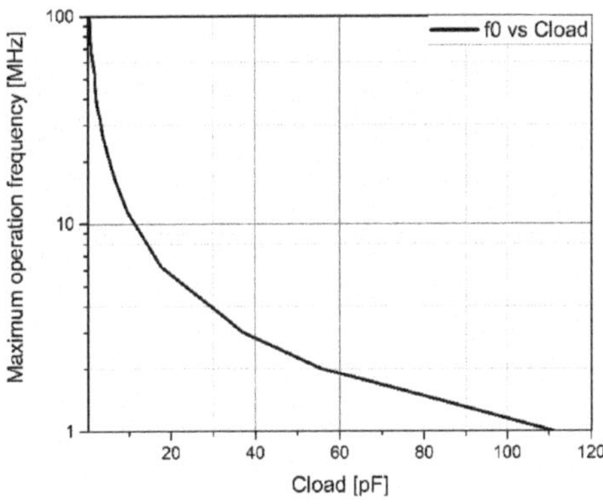

Figure 4. Maximum transmitter operating frequency as a function of load capacitance (simulation results).

2.3. Front-End Amplifier

The operation of PMUTs as a sensor is governed by the direct piezoelectric effect, where an electrical charge is generated when the acoustic wave arrives at the PMUT. In order to measure this electric charge, it was necessary to design a low-noise amplifier (LNA) at the first stage of the front-end PMUT receiver circuit that was a tradeoff between power supply, noise performance, gain, bandwidth, dynamic range, and die size. Transimpedance amplifier topologies (TIAs) are widely used to amplify the weak signal generated by relatively high-impedance CMUTs [12,16]. Nevertheless, piezoelectric transducers (PZTs and PMUTs) with similar sizes present much lower equivalent impedance around their resonance frequency, whereas this topology suffers from a low noise/power tradeoff when it is chosen as a front-end amplifier for PMUT devices. A capacitive feedback voltage amplifier (CFVA) with a split-capacitor feedback network has been a very common topology in readout integrated circuits (ICs) for PZT transducers [17,18] (in order to sense the transducer's voltage rather than its electric charge). Nevertheless, this topology suffers from a large area due to the use of several capacitors. This work presents a comparative analysis of two of the most used front-end amplifier configurations in sensors based on PMUTs: a voltage amplifier (VA) based on capacitive integration [11,13,15] and a charge-sensitive amplifier (CSA) [3,19].

2.3.1. Voltage Amplifier (VA) Description

Figure 5a shows a scheme of a front-end amplifier based on capacitive integration. The accumulated electric charge on the input equivalent capacitance (the parallel between C_{PMUT}, the electric capacitance of the PMUT; C_{in}, the input capacitance of the voltage amplifier; and C_p, all of the parasitic capacitances) is converted into voltage and is later amplified by the designed voltage amplifier. This relationship can be expressed by Equation (1), where A_0 is the open-loop amplifier voltage gain and Q_E is the electric charge generated by the PMUT:

$$V_{out} = A_0 \frac{Q_E}{C_{PMUT} + C_P + C_{in}} \tag{1}$$

To find a tradeoff between power consumption, gain, bandwidth, and area, a single-ended input self-biased push–pull configuration was selected as the VA topology (see Figure 5c). With this configuration, a smaller possible area could be used for the amplifier, which allows for a pitch-matched circuit with a single PMUT.

The input-referred noise current of this topology was derived from a small-signal equivalent model considering all parasitic capacitance (see Equation (2)):

$$i_{n,in}(s) = \frac{sC_{in}(C_f + C_{in})(i_{n,n} + i_{n,p})[1 + \frac{s(C_f + C_{out})}{g_{ds,n} + g_{ds,p}}]}{A_0[C_f(g_{m,n} + g_{m,p}) + C_{in}(g_{ds,n} + g_{ds,p})](1 - \frac{sC_f}{g_{m,n} + g_{m,p}})[1 + \frac{s(C_{in}C_f + C_{out}C_f + C_{out}C_{in})}{C_f(g_{m,n} + g_{m,p}) + C_{in}(g_{ds,n} + g_{ds,p})}]} \tag{2}$$

where $i_{n,n}$ and $i_{n,p}$ are the square root of the mean square current noise for nMOS and pMOS transistors; $g_{m,n}$ and $g_{m,p}$ are the transconductance for the M1 and M2 transistors, respectively; $g_{ds,n}$ and $g_{ds,p}$ are the channel conductance of M1 and M2, respectively; C_f is the feedback capacitance (in this topology, the sum of gate-to-drain capacitances); and C_{out} is the output capacitance of the circuit (approximately the load capacitance). Due to the small current generated by the PMUT (nanoamperes), the minimization of amplifier noise is a must. In order to minimize the input current noise and maximize the transimpedance gain, the pMOS M1 and nMOS M2 transistors were correctly sized. To maximize the dynamic range, the operation point of the amplifier was fixed to $V_{DD}/2$ by a pMOS M3 transistor connected between the input and output (operating in the subthreshold region, which functions as very high-impedance resistance) [20]. A source follower buffer was added to match the output impedance to 50 Ω for testing.

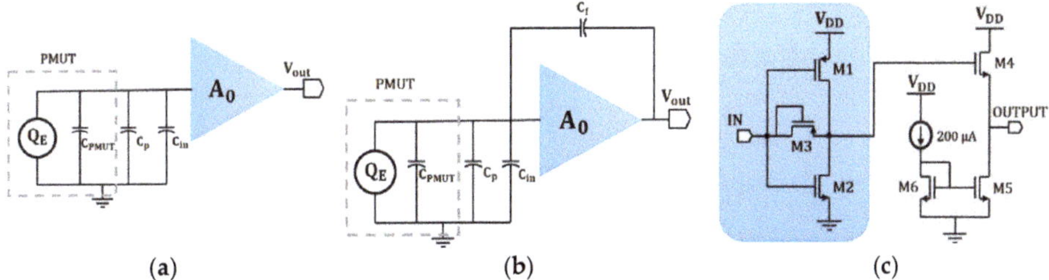

Figure 5. PMUT front-end amplifier configurations: (**a**) a voltage amplifier (VA) based on capacitive integration, (**b**) a CSA amplifier, and (**c**) a proposed schematic for the open-loop voltage amplifier, A_0, for the VA and CSA receivers.

2.3.2. Charge-Sensitive Amplifier (CSA) Description

The second approach is the charge-sensitive amplifier shown in Figure 5b, where the output voltage, given by Equation (3), only depends on the feedback capacitance (C_f), considering that the input impedance of the amplifier is much smaller than the equivalent impedance given by PMUT and the parasitic elements:

$$V_{out} = \frac{Q_E}{C_f}. \tag{3}$$

Since the input impedance is proportional to $1/(C_{in} + C_f{*}A_0)$, it is necessary to choose an amplifier topology that guarantees a high open loop gain A_0 (with minimum dimensions) to decrease Cin and the area. This reduction in the input capacitance impedance allows for the use of a low-feedback capacitance, C_f, thus achieving a maximum charge transfer and SNR. Therefore, the same single-ended input self-biased push–pull configuration used in the VA was the right selection.

In order to select the C_f, we took into account the input referred noise, the charge transfer from the PMUT to the feedback capacitance, and the gain. The input-referred noise current of this configuration is given by the following expression:

$$i_{n,in}(s) = \frac{s(C_f + C_{in})\left(i_{n,n} + i_{n,p}\right)}{\left(g_{m,n} + g_{m,p}\right)\left(1 - \frac{sC_f}{g_{m,n} + g_{m,p}}\right)} \tag{4}$$

This equation shows that the input noise current is directly proportional to C_f, and therefore the feedback capacitance must be selected to be as low as possible. In order to perform good signal processing, at least 85% of the charge transfer must be guaranteed. Considering a $C_{PMUT} = 1.2$ pF (equivalent to one row of six 80-µm PMUTs [10]), a parametric AC analysis was performed to compute the charge transfer. Figure 6 shows that for a C_f higher than 400 fF, the amplifier behaves as a CSA and guarantees at least an 85% charge transfer. Therefore, in order to maximize the output voltage given by Equation (3) and to find a tradeoff between noise performance and charge transfer, 400 fF was selected as the feedback capacitance. In this case, a pMOS M3 transistor was used to fix the operation point and was dimensioned (without affecting the bias) to avoid any leakage current that could charge C_f, leading to saturation of the amplifier. The same source-follower buffer was used to match the output impedance to 50 Ω.

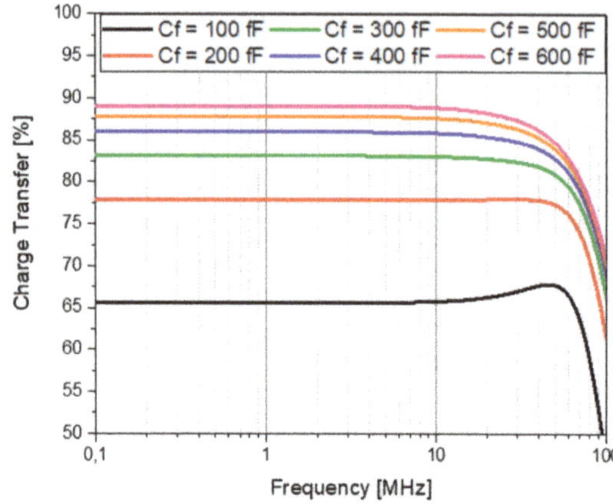

Figure 6. Charge transfer obtained for different feedback capacitances (C_f).

In Figure 7, the final layout of the LNA based on a VA is shown together with an optical image of the PMUT array, demonstrating its capabilities for a minute area and a pitch-matched system (for the CMOS receiver).

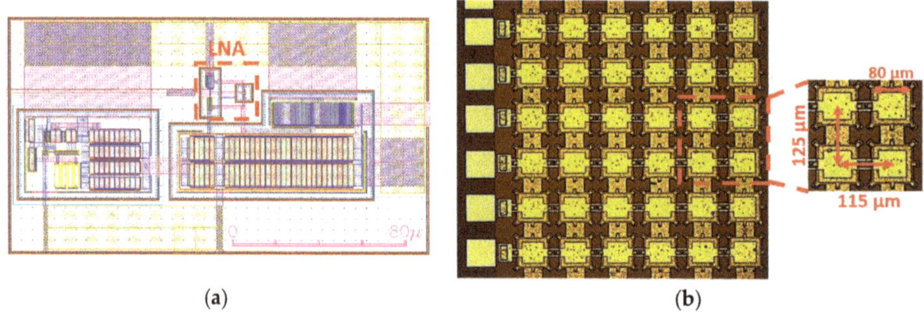

(**a**)   (**b**)

Figure 7. (**a**) Layout of the LNA (VA topology) with an output buffer. The LNA dimensions are 23 μm of width and 25 μm of length. (**b**) Optical image of a 6 × 6 (square) 80-μm/side PMUT array.

2.4. Front-End Amplifier Simulation Results

All simulations were done with 14 pF and 50 Ω as the load impedance (input impedance of the oscilloscope). The PMUT was modeled as a current source in parallel with its electrical capacitance. The assumed operation frequency was the resonance frequency of the PMUT array in liquid (3 MHz), which was given by Reference [10].

2.4.1. Transimpedance Gain and Input Impedance

Figure 8 shows the transimpedance gain of both approaches (the voltage amplifier and charge-sensitive amplifier) and the voltage open-loop gain A_0. It can be seen that the VA amplifier had a higher transimpedance gain ($\approx$23 dB Ω), since the output voltage of the VA depended not only on the amplifier open-loop gain A_0, but also on the integration of the current in the input capacitance. The CSA transimpedance gain depended only on the value of the feedback capacitance (it needs to be lower than the total VA input capacitance). However, the VA amplifier frequency response was limited by the open loop bandwidth (21.7 MHz). In the case of the CSA, its frequency behavior was limited by its input impedance and the equivalent impedance of PMUTs, plus the parasitic capacitance. At the frequency where both impedances were similar (around 140 MHz), the electric charge generated by the PMUT was divided, and therefore the CSA stopped behaving like a charge-sensitive amplifier.

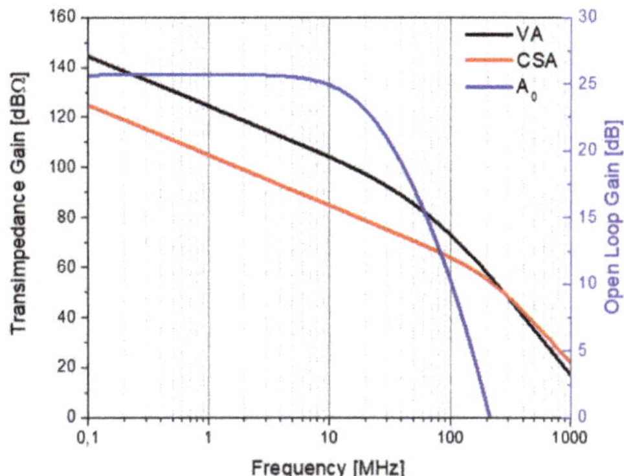

Figure 8. Frequency response of both amplifiers (VA and CSA) (left axis) along with the push–pull open-loop response of A_0 (right axis).

Figure 9 shows the dependency of the transimpedance gain in both cases (VA and CSA amplifiers), with the PMUTs and parasitic capacitances. As expected given Equations (1) and (3), the VA had a steeper slope (ΔTIGain = 7.5 dBΩ), since it amplified the generated input voltage through the integration of the PMUT current in all of the capacitances present in the input of the amplifier. Instead, the transimpedance gain of the CSA had a linear dependency with the lower slope (ΔTIGain = 0.8 dBΩ), but still the VA had more gain.

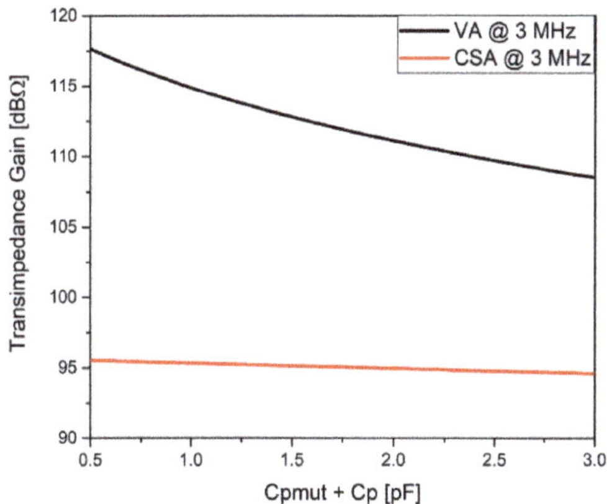

Figure 9. Dependence of the transimpedance gain at 3 MHz, with equivalent capacitance given by the PMUT plus parasitic capacitances.

2.4.2. Noise Performance and Dynamic Range

To compare the noise performance of both amplifiers, the output noise voltage ($V_{n,out}$) and the input noise current ($I_{n,in}$), defined by Equations (2) and (4), were taken into account. The presence of the feedback capacitor in the CSA amplifier caused its output impedance to be smaller than the VA output impedance, causing the output voltage noise to be smaller (see Figure 10). However, due to the big difference between their transimpedance gains, the input noise current was lower for the VA than for the CSA (0.08 pA/$\sqrt{\text{Hz}}$ and 0.12 pA/$\sqrt{\text{Hz}}$, respectively, at a 3-MHz frequency).

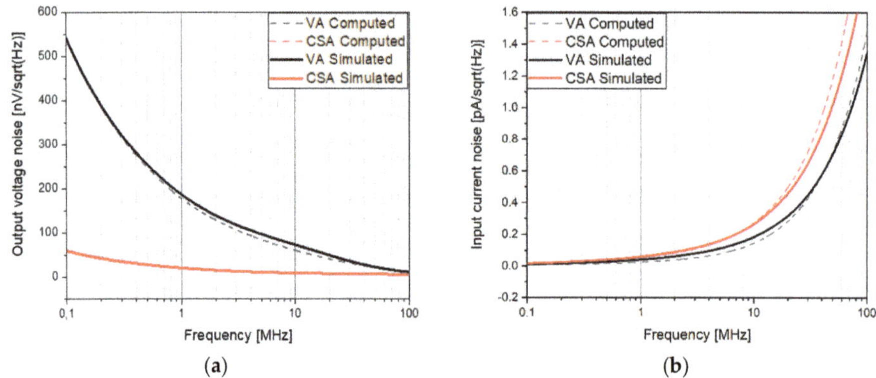

Figure 10. Computations using Equations (2) and (4) and the simulated RX amplifiers' noise spectral density. (**a**) Output voltage noise and (**b**) input current noise.

Figure 11 shows the simulated dynamic range for a 3-MHz input signal. A maximum input current of 100 nA and 1 μA for the VA and CSA were obtained, respectively. This great difference was expected due to the big difference between their gains. In these results, the CSA provided a dynamic range 10 times higher than the VA amplifier, allowing it to amplify PMUT signals 10 times larger than those amplified by the VA.

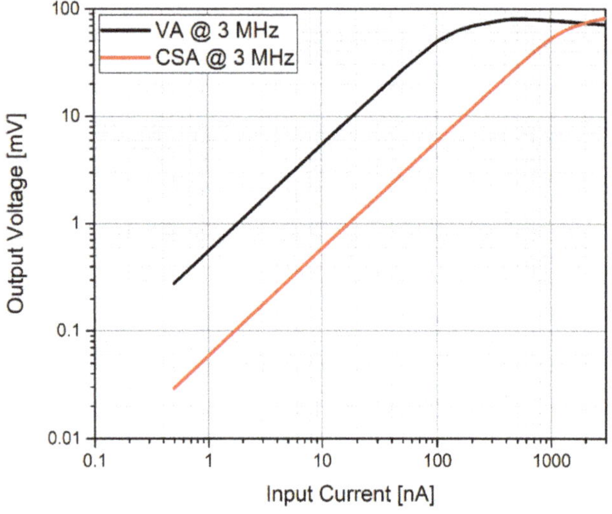

Figure 11. Simulated dynamic range at 3 MHz.

3. Electrical Characterization

3.1. High-Voltage Transmitter

The HV transmitter circuit was tested using a signal generator and an oscilloscope. The circuitry, operated from two power supplies (3.3 V and 32 V) and with a 3.3-V-squared input signal, generated output pulses up to 32 V. Figure 12 shows the input and output waveforms of this circuit, which achieved a rise/fall time of 47.5 ns. Using 14 pF as the load capacitance from the oscilloscope plus 3 pF due to the bonding pad, printed circuit board (PCB) pad, and connectors, this value differed by

6.36 ns compared to the simulation result. This difference could have been due to the estimation of the load capacitance (parasitic capacitance) as well as due to process variations. For this estimated load capacitance, the HV transmitter operated at a maximum frequency of 5 MHz, achieving a 10–90% rise time and a 50–50% latency of 47.5 ns and 31 ns, respectively.

The static power consumption was negligible, and the dynamic power consumption, defined as $C_L V^2 f$, was 3.68 mW for a 1.2-pF load capacitance.

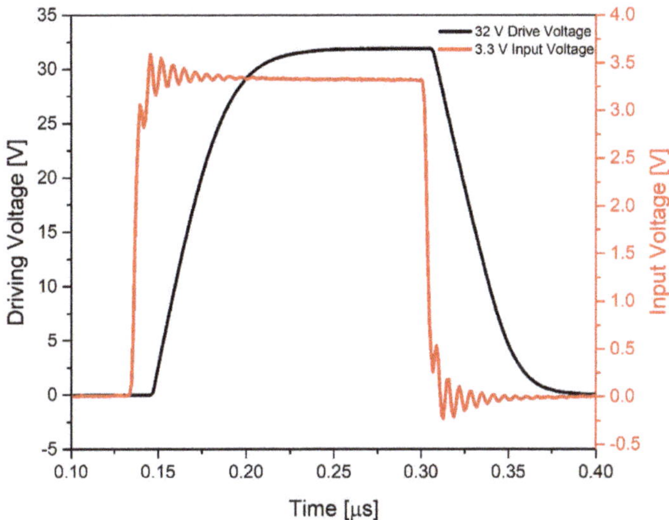

Figure 12. Waveform of the HV transmitter input and output.

3.2. Front-End Amplifiers

The RX amplifiers were characterized in terms of voltage–voltage gain for the VA, transimpedance gain for the CSA, and input current noise and output dynamic range for both of them.

Figure 13 shows the measured and postlayout simulation voltage–voltage gain of the voltage amplifier (VA). A measured bandwidth of about 22 MHz was achieved. Compared to the postlayout simulation and considering the technology variations, the simulated values were close to the experimental ones.

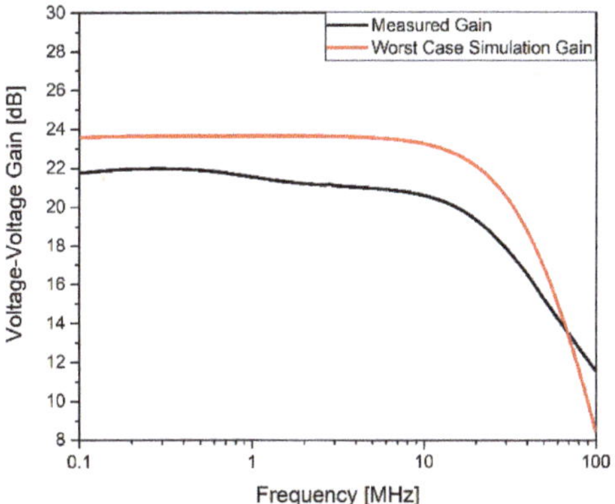

Figure 13. Transfer function of the VA amplifier.

In order to compute the gain of the CSA, we employed the charge injection method. A 100-fF capacitor connected in series at the CSA input was used as a test injection capacitor (C_{TEST}). Small voltage steps (ΔV) applied to the C_{TEST} produced charge pulses at the CSA input with a value of C_{TEST} $*\Delta V$. Figure 14 shows the output voltage measured at the buffer output for an input charge sweep from 2.5 fC to 18 fC (corresponding to voltage steps from 25 mV to 180 mV). According to Equation (3), the slope of this graph must be inversely proportional to the feedback capacitance, giving 472 fF with a 100% charge transfer. Taking into account this and using the designed feedback capacitance (Cf = 410 fF in the layout), the obtained charge transfer was 87%, close to the simulated results.

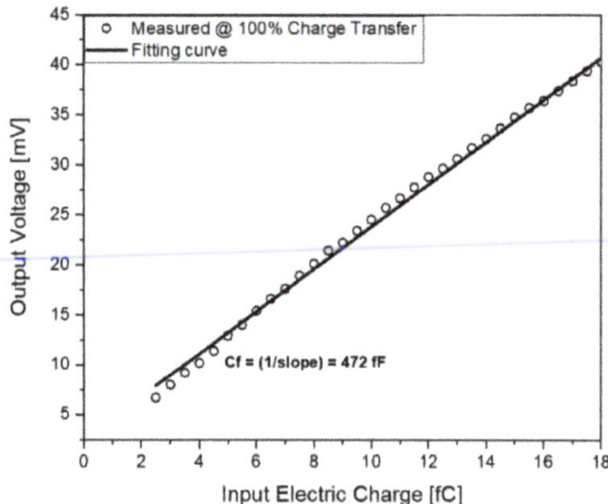

Figure 14. Tested results of the output voltage of the CSA.

The noise performance of both the VA and CSA amplifiers was measured by opening the input and measuring the output noise. The input noise was computed by dividing the measured output noise by the corresponding transfer function. The resulting spectral density noise is shown in Figure 15, which was in good agreement with the simulations. The measured output-referred voltage noise density was 60.68 nV/$\sqrt{\text{Hz}}$ and 19.01 nV/$\sqrt{\text{Hz}}$ at 3 MHz for the VA and CSA amplifiers, respectively. At the same frequency, the measured input-referred current noise was 68 fA/$\sqrt{\text{Hz}}$ and 147 fA/$\sqrt{\text{Hz}}$ for the VA and CSA, respectively. Taking into account the PMUT bandwidth (1 MHz in liquid), an input-referred integrated noise of 59.18 pA$_{rms}$ and 142.7 pA$_{rms}$ was obtained for the VA and CSA, respectively.

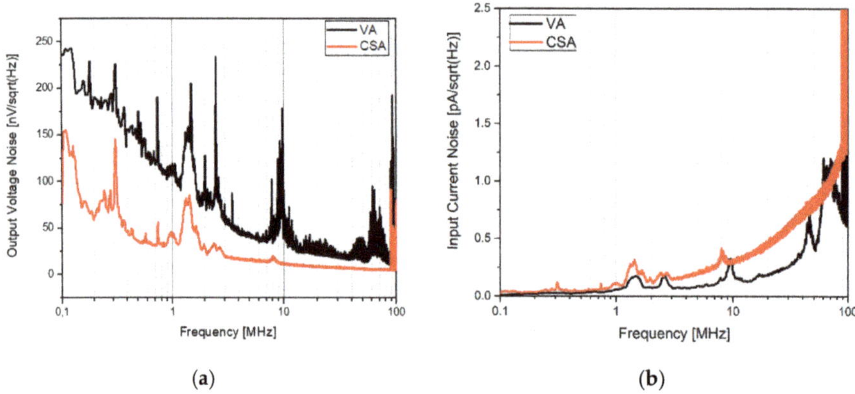

Figure 15. Measured spectral density noise: (**a**) output-referred voltage noise and (**b**) input-referred current noise.

Considering that the reception sensitivity of the PMUT array [10] was 5.9 V/MPa, the measured noise equivalent pressure was 0.84 mPa/$\sqrt{\text{Hz}}$ and 1.32 mPa/$\sqrt{\text{Hz}}$ at 3 MHz. These values are very low for imaging applications, where the received echoes are in the order of kPa, so a good signal-to-noise ratio can be guaranteed.

Figure 16 shows the measurements of the VA and CSA output voltage for different input current levels at 3 MHz. The measured transimpedance gain was 109.22 dBΩ and 99.57 dBΩ, respectively. An input-measured dynamic range of 69 dB and 71 dB was achieved, where the previous measured integrated input noise was taken as the minimum detectable current.

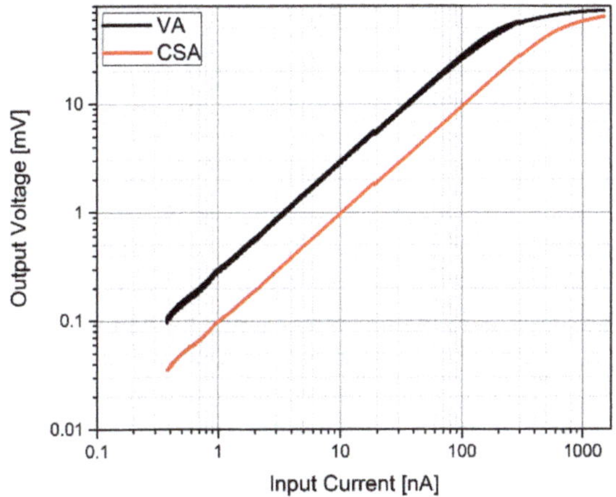

Figure 16. Dynamic range measurements: the VA amplifier (black) and the CSA amplifier (red).

4. Acoustic Characterization

To validate the analog front-end CMOS in its intended application, an acoustic characterization was done using fluorinert (FC-70, where the mass density is $\varrho = 1940$ kg m^{-3} and the sound velocity is c = 690 m/s) as a propagation medium. The set-up is shown in Figure 17, where a grating phantom was introduced in a small pool, confining the FC-70 to approximately 2.1 mm over the 6 × 6 PMUT array surface. One row of this array was excited by the designed HV transmitter (four cycles at 3 MHz with a 32-V peak-to-peak), and another row was dedicated to reception using the proposed VA as an LNA front-end amplifier. In order to obtain a 2D image, the grating phantom was shifted in the × direction and y direction, with steps of 50 μm and 100 μm, respectively. Figure 18b shows an image of a section of this phantom (see Figure 18a). The blue bars correspond with the slots, with a width close to 1.1 mm and 1 mm, as was expected (physical width), demonstrating the capability of the system to make an ultrasonic image.

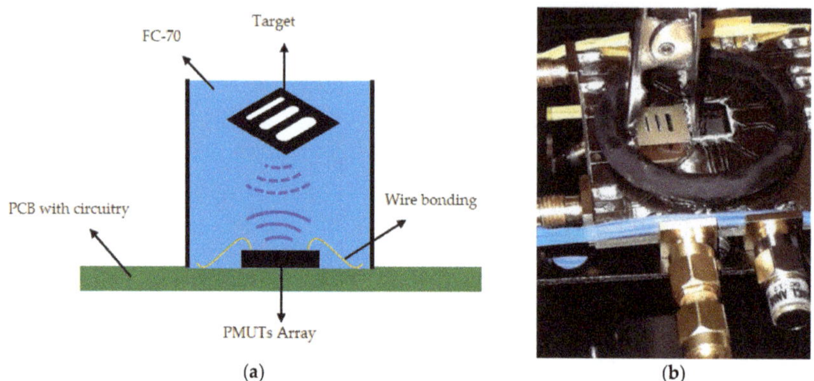

Figure 17. Set-up for acoustic characterization using an FC-70: (**a**) the schematic set-up; (**b**) an optical image of the experimental set-up using a grating phantom.

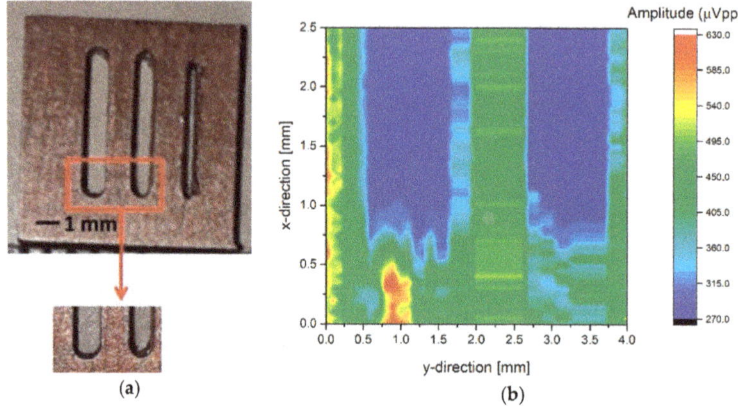

Figure 18. (**a**) Grating phantom and (**b**) an ultrasonic image of a phantom section.

5. Discussion

In this work, we present the design and characterization of two functional blocks of an imaging ultrasound system: an HV transmitter and a front-end receiver amplifier. In addition, we compared two of the most used front-end amplifiers in sensor-based PMUTs.

In this section, we will compare these proposals to prior work, taking into account all of the obtained results.

5.1. HV Transmitter

Table 2 summarizes and compares our transmitter to prior reported works. The performance of the HV transmitter was measured in terms of its speed, area, power consumption, and load capacitance. In order to compare different systems, we defined a figure-of-merit (FOM) that takes into account all of the key parameters that intervene in the transmitter's performance (see Equation (5)):

$$\text{FOM}_{TX}\left[\frac{mA}{mm^2}\right] = \frac{V_{MAX}[V] * C_L[pF]}{t_{rise}[ns] * area[mm^2]}, \tag{5}$$

where C_L corresponds to the nominal load capacitance, V_{MAX} is the maximum pulsed output voltage, and t_{rise} is the rise time. Our designed HV transmitter achieved the best FOM_{TX}

(1018 mA/mm^2), which translated to the best electrical performance with the highest integration density. This performance was attributable in part to the small area that the transmitter has.

Table 2. HV transmitter performance summary and comparisons.

Parameter	This Work	[13] (2019)	[21] (2017)	[3] (2016)	[16] (2016)
Process technology	0.13 μm HV CMOS	0.18 μm CMOS	0.18 μm HV BCD8-SOI	0.18 μm HV CMOS	TSMC 0.18 μm HV CMOS
Transducer	PMUT	PMUT	CMUT	PMUT	CMUT
Pulsed output voltage (V)	32	5/13.2	100	24	30
Nominal operation frequency (MHz)	3	5	10	14	5
Nominal load capacitance (pF)	1.2	N/A	9.2	2	2
Area (mm^2)	0.013	N/A	0.09 [1]	0.017 [1]	0.016
Rise time (ns)	2.9	N/A	14	10.56 [2]	6.6
FOM$_{TX}$ (mA/mm^2)	1018	-	730	267	568

[1] These areas were estimated using a chip micrograph. [2] Obtained as $t_{rise} = 2.2RC_L$, where $R = Vmax/Imax$.

5.2. Front-End Receiver Amplifier

To compare both presented LNA topologies to the state-of-the-art topologies, we also defined a figure-of-merit that includes the main parameters of the LNA design. Taking into account the gain, input referred noise, power consumption, operation frequency, and area, we defined the following FOMs:

$$\text{FOM}_{RX\_1}\left[\frac{\text{MHz}}{\text{V}^2*\text{A}*\mu\text{m}^2}\right] = \frac{\text{Gain}[\frac{V}{V}] * \sqrt{\text{frequency}[\text{MHz}]}}{\text{input voltage noise }[\frac{\text{nV}}{\sqrt{\text{HZ}}}]*\text{Power Consumption }[\text{mW}]*\text{area }[\text{mm}^2]}, \quad (6)$$

$$\text{FOM}_{RX\_2}\left[\frac{\text{Hz}}{\text{mA}^3*\mu\text{m}^2}\right] = \frac{\text{Transimpedance gain}[\frac{V}{A}] * \sqrt{\text{frequency}[\text{MHz}]}}{\text{input current noise }[\frac{\text{pA}}{\sqrt{\text{HZ}}}]*\text{Power Consumption }[\text{mW}]*\text{area }[\text{mm}^2]}, \quad (7)$$

where Equation (6) was used to compare the VA and CFVA topologies and Equation (7) was used to compare the CSA and TIA topologies. Table 3 summarizes the comparisons between the performance of the designed LNAs and that of the other amplifiers reported as being state-of-the-art, demonstrating competitive results. Taking the best computed FOM$_{RX_1,2}$, the designed VA and CSA topologies achieved ~2.5× and ~45× higher improvement, respectively, guaranteeing the best electrical performance with a minimum area. These competitive FOM values were possible since our two LNAs (based on a self-biased push–pull amplifier) have a smaller area (almost by 10 times), which makes these designs good candidates to implement a pitch-matched system, where the LNA can be implemented just below its ultrasound transducer.

Finally, comparing the two presented LNA implementations in this work, the VA amplifier provided a remarkable improvement in terms of transimpedance gain and input-referred noise, achieving a higher signal-to-noise ratio. Nevertheless, the great dependency of the VA gain on the parasitic capacitance means that the CSA is the more useful option when the amplifier is not fully integrated with the PMUTs, since all of the parasitic capacitances are difficult to efficiently control, providing different gains for the same devices. The VA is more adequate when the amplifier can be directly connected to the sensor, as happens when there is a monolithic integration of the CMOS and PMUTs.

Table 3. Measured LNA performance summary and comparisons.

Parameter	This Work		[13] (2019)	[18] (2018)	[16] (2016)	[11] (2015)	[17] (2015)
Topology	VA	CSA	VA	CFVA	TIA	VA	CFVA
Process technology	0.13 µm HV CMOS	0.13 µm HV CMOS	0.18 µm CMOS	0.18 µm HV-BCD	TSMC 0.18 µm HV CMOS	0.18 µm HV CMOS	0.18 µm CMOS
Transducer	PMUT	PMUT	PMUT	PZT	CMUT	PMUT	PZT
Power supply (V)	1.5	1.5	1.5	1.8	1.8	1.8	1.8
Power consumption (mW)	0.3	0.3	0.08	0.79	1.4	N/A	0.135
Area (10^{-4} mm^2)	6	9	N/A	30 [1]	280	310	60
Voltage–voltage gain (dB)	21.8	N/A	29/30/42/53 [2]	18	N/A	N/A	−12/6/24
Transimpedance gain (dBΩ)	109.22 at 3 MHz	99.57 at 3 MHz	N/A	N/A	116/113.5 110/104	N/A	N/A
Bandwidth (MHz)	22	N/A	10	20	10.2/10.8 10.6/10.5	N/A	9.8
Input current noise (pA/$\sqrt{\text{Hz}}$)	0.08 at 3 MHz	0.15 at 3 MHz	N/A	N/A	0.41 @ 5 MHz	N/A	N/A
Input voltage noise (nV/$\sqrt{\text{Hz}}$)	7.1 at 3 MHz	N/A	N/A	7.9 at 5 MHz	N/A	11 at 0.22 MHz	5.9 at 4 MHz
Input dynamic range (dB)	69	71	90	75	N/A	N/A	81
FOM$_{RX_1}$ (MHz/V^{2}Aµm^2)	16674	N/A	N/A	949	N/A	N/A	6633 [3]
FOM$_{RX_2}$ (Hz/mA^{3}µm^2)	N/A	4.1*10^9	N/A	N/A	0.09*10^9 [3]	N/A	N/A

[1] This area was estimated from a chip micrograph. [2] Including a TGC amplifier as a second stage. [3] Computed considering its higher gain.

6. Conclusions

A CMOS circuit was designed and implemented using the novel CMOS-MEMS monolithic integration platform from Silterra in order to drive a 6 × 6 PMUT array. The system presented with competitive performance in comparison to state-of-the-art highlighting area and noise performance. A comprehensive comparison between two RX front-end amplifiers was done, showing that the VA amplifier based on capacitive integration provided remarkable improvements in general terms but was more sensitive to parasitic elements, which will be controlled with monolithic integration. This system was successfully applied in imaging experiments.

Author Contributions: conceptualization, I.Z., A.U., and N.B.; methodology, I.Z.; electrical experiments, I.Z. and A.U.; acoustic experiments, E.L.; writing—original draft preparation, I.Z.; writing—review and editing, A.U. and N.B.; supervision, A.U. and N.B.; project administration, N.B. All authors have read and agreed to the published version of the manuscript.

Acknowledgments: This work has been partially supported by project TEC2015-66337-R (MINECO/ FEDER). Special acknowledgements are made to E.M.F., C.B.F., and all the other members of the SilTerra's MEMS & SENSORS technology development team for support in fabricating the wafers.

Conflicts of Interest: The authors declare no conflicts of interest.

References

1. Janjic, J.; Tan, M.; Daeichin, V.; Noothout, E.; Chen, C.; Chen, Z.; Chang, Z.Y.; Beurskens, R.H.S.H.; van Soest, G.; van der Steen, A.F.W.; et al. A 2D Ultrasound Transducer with Front-End ASIC and Low Cable Count for 3D Forward-Looking Intravascular Imaging: Performance and Characterization. *IEEE Trans. Ultrason. Ferroelectr. Freq. Control* **2018**, *65*, 1832–1844. [CrossRef] [PubMed]
2. Kim, H.; Kim, S.; Lee, S.; Kang, B.B.; Jo, Y.; Kook, G.; Oh, C.; Kim, M.K.; Kim, H.; Lee, H.J. MEMS Transducers for Non-Invasive Ultrasound Brain Stimulation. In Proceedings of the 2019 20th International Conference on Solid-State Sensors, Actuators and Microsystems & Eurosensors XXXIII (TRANSDUCERS & EUROSENSORS XXXIII), Berlin, Germany, 23–27 June 2019; pp. 242–245.

3. Tang, H.-Y.; Lu, Y.; Jiang, X.; Ng, E.J.; Tsai, J.M.; Horsley, D.A.; Boser, B.E. 3-D Ultrasonic Fingerprint Sensor-on-a-Chip. *IEEE J. Solid-State Circuits* **2016**, *51*, 2522–2533. [CrossRef]

4. Przybyla, R.J.; Tang, H.-Y.; Shelton, S.E.; Horsley, D.A.; Boser, B.E. 12.1 3D Ultrasonic Gesture Recognition. In Proceedings of the 2014 IEEE International Solid-State Circuits Conference Digest of Technical Papers (ISSCC), San Francisco, CA, USA, 9–13 February 2014; pp. 210–211.

5. Qiu, Y.; Gigliotti, J.V.; Wallace, M.; Griggio, F.; Demore, C.E.M.; Cochran, S.; Trolier-McKinstry, S. Piezoelectric micromachined ultrasound transducer (PMUT) arrays for integrated sensing, actuation and imaging. *Sensors* **2015**, *15*, 8020–8041. [CrossRef] [PubMed]

6. Jiang, X.; Tang, H.-Y.; Lu, Y.; Ng, E.J.; Tsai, J.M.; Boser, B.E.; Horsley, D.A. Ultrasonic Fingerprint Sensor with Transmit Beamforming Based on a PMUT Array Bonded to CMOS Circuitry. *IEEE Trans. Ultrason. Ferroelectr. Freq. Control* **2017**, *64*, 1401–1408. [CrossRef] [PubMed]

7. Lu, Y.; Tang, H.; Fung, S.; Wang, Q.; Tsai, J.M.; Daneman, M.; Boser, B.E.; Horsley, D.A. Ultrasonic fingerprint sensor using a piezoelectric micromachined ultrasonic transducer array integrated with complementary metal oxide semiconductor electronics. *Appl. Phys. Lett.* **2015**, *106*, 263503. [CrossRef]

8. Jiang, X.; Lu, Y.; Tang, H.Y.; Tsai, J.M.; Ng, E.J.; Daneman, M.J.; Boser, B.E.; Horsley, D.A. Monolithic ultrasound fingerprint sensor. *Microsyst. Nanoeng.* **2017**, *3*, 1–8. [CrossRef] [PubMed]

9. Pandian, M.S.; Ferrer, E.M.; Tay, W.S.; Madhaven, V.; Kantimahanti, A.K.; Sobreviela, G.; Uranga, A.; Barniol, N. Thin Film Piezoelectric Devices Integrated on CMOS. In Proceedings of the 2016 Symposium on Piezoelectricity, Acoustic Waves and Device Applications, SPAWDA 2016, Xi'an, China, 21–24 October 2016.

10. Ledesma, E.; Zamora, I.; Torres, F.; Uranga, A.; Tzanov, V.; Bhd, S. ALN Piezoelectric Micromachined Ultrasonic Transducer Array Monolothically Fabricated on Top of Pre-Processed CMOS Substrates. In Proceedings of the 2019 20th International Conference on Solid-State Sensors, Actuators and Microsystems & Eurosensors XXXIII (TRANSDUCERS & EUROSENSORS XXXIII), Berlin, Germany, 23–27 June 2019; pp. 655–658.

11. Przybyla, R.J.; Tang, H.-Y.; Guedes, A.; Shelton, S.E.; Horsley, D.A.; Boser, B.E. 3D Ultrasonic Rangefinder on a Chip. *IEEE J. Solid-State Circuits* **2015**, *50*, 320–334. [CrossRef]

12. Chen, K.; Lee, H.-S.; Chandrakasan, A.P.; Sodini, C.G. Ultrasonic Imaging Transceiver Design for CMUT: A Three-Level 30-Vpp Pulse-Shaping Pulser With Improved Efficiency and a Noise-Optimized Receiver. *IEEE J. Solid-State Circuits* **2013**, *48*, 2734–2745. [CrossRef]

13. Lee, J.; Lee, K.; Eovino, B.E.; Park, J.H.; Lin, L.; Yoo, H.; Yoo, J. 11.1 A 5.37mW/Channel Pitch-Matched Ultrasound ASIC with Dynamic-Bit-Shared SAR ADC and 13.2V Charge-Recycling TX in Standard CMOS for Intracardiac Echocardiography. In Proceedings of the IEEE International Solid-State Circuits Conference, San Francisco, CA, USA, 17–21 February 2019; pp. 2018–2020.

14. Tang, H.-Y.; Seo, D.; Singhal, U.; Li, X.; Maharbiz, M.M.; Alon, E.; Boser, B.E. Miniaturizing Ultrasonic System for Portable Health Care and Fitness. *IEEE Trans. Biomed. Circuits Syst.* **2015**, *9*, 767–776. [CrossRef] [PubMed]

15. Zamora, I.; Ledesma, E.; Uranga, A.; Barniol, N. Fully Integrated CMOS-PMUT Transceiver. In Proceedings of the 2018 25th IEEE International Conference on Electronics Circuits and Systems, Bordeaux, France, 9–12 December 2018; pp. 149–152.

16. Chen, K.; Lee, H.S.; Sodini, C.G. A Column-Row-Parallel ASIC Architecture for 3-D Portable Medical Ultrasonic Imaging. *IEEE J. Solid-State Circuits* **2016**, *51*, 738–751.

17. Chen, C.; Chen, Z.; Chang, Z.Y.; Pertijs, M.A.P. A Compact 0.135-mW/Channel LNA Array for Piezoelectric Ultrasound Transducers. In Proceedings of the ESSCIRC Conference 2015—41st European Solid-State Circuits Conference (ESSCIRC), Graz, Austria, 14–18 September 2015.

18. Kang, E.; Ding, Q.; Shabanimotlagh, M.; Kruizinga, P.; Chang, Z.Y.; Noothout, E.; Vos, H.J.; Bosch, J.G.; Verweij, M.D.; De Jong, N.; et al. A Reconfigurable Ultrasound Transceiver ASIC with 24 × 40 Elements for 3-D Carotid Artery Imaging. *IEEE J. Solid-State Circuits* **2018**, *53*, 2065–2075. [CrossRef]

19. Tang, H.; Lu, Y.; Fung, S.; Tsai, J.M.; Daneman, M.; Horsley, D.A.; Boser, B.E. Pulse-Echo Ultrasonic Fingerprint Sensor on a Chip. In Proceedings of the 2015 Transducers—2015 18th International Conference on Solid-State Sensors, Actuators and Microsystems (TRANSDUCERS), Anchorage, AK, USA, 21–25 June 2015; pp. 674–677.

20. Riverola, M.; Sobreviela, G.; Torres, F.; Uranga, A.; Barniol, N. Single-resonator dual-frequency BEOL-embedded CMOS-MEMS oscillator with low-power and ultra-compact TIA core. *IEEE Electron Device Lett.* **2017**, *38*, 273–276. [CrossRef]

21. Sautto, M.; Savoia, A.S.; Quaglia, F.; Caliano, G.; Mazzanti, A. A comparative analysis of CMUT receiving architectures for the design optimization of integrated transceiver front ends. *IEEE Trans. Ultrason. Ferroelectr. Freq. Control* **2017**, *64*, 826–838. [CrossRef] [PubMed]

© 2020 by the authors. Licensee MDPI, Basel, Switzerland. This article is an open access article distributed under the terms and conditions of the Creative Commons Attribution (CC BY) license (http://creativecommons.org/licenses/by/4.0/).

 *sensors*

Article

A Highly Linear CMOS Image Sensor Design Based on an Adaptive Nonlinear Ramp Generator and Fully Differential Pipeline Sampling Quantization with a Double Auto-Zeroing Technique

Chuangze Li [1], Benguang Han [2], Jie He [1], Zhongjie Guo [3] and Longsheng Wu [1,*]

[1]   Xi'an Microelectronic Technology Institute, No. 198 Taibai South Road, Yanta District, Xi'an 710071, China; li_chuang_ze@163.com (C.L.); hejiexian@126.com (J.H.)
[2]   School of Microelectronics, Xi dian University, No. 2 Taibai South Road, Yanta District, Xi'an 710071, China; bghan@xidian.edu.cn
[3]   Department of Electronic Engineering, Xi'an University of Technology, No. 5, Jinhua South Road, Beilin District, Xi'an 710054, China; zjguo@xaut.edu.cn
*   Correspondence: wls771@163.com; Tel./Fax: +86-29-8860-9000

Received: 28 January 2020; Accepted: 12 February 2020; Published: 14 February 2020

Abstract: For a complementary metal-oxide-semiconductor image sensor with highly linear, low noise and high frame rate, the nonlinear correction and frame rate improvement techniques are becoming very important. The in-pixel source follower transistor and the integration capacitor on the floating diffusion node cause linearity degradation. In order to address this problem, this paper proposes an adaptive nonlinear ramp generator circuit based on dummy pixels used in single-slope analog-to-digital converter topology for a complementary metal-oxide-semiconductor (CMOS) image sensor. In the proposed approach, the traditional linear ramp generator circuit is replaced with the new proposed adaptive nonlinear ramp generator circuit that can mitigate the nonlinearity of the pixel unit circuit, especially the gain nonlinearity of the source follower transistor and the integration capacitor nonlinearity of the floating diffusion node. Moreover, in order to enhance the frame rate and address the issue of high column fixed pattern noise, a new readout scheme of fully differential pipeline sampling quantization with a double auto-zeroing technique is proposed. Compared with the conventional readout structure without a fully differential pipeline sampling quantization technique and double auto-zeroing technique, the proposed readout scheme cannot only enhance the frame rate but can also improve the consistency of the offset and delay information of different column comparators and significantly reduce the column fixed pattern noise. The proposed techniques are simulated and verified with a prototype chip fabricated using typical 180 nm CMOS process technology. The obtained measurement results demonstrate that the overall nonlinearity of the CMOS image sensor is reduced from 1.03% to 0.047%, the efficiency of the comparator is improved from 85.3% to 100%, and the column fixed pattern noise is reduced from 0.43% to 0.019%.

Keywords: CMOS image sensor; linearity; adaptive nonlinear ramp; fully differential pipeline; double auto-zeroing; high framerate; fixed pattern noise; floating diffusion; readout scheme; ramp generator circuit

---

1. Introduction

Highly linear complementary metal-oxide-semiconductor (CMOS) image sensors (CISs) have a wide range of applications in time-of-flight (ToF) ranging, medical imaging, space remote sensing imaging and scientific imaging [1–6]. The most critical element of a CMOS image sensor (CIS) is the column-parallel analog-to-digital converter (ADC). Various architectural designs of column-parallel

ADCs have been proposed, including Cyclic-ADC, SAR-ADC, pipeline-ADC, single-slope ADC (SS_ADC) and multi-slope ADC [7–12]. The single-slope ADC structure is the most popular for column-parallel ADCs in CISs due to its simple circuit topologies and small layout area. The ramp generator circuit and the high-speed comparator in the SS-ADC structure are the most critical modules in determining the performance of the CIS.

Many research studies have been conducted to improve the linearity of CIS. According to the causes of nonlinearity of CISs, the solutions adopted in the literature are categorized into four types. The first type is employing the generated linear ramp signal buffered by the source follower (SF) transistor [8] to eliminate the gain nonlinearity of the SF transistor in pixel. However, the disadvantage of this method is that the nonlinearity of the integration capacitor (C_{FD}) caused by the floating diffusion (FD) node still exists. The second type is to utilize the regulating output voltage follower method to calibrate the nonlinearity outside the pixel structure [2,3,13,14]. The third type is utilizing the unit-gain analog buffer instead of the SF in the pixel circuit [1–4] to reduce the gain nonlinearity of the SF transistor. However, this method increases the pixel area and reduces the fill factor. The fourth type is utilizing the off-chip high-precision ADC and digital-to-analog converter (DAC) to calibrate the nonlinearity of the CMOS image sensor system [1–3,15]. The advantages of this correction method are high linearity and high precision. However, the method is complicated and has a high cost.

The SS-ADC in the conventional CIS has different crossing point voltages and different offset voltages of the column-level comparator. In order to address this issue, a fully differential comparator with pipeline sampling quantization based on the double auto-zeroing (AZ) technique was proposed to improve the frame rate and reduce the column fixed pattern noise (FPN) in the spatial domain. In the literature [6], a classic CIS composed of a programmable gain amplifier (PGA) and an SS-ADC has been proposed for the column-level readout circuit. The advantage of this readout structure is that the dynamic range (DR) of the CIS can be extended, and the gain of the PGA can suppress the noises of the next stages (for example, the readout noise of the SS-ADC). However, the PGA itself has high noise, high power consumption, and the readout circuit sampling and quantization work in sequence without concurrent execution. Therefore, the comparator in the SS-ADC will have a fixed idle time, which reduces the efficiency of the comparator. The general readout scheme leads to the problem of large column FPN caused by different comparator crossing voltage levels. In order to address the above-mentioned problems, this paper proposes an adaptive nonlinear ramp generator circuit design and a readout circuit with fully differential pipeline sampling quantization based on the double AZ technique.

The remainder of this paper is organized as follows. Section 2 introduces the architecture of highly linear CIS. Section 3 describes the proposed adaptive nonlinear ramp generator technique and the fully differential pipeline sampling quantization based on the double AZ technique. The principle analysis of highly linear CIS with linear and nonlinear ramp generators is presented in Section 4. The simulation and experimental results of the fabricated CIS are discussed in Section 5, followed by conclusions in Section 6.

2. Image Sensor Architecture

The CIS proposed in this paper consists of a typical 5T pixel [6] array, an adaptive nonlinear ramp generator, a readout circuit with fully differential pipeline sampling quantization based on the double AZ technique, row/column decoder and driver, timing sequence controller, phase-locked loop (PLL), charge pump, temperature sensor, and high-speed, high-precision, low-power LVDS serial data transfer circuit. The overall architecture of the proposed highly linear CIS is shown in Figure 1.

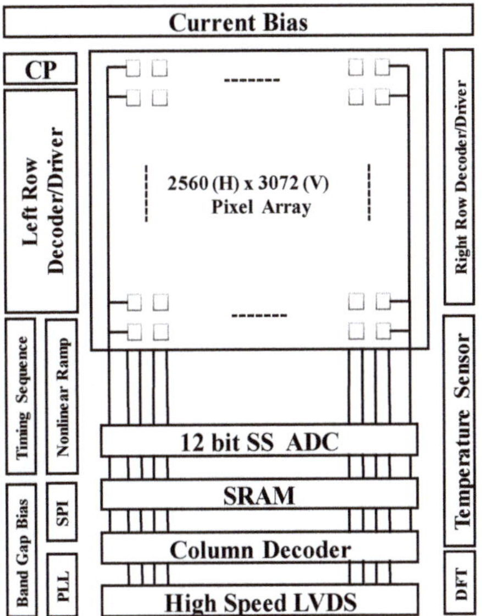

Figure 1. System architecture of the proposed highly linear complementary metal-oxide-semiconductor (CMOS) image sensor.

The central part of Figure 1 is a 2560 × 3072 active pixel array. The active pixel array is distributed around the dummy pixel. The dummy pixel can be used to generate an integration capacitor for an adaptive nonlinear ramp that is highly consistent with the characteristics of floating diffusion node capacitance. The left part of Figure 1 includes bandgap reference voltage generator circuit and current bias circuit, charge pump module, PLL module, timing sequence driving circuit, serial peripheral interface (SPI) and other modules. The right part of Figure 1 includes the temperature sensor, design for testability (DFT) circuit and the right row decoder driver circuit.

3. Proposed Techniques

The essential blocks of column-parallel SS-ADC in CIS is the ramp generator circuit and the high-speed readout circuit. This paper proposes an adaptive [16,17] nonlinear ramp generator and a fully differential pipeline sampling quantization scheme based on the double AZ technique. The two proposed techniques are dedicated for enhancing the performance of system linearity, improving the frame rate and reducing the column FPN of the CIS.

3.1. Nonlinear Ramp Generation Technique Based on Dummy Pixel Array

Figure 2 shows the system signal process flow diagram of CIS using the nonlinear ramp generation technique based on the dummy pixel and fully differential pipeline sampling quantization technique.

Figure 3a,b shows the system digital number (DN) outputs using the linear and the nonlinear ramp generators, respectively.

A typical SS-ADC utilizes the linear ramp generation technique, which cannot eliminate the nonlinearity of the pixel. The proposed nonlinear ramp generation technique based on dummy pixels can not only eliminate the gain nonlinearity of the SF, which changes with the input voltage but can also reduce the nonlinearity of C_{FD} in the FD node.

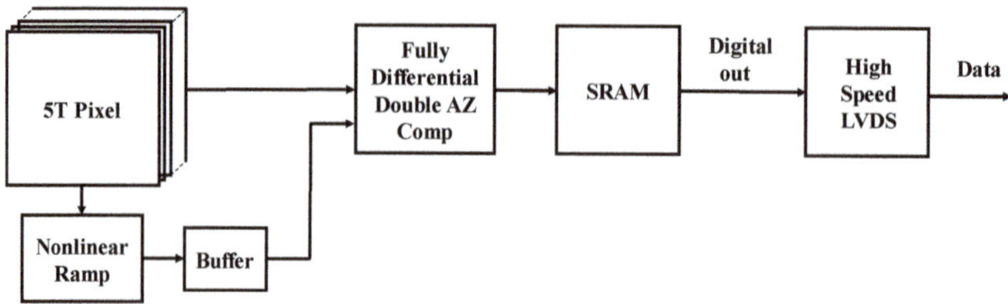

Figure 2. System signal process flow diagram of a highly linear CMOS image sensor based on nonlinear ramp generator.

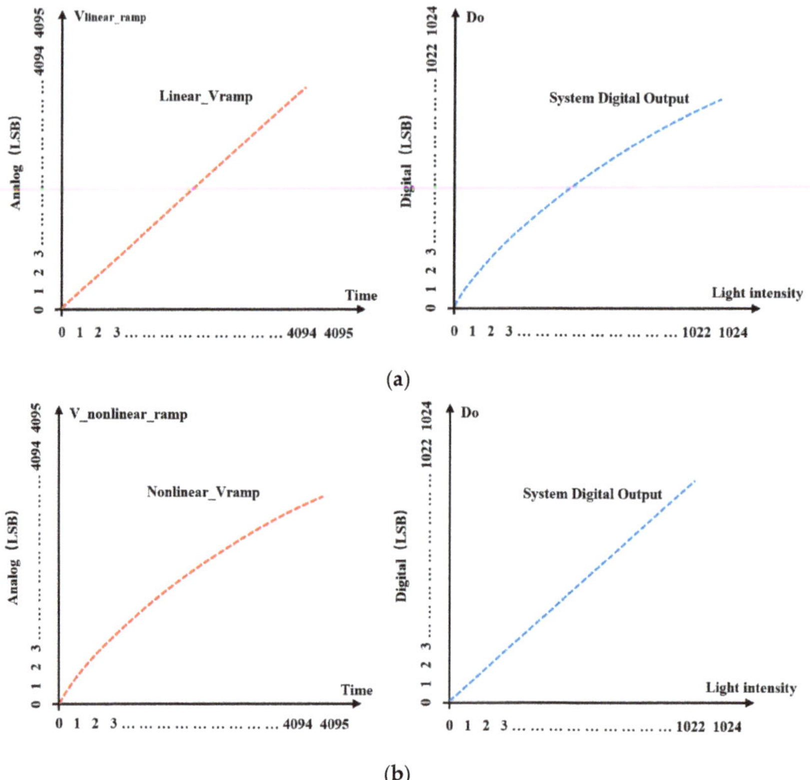

Figure 3. System digital number (DN) output values with linear and nonlinear ramp generators: (**a**) system DN output with linear ramp generator; (**b**) system DN output with a nonlinear ramp generator.

The proposed technique utilizes the multiple dummy pixel units surrounded by active pixel array, which are connected in parallel to form the C_{FD} of the ramp generator circuit. That is, the equivalent C_{FD} is generated by the parallel connection of multiple FD nodes of pixel units, and then a discrete sampling negative feedback technique is utilized to adaptively adjust the level of the current source. This technique produces an adaptively current source that creates a discharge path between the FD node's capacitance and the ground, thereby generating a nonlinear ramp signal. Then the nonlinear signal is buffered by the SF transistor that has the same characteristics as the active pixel. The buffered ramp signal is then compared with the pixel output voltage by the fully differential pipeline sampling quantization comparator based on the double auto-zeroing technique. When the sampled active pixel output voltage signal V_{PIX} is equal to the nonlinear ramp signal V_{ramp}, the comparator toggles while the counter stops counting and stores the counter value into the static random-access memory (SRAM). Figure 4 shows the schematic diagram of a nonlinear ramp using the dummy pixel based on a typical voltage mode 5T pixel structure.

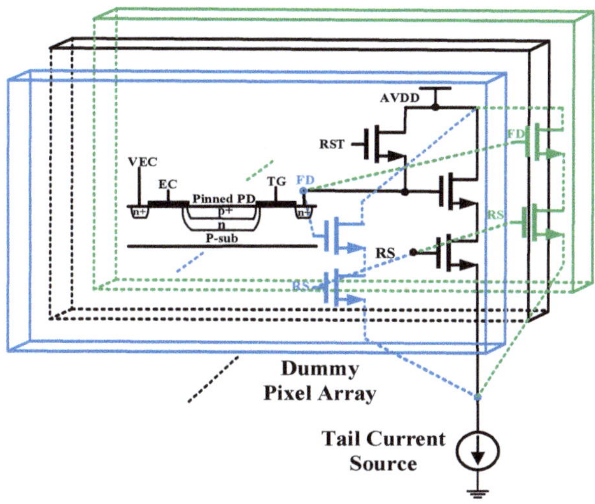

Figure 4. Schematic diagram illustrating the implementation of the integration capacitor by a floating diffusion node based on a dummy pixel array.

The schematic diagram of the adaptive nonlinear ramp generator based on the dummy pixel array is shown in Figure 5. The left part of Figure 5 shows the discrete negative feedback control circuit that generates the corresponding tail current source of nonlinear ramp generator. The right part of Figure 5 shows the circuit of C_{FD} based on a dummy pixel. The C_{FD} produced by the dummy pixel unit and the operational amplifier circuit constitutes the entire nonlinear ramp generator circuit, which is sampled and corrected by the adaptive discrete negative feedback control technique. The nonlinearity characteristic of the generated ramp signal is the same as the output of the pixel unit. Therefore, this method can reduce the nonlinearity caused by the capacitance of the FD node and SF's variable gain in the pixel. Hence, the system linearity of the CIS is improved.

In the 5T pixel, the row selection transistor and the transfer gate transistor (including EC and TG) are connected to the power supply and analog ground, respectively, while the reset transistor NM2 is controlled by the timing signal (Ramp_Adj_P).

The timing of the adaptive nonlinear ramp generator is divided into three phases. During the initial phase, both the pulse Ramp_Init (driving the switch K9) and the Ramp_RST (driving the switch K6–K9) are set low in the initial state of the adaptive nonlinear ramp generator circuit, while both are held high for other phases. The non-overlapped clocks Ramp_Adj_P and Ramp_Adj_N are driving the switches K1–K5 and NM2 to correct and generate the ramp, respectively. The second phase adaptively corrects the slope of the ramp signal. The third phase is the ramp generation phase. The timing diagram of the adaptive nonlinear ramp generator circuit is shown in Figure 6.

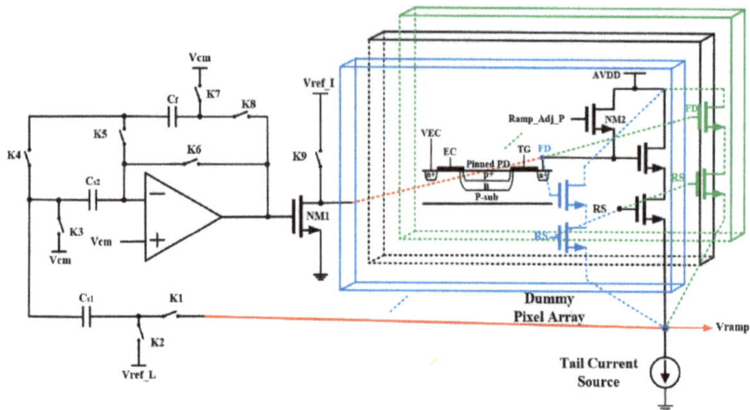

Figure 5. Schematic diagram of adaptive nonlinear ramp generator based on a dummy 5T pixel.

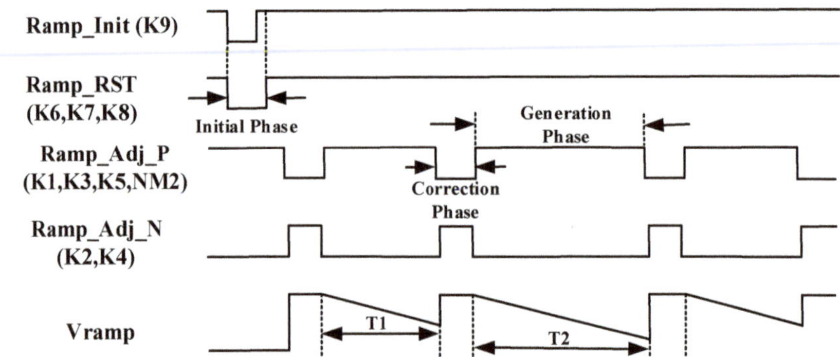

Figure 6. Timing diagram of adaptive nonlinear ramp generator.

3.2. Fully Differential Pipeline Sampling Quantization Based on Double Auto-Zeroing Technique

Generally, the readout circuit for the CMOS image sensor consists of a programmable gain amplifier (PGA) and an SS-ADC. The sampling and the quantization phases are implemented by the PGA module and the SS-ADC, respectively. The sampling and the quantization phases usually work in sequential order. One disadvantage of this design used in the classic CIS is the inefficient use of the SS-ADC. Another disadvantage is that the comparator of SS-ADC has different crossing common-mode voltages that cause the offset and the delay information to vary in different columns, resulting in a large column FPN. In order to overcome the shortcomings of sampling and quantization phases, which cannot operate simultaneously for conventional SS-ADC, a readout circuit of fully differential pipeline sampling quantization based on a double auto-zeroing technique is proposed in this section. The proposed circuit realizes sampling and quantization executions concurrently to improve the frame rate and reduce the column FPN.

In the proposed design, a fully differential analog comparator with pipeline sampling quantization based on the double AZ technique is employed to improve the frame rate and reduce the influence of column FPN on the performance of CIS. Figure 7 shows the double AZ technique comparator circuit for fully differential pipeline sampling.

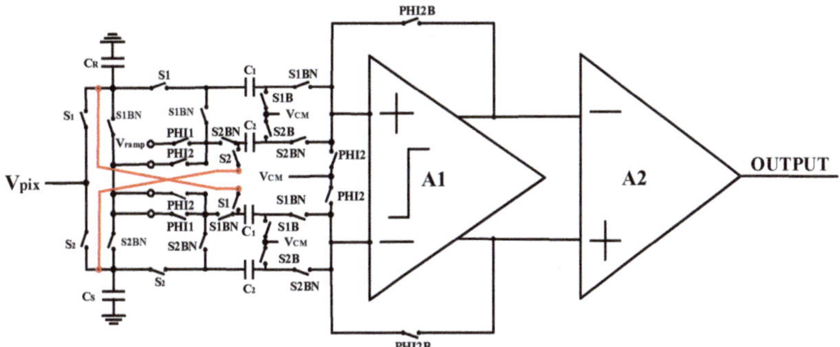

Figure 7. Fully differential comparator with pipeline sampling quantization based on the double auto-zeroing technique.

In Figure 7, C_R and C_S are the reset and the signal sampling capacitors, respectively. S1 and S2 represent the sampling switch control signals in the signal and the reset phases, respectively. S1B, S1BN, and S2B, S2BN are the control switches related to the sampling reset and signal, respectively. Clock signal PHI1, PHI2 and PHI2B are utilized to realize double reset operation in the quantization phase. Where PHI2 and PHI2B are driven by a pair of non-overlapping clocks, and PHI2B is a control signal of the comparator in the reset phase ahead of PHI2 in phase by a constant time. Figure 8 shows a detailed working timing diagram for the fully differential comparator with pipeline sampling quantization based on the double AZ technique.

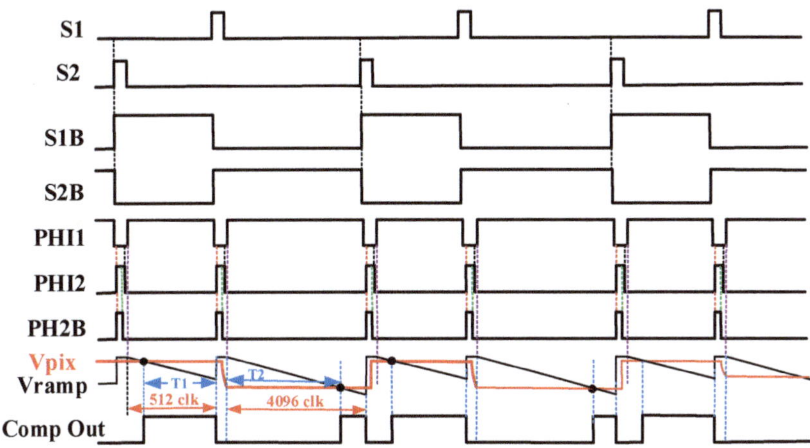

Figure 8. Working timing diagram of fully differential comparator with pipeline sampling quantization based on the double auto-zeroing technique.

The column-parallel readout timing of the CIS is illustrated in Figure 9, and the whole frame time is defined as:

$$T_{frame} = T_{FOT} + T_{row} \times N_{row} \tag{1}$$

where T_{FOT} is the frame overhead time, T_{row} is the row readout time, and N_{row} is the number of rows of CIS. For the readout circuit structure of sequential sampling quantization, the time of row readout can be defined as:

$$T_{row\_seq} = T_{sample} + \max(T_{ADC}, T_{LVDS}) \tag{2}$$

where T_{row_seq} is the row sequential sampling time, T_{sample} is the row sampling time, T_{ADC} is the ADC conversion time and T_{LVDS} is the LVDS output time. According to the column-parallel pipeline signal processing, the row readout time can be expressed as:

$$T_{row} = \max(T_{sample}, T_{ADC}, T_{LVDS}) \tag{3}$$

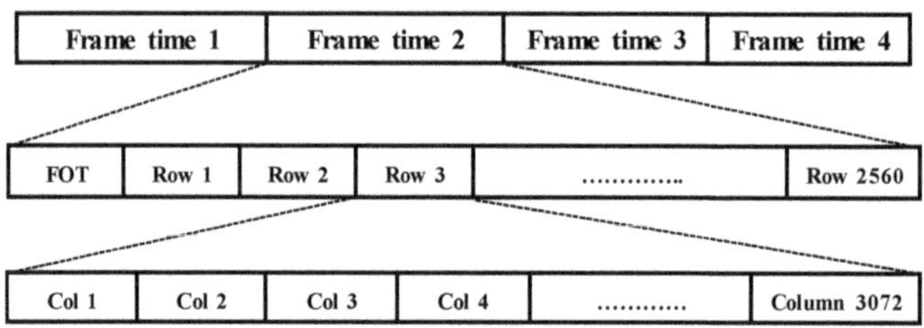

Figure 9. Readout timing of the column-parallel signal processing.

It can be observed from the above equations that the pipeline sampling structure will reduce the row readout time. Thus, the frame rate will be improved.

The proposed readout circuit combines the advantages of both pipeline sampling quantization and double AZ techniques. The proposed method saves two sampling times (the time of the correlated double sampling), avoids the waiting time as in the conventional readout scheme, increasing the efficiency of the comparator up to 100%, and improves the frame rate from 75 to 86 fps under the clock frequency of 400 MHz compared with the classic sequential sampling and quantization method. Since a double reset operation is performed in the quantization phase (double AZ technique), the values of signal and reset phases are all at the same constant crossing detector voltage level. Thus, the column FPN is significantly reduced and the inter-columns consistency of the CIS is effectively improved.

4. Analysis of Proposed Techniques

This section presents the analysis of the eliminating principle of nonlinearity when using the linear and the nonlinear ramp generators. Usually, the principle of a typical SS-ADC structure consisting of a high-speed comparator, a ramp generator circuit, and a digital counter is implemented by converting the counter time into the digital code, assuming that the photocurrent of the pinned photodiode is highly linear to the illumination intensity [1–3]. Equivalently, the analysis in this section uses time t to represent the digital output code of SS-ADC while conducting the relationship between the digital output code of an SS-ADC and the photocurrent when exploring both the linear and the nonlinear ramp generators.

4.1. Linearity Analysis of CMOS Image Sensor with Linear Ramp

Using a linear ramp, the ramp output voltage V_{ramp} is defined as:

$$V_{ramp} = \frac{I_{ramp}}{C} \times t \tag{4}$$

where I_{ramp} is the discharging current of the ramp circuit, the integral capacitance C is a constant value, and t is the integration time of the ramp circuit. Since the illumination intensity changes linearly based on the above assumption, the output photocurrent of the pinned photodiode I_{pd} is also changing linearly. When the voltage of the FD node voltage discharges for an integration time T_{int} with the photocurrent I_{pd}, the resulting voltage V_{FD} is defined as:

$$V_{FD} = \frac{I_{pd}}{C_{FD}(V_{FD})} \times T_{int} \tag{5}$$

where, $C_{FD}(V_{FD})$ is the total parasitic capacitance of the FD node, which is related to the voltage of the FD node, and is also one of the nonlinear causes of the pixel. The complete capacitance of the FD node is defined as [2,3]:

$$C_{FD} = C_{RST\_OV} + C_{TX\_OV} + C_{SF\_OV} + C_{FD\_VERTICAL} + C_{METAL} \tag{6}$$

where, C_{METAL}, C_{TX_OV}, and C_{RST_OV} are the parasitic capacitances related to the size of the metal wire, the TX transistor and the reset transistor, respectively. Once the sizes of the TX and the reset transistors in the pixel are determined, the corresponding parasitic capacitances will remain unchanged, while $C_{FD\_VERTICAL}$, and C_{SF_OV} are related to the overlap parasitic capacitances of the floating diffusion node that changes with the voltage of the FD node. The output voltage of pixel V_{PIX} is equal to the voltage of the floating diffusion node buffered by the SF, as shown below [2,3]:

$$V_{PIX} = \frac{I_{pd}}{C_{FD}(V_{FD})} \times T_{int} \times G_{SF}(V_{FD}) \tag{7}$$

When the comparator of a single-slope ADC utilizes a linear ramp, the comparator toggles if the voltage V_{ramp} is equal to V_{PIX}, so depending on time t, Equation (5) can be obtained as:

$$t = \frac{I_{pd}}{I_{ramp}} \times \frac{C}{C_{FD}(V_{FD})} \times G_{SF}(V_{FD}) \times T_{int} \tag{8}$$

where, $G_{SF}(V_{FD})$ is the gain of SF in the pixel. It can also be seen from Equation (8) that, due to the nonlinearity of integral capacitance of the FD node and the gain nonlinearity of the SF, nonlinearity exists between the integration time t and the photocurrent I_{pd} when T_{int} is constant. That is, the integration time of the linear ramp generator has a nonlinear relationship with the integration time of the FD node in the pixel is constant (which can be equivalent to a linear change in the integration time when the illumination intensity is constant). It can be further observed that there is a nonlinear relationship between the digital output code of the SS-ADC and the linear ramp generator. Thus, the overall illumination intensity is not linear between the digital code (by representing the time t) of the SS-ADC and the illumination intensity I_{pd}. The nonlinear relationship reduces the system linearity of the CMOS image sensor.

4.2. Linearity Analysis of CMOS Image Sensor with Adaptive Nonlinear Ramp

When using the adaptive nonlinear ramp proposed in this paper, the nonlinear ramp generator voltage is defined as:

$$V_{ramp} = \frac{I_{ramp}}{K \times C_{FD}(V_{FD})} \times G_{SF}(V_{FD}) \times t \tag{9}$$

where K is the scale factor, I_{ramp} is the discharging current of the nonlinear ramp, and $C_{FD}(V_{FD})$ is the nonlinear capacitance produced by the dummy pixel, which is related to the voltage of the FD node. When the output voltage of pixel V_{PIX} is equal to the ramp generator voltage V_{ramp}, the output of the comparator is toggled, and the counter stops counting. At this moment, the output value of the digital counter represents the pixel signal. There is a linear relationship between the digital number (DN) and the ramp integration time t when using a nonlinear ramp. The integration time t is:

$$t = K \times \frac{I_{pd}}{I_{ramp}} \times \frac{C_{FD}(V_{ramp\_FD})}{C_{FD}(V_{FD})} \times \frac{G_{SF}(V_{ramp\_FD})}{G_{SF}(V_{FD})} \times T_{int} \qquad (10)$$

where V_{ramp_FD} is the voltage of the FD node of the adaptive nonlinear ramp generator. According to Equation (10), when the voltage values V_{ramp} and V_{PIX} are equal, $C_{FD}(V_{ramp\_FD})$, $C_{FD}(V_{FD})$, $G_{SF}(V_{ramp\_FD})$ and $G_{SF}(V_{FD})$ are all also equal. Thus, the two variables in Equation (10) cancel each other out, thereby eliminating the nonlinear error caused by the FD node integral capacitance and the gain nonlinearity of the SF transistor. The linear relationship between the DN and the input derived from the above analysis is given as:

$$t = K \times \frac{I_{pd}}{I_{ramp}} \times T_{int} \qquad (11)$$

where K is the number of dummy pixel units used for producing the nonlinear ramp integration capacitance.

5. Results

5.1. Simulation Results of Nonlinear Ramp Generation Circuit Based on Dummy Pixel

Figure 10 compares the simulation results of nonlinear and linear ramp generators based on the dummy pixel. The red and gray curves represent linear and nonlinear ramp signals, respectively. It can be seen from the simulation results that the nonlinear ramp signal already contains the nonlinear components of the pixel structure, which is the nonlinearity of the C_{FD} of the FD node and the gain nonlinearity of the SF.

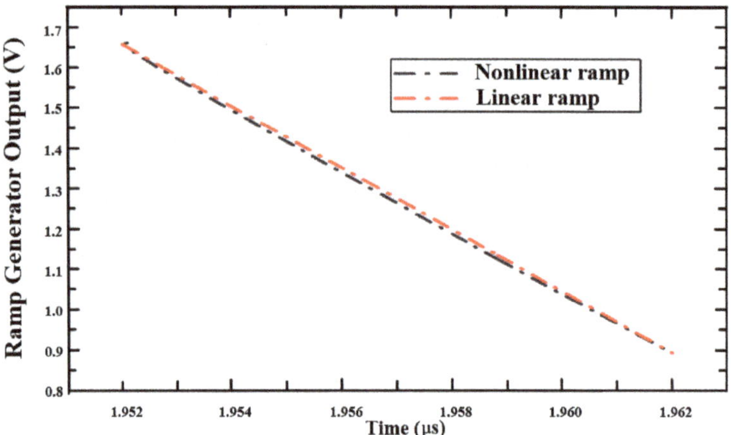

Figure 10. Comparison of nonlinear and linear ramp signals.

Figure 11 shows the nonlinearity of CIS along with the linear change of photocurrent (I_{pd}) when using the linear and the adaptive nonlinear ramp techniques. Figure 11 includes the nonlinearity characteristics of typical CIS using the linear and the adaptive nonlinear ramp techniques. It can be seen from the simulation results that the nonlinearity is significantly reduced between the output digital code and the photocurrent (light intensity). The maximum differential nonlinearity (DNL) is 0.044 LSB, and the maximum integral nonlinearity (INL) is 0.054 LSB.

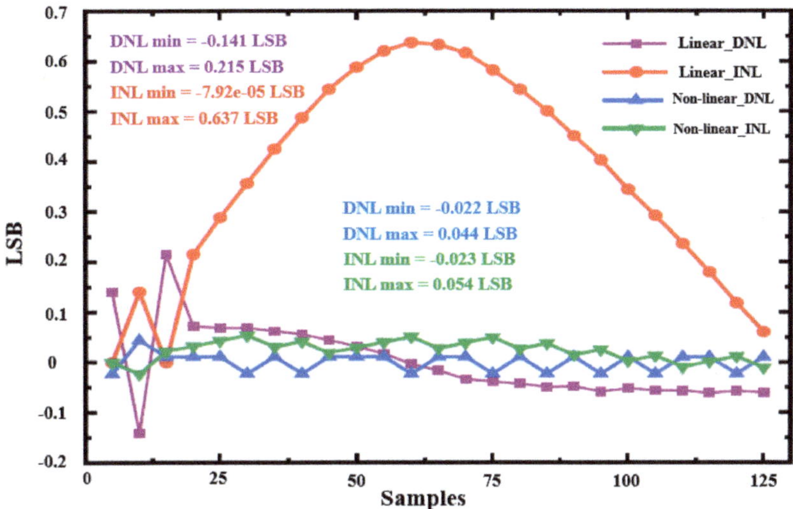

Figure 11. Simulation results of the linearity of CMOS image sensor (CIS) using the linear ramp and the adaptive nonlinear single-slope analog-to-digital converter (ADC) techniques.

5.2. Simulation Results of Fully Differential Pipeline Sampling Quantization with Double Auto-Zeroing Technique

Figure 12 shows the simulation results of the operation timing of the fully differential comparator with pipeline sampling quantization based on the double AZ technique. It is easy to observe the correct function of the proposed fully differential comparator from the operation timing diagram. The crossing detector voltage level of the input positive level VCP is always the same as the negative level VCN for the comparator.

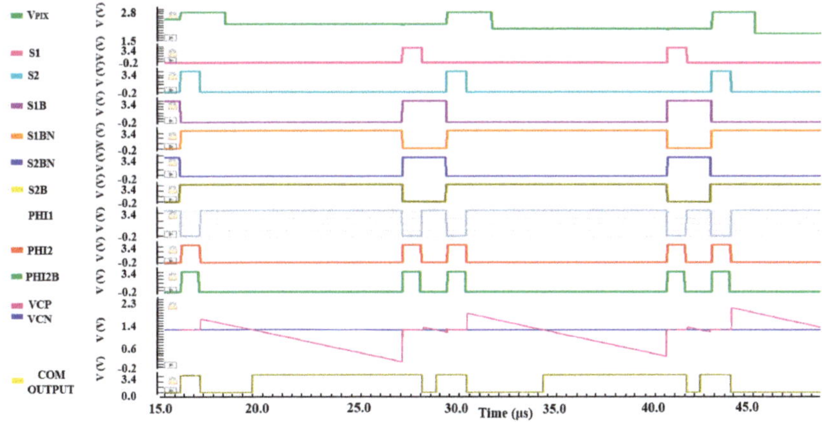

Figure 12. Operation timing of the fully differential pipeline sampling quantization with double auto-zeroing technique.

5.3. Experimental Result

The size of the pixel array in the proposed design is 2560 × 3072. Figure 13 shows the sample image taken by a typical global shutter pixel 5T structure with an adaptive nonlinear ramp generator and the fully differential comparator based on pipeline sampling quantization with the double AZ technique.

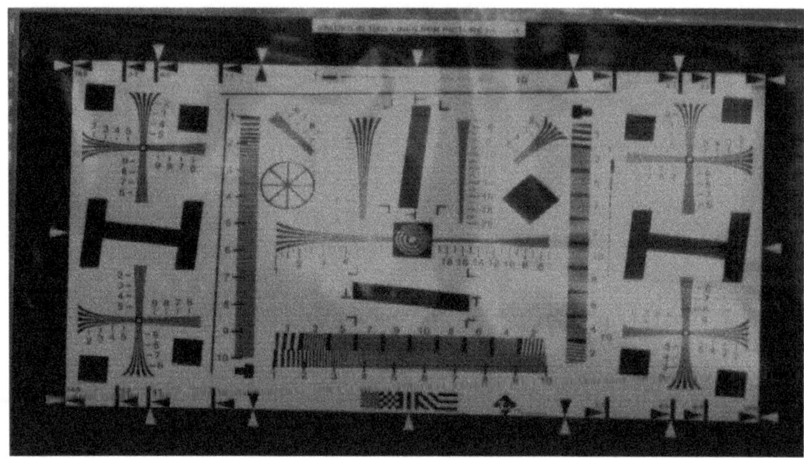

Figure 13. Sample image taken from the proposed highly linear CMOS image sensor.

Table 1 compares the performance of the highly linear CMOS image sensor proposed in this paper with those reported in the literature.

Table 1. Performance comparison of the proposed highly linear CMOS image sensor with the literature.

Reference	This Work	[2]	[3]	[3]	[8]
Process (nm)	180	180	180	180	40
Linearity improvement techniques	Adaptive nonlinear ramp based on dummy pixel	Pixel optimization and calibration	Analog buffer	Off-chip digital calibration	Linear ramp and source follower buffer
Array size	2560 × 3072	128 × 160	128 × 128	128 × 128	NA
Global/Rolling shutter	GS/RS	RS	RS	RS	RS
ADC Architecture	12/14 bit SS-ADC	12 bit SS-ADC	10 bit SS-ADC	10 bit SS-ADC	12 bit SS-ADC
Frame rate (fps)	86 (400 MHz clock)	NA	60 (12.5 MHz clock)	60	NA
Digital CDS	Y	Y	Y	Y	Y
Pixel size	6.5 μm × 6.5 μm	12 μm × 10 μm	10 μm × 10 μm	10 μm × 10 μm	NA
Fill factor	100% (BSI)	40% (FSI)	47% (FSI)	47% (FSI)	100% (3D BSI)
Pixel type	5T	CTIA	4T	4T	NA
Conversion gain	17.4 μV/e$^-$	40 μV/e-	56.8 μV/e-	45.3 μV/e-	NA
Read noise	7.9 e$^-$	16.4 (gain = 8)	4.12 (gain = 8)	4.17 (gain = 8)	261.5 μVrms
Full well capacity	91.7 ke$^-$	30.613 ke$^-$	17.27 ke$^-$	20.96 ke$^-$	NA
Dynamic range(dB)	81.3	65	72.4	74	71.8
Column FPN (%)	0.019	NA	NA	NA	0.028
SNR (dB)	49.4	44.2	42.1	42.9	NA
Nonlinearity (%)	0.047	0.095	0.058	0.06	NA
Dark current	8.3 pA/cm^2@23 °C	NA	5.6pA/cm^2	NA	NA
Per column Power	96.3 μW	NA	NA	NA	66.8 μW

6. Conclusions

In this paper, a novel adaptive nonlinear ramp generator design technique based on dummy pixels and a readout scheme of fully differential pipeline sampling quantization with a double auto-zeroing technique are proposed. The proposed ramp generator design can significantly improve the linearity of the pixel, which eliminates the nonlinearity caused by the capacitance of the floating diffusion node of the pixel and the gain nonlinearity of the SF transistor.

The proposed readout circuit improves the frame rate by 14% in the fully differential comparator with the pipeline sampling quantization technique, while increasing the efficiency of the comparator up to 100%. Moreover, the readout chain with the fully differential comparator based on the double auto-zeroing technique reduces the fixed pattern noise to 0.019%. A wide dynamic range of 81.3dB is also achieved by the proposed design without using the column amplifier gain stages. The resulting readout circuit noise is lowered to 7.9e-, and the system linearity is reduced to 0.047% using the adaptive nonlinear ramp technique.

Author Contributions: C.L. and B.H. conceived the ideas and innovations, performed the simulation and analyzed the data; J.H. and Z.G. contributed a lot to the principle verification and results and discussion of this paper; L.W. provided supervision and guidance in this work; C.L. wrote the paper. All authors have read and agreed to the published version of the manuscript.

Funding: This research was funded by National Major Science and Technology Industry Project grant number 2017ZX01006101-001 and Scientific Research Project of the Shaanxi Education Department grant number 19JC029. The APC was funded by National Major Science and Technology Industry Project.

Conflicts of Interest: The authors declare no conflict of interest.

References

1. Fei, W. *Linearity Research of ACMOS Image Sensor*; Delft University of Technology: Delft, The Netherlands, 2018.
2. Wang, F.; Theuwissen, A.J.P. Pixel optimizations and digital calibration methods of a CMOS image sensor targeting high linearity. *IEEE Trans. Circuit Syst.* **2018**, *66*, 930–940. [CrossRef]
3. Wang, F.; Han, L.; Theuwissen, A.J.P. Development and Evaluation of a Highly Linear CMOS Image sensor with a digitally assisted linearity calibration. *IEEE J. Solid State Circuit* **2018**, *53*, 2970–2981. [CrossRef]
4. De Sá, L.B.; Dias, M.H.C.; Dupret, A.; Filho, A.C.D. A 99.95% linearity readout circuit with 72 dB dynamic range for active pixel sensors. *Int. J. Circuit Theory Appl.* **2018**, *46*, 1580–1592. [CrossRef]
5. Wang, F.; Theuwissen, A.J.P. Techniques for pixel-level linearity optimization. In Proceedings of the Workshop CMOS Image Sensors High Performance Applications, Paris, France, 21–22 November 2017; pp. 1–20.
6. Xiaoliang, G. *The Design of a Global Shutter CMOS Image Sensor in 110nm Technology*; Delft University: Delft, The Netherlands, 2012.
7. Shoji, K. Column-parallel adcs for cmos image sensors and their fom-based evaluations. *Ieice Trans. Electron.* **2018**, *E101*, 444–456.
8. Qiyuan, L.; Edward, A.; Kinyua, M.; Soenen, E.G.; Silva-Martinez, J. A low power digitizer for back-illuminated 3-D-stacked CMOS image sensor readout with passing window and double auto-zeroing techniques. *IEEE J. Solid State Circuits* **2017**, *52*, 1591–1604.
9. Samuel, S.I.; Blanca, P.G.; Espejo-Meana, S.; Ragel-Morales, A.; Ceballos-Cáceres, J.; Muñoz-Díaz, M.; Carranza-González, L.; Arias-Drake, A.; Mora-Gutiérrez, J.M.; Lagos-Florido, M.A. An adaptive approach to on-chip cmos ramp generation for high resolution single slope ADCs. In Proceedings of the 21st European Conference on Circuit Theory and Design, Dresden, Germany, 8–12 September 2012; Volume 12.
10. Snoeij, M.F.; Gan, P.D.; Theuwissen, J.P.; Makinwa, K.A.A.; Huijsing, J.H. A Cmos Image Sensor with a column-level multiple-ramp single-slope adc. In Proceedings of the 2007 IEEE International Solid-State Circuits Conference. Digest of Technical Papers, San Francisco, CA, USA, 11–15 February 2007.
11. Lim, S.; Cheon, J.; Ham, S.; Han, G. A new correlated double sampling and single slope ADC circuit for CMOS image sensors. *Int. SoC Des. Conf.* **2004**, *10*, 129–131.
12. Chen, Y.; Theuwissen, J.P.; Chae, Y.C. Column parallel single slope ADC with digital correlated multiple sampling for low noise cmos image sensors. *Procedia Eng.* **2011**, *25*, 1265–1268. [CrossRef]

13. Masood, T.; Jafar, S. An ultra-linear CMOS image sensor for a high—Accuracy imaging system. *Int. J. Circuit Theory Appl.* **2018**, *3*, 1593–1605.

14. Masood, T. A highly linear and high-accurate CMOS image sensor. *Analog Integr. Circuits Signal Proc.* **2019**, *10*, 91–96.

15. Sudhir, M.G.; Hyun, J.S.; Hon, S.P.W. Image Sensor with Dummy Pixel Array. US6344877B1, 12 February 2002.

16. Benoit, P.; Sanchez-Sinencio, E. Auto-calibrating analog timer for on-chip testing. In Proceedings of the International Test Conference 1999 Proceedings, Atlantic City, NJ, USA, 30 September 1999; Volume 9.

17. Benoit, P.; Sanchez-Sinencio, E. On-chip ramp generators for mixed-signal BIST and ADC self-test. *IEEE J. Solid State Circuits* **2003**, *38*, 263–273.

© 2020 by the authors. Licensee MDPI, Basel, Switzerland. This article is an open access article distributed under the terms and conditions of the Creative Commons Attribution (CC BY) license (http://creativecommons.org/licenses/by/4.0/).

Article

Temperature Hysteresis Mechanism and Compensation of Quartz Flexible Accelerometer in Aerial Inertial Navigation System

Chunxi Zhang, Xin Wang, Lailiang Song and Longjun Ran *

School of Instrumentation and Optoelectronic Engineering, Beihang University, Beijing 100191, China; zhangchunxi@buaa.edu.cn (C.Z.); Vong_Hsin@163.com (X.W.); songlailiang@buaa.edu.cn (L.S.)
* Correspondence: rlj_buaa@163.com; Tel.: +86-131-2020-9442

Abstract: Strap-down inertial navigation systems (INSs) with quartz flexible accelerometers (QFAs) are widely used in many conditions, particularly in aerial vehicles. Temperature is one of the significant issues impacting the performance of INS. The variation and the gradient of temperature are complex under aerial conditions, which severely degrades the navigation performance of INS. Previous work has indicated that parts of navigation errors could be restrained by simple temperature compensation of QFA. However, the temperature hysteresis of the accelerometer is seldom considered in INS. In this paper, the temperature hysteresis mechanism of QFA and the compensation method would be analyzed. Based on the fundamental model, a comprehensive temperature hysteresis model is proposed and the parameters in this model were derived through a temperature cycling test. Furthermore, the comparative experiments in the laboratory were executed to refine the temperature hysteresis model and to verify the effectiveness of the new compensation method. Applying the temperature hysteresis compensation in flight condition, the result shows that the position error (CEP) is restrained from 1.54 nmile/h to 1.29 nmile/h. The proposed temperature hysteresis compensation method improves the performance of INS effectively and feasibly, which could be promoted to other applications of INS in similar temperature changing environment correspondingly.

Keywords: temperature compensation; hysteresis; quartz flexible accelerometer; aerial inertial navigation system; thermal effect; creep effect

Citation: Zhang, C.; Wang, X.; Song, L.; Ran, L. Temperature Hysteresis Mechanism and Compensation of Quartz Flexible Accelerometer in Aerial Inertial Navigation System. *Sensors* **2021**, *21*, 294. https://doi.org/10.3390/s21010294

Received: 11 November 2020
Accepted: 30 December 2020
Published: 4 January 2021

Publisher's Note: MDPI stays neutral with regard to jurisdictional claims in published maps and institutional affiliations.

Copyright: © 2021 by the authors. Licensee MDPI, Basel, Switzerland. This article is an open access article distributed under the terms and conditions of the Creative Commons Attribution (CC BY) license (https://creativecommons.org/licenses/by/4.0/).

1. Introduction

INS is one of the most important measurements in the aerial vehicles. It provides the attitude information to the control system, which can guide an aerial vehicle without external information, such as GPS. Therefore, the performance of INS is vital to aircraft. With the rapid development of inertial technology, the performance of gyroscopes has improved tremendously. Nevertheless, the performance of QFA is not comparable with that of gyroscopes. In an aerial platform, vibration and temperature issues are two of the main factors degrading the performance of the QFA [1–4].

The QA-3000 manufactured by Honeywell (Morristown, NJ., USA) and the GJN096 manufactured by China Aerospace Science and Industry Corporation (Beijing, China) whose performances are similar cover the medium and high classes of INS. The performances of these two QFAs are shown in Table 1. The thermal issue is the main factor degrading the performance of QFA. Taking the Honeywell QA-3000-030 as an example, the temperature coefficient of bias is typical 15 μg/°C and the temperature coefficient of scale factor is typical 120 ppm/°C, which means when temperature varies from −20 °C to +70 °C, the bias of QFA will drift for nearly 1.35 mg and the scale factor will drift for 10,800 ppm at most, if there is no effective temperature compensation for QFA. Therefore, improving the temperature adaptability of QFA is vital for INS.

Table 1. Main performance characteristics of QA3000 and GJN06.

	QA3000-030	GJN096-D
Input range [g]	±60	±70
Bias repeatability [µg]	<40	<60
Bias temperature sensitivity [µg/°C]	15	50
Scale factor repeatability [ppm]	<80	<80
Scale factor temperature sensitivity [ppm/°C]	120	50
Operating temperature [°C]	−28~+78	−48~+80
Shock [g]	100	200
Resolution/Threshold [µg]	<1	<5
Bandwidth [Hz]	>300	<2000

Data are from their respective product descriptions.

Lots of work has been conducted to improve the temperature experiment performance of QFA. Temperature is used to build a general temperature model through simple linear regression and the result shows that the general performance is improved in pure inertial navigation [5,6]. Establishing the temperature model by linear regression is a common strategy. However, the temperature model built by simple linear regression only slightly improves performance [7,8]. In order to optimize the temperature model, an improved linear regression algorithm that focuses on determining the order of the model is proposed. The simulation result indicates that the temperature performance of compensated bias is better compared with a simple regression algorithm [9]. Nevertheless, considering the high performance of INS, these simple linear regression models which only considers thermal effect do not meet the demand for temperature performance. Consequently, a temperature-dependent model based on a neural network is proposed. The drifts of bias and scale factor are well compensated in a wide temperature range [10–12]. But the repeatability of this temperature compensation method remains suspicious.

The proposed temperature hysteresis compensation method in this paper is motivated by studying the viscoelasticity of a macromolecular compound in QFA, and it might be a new avenue to support the improvement of the INS. The main contributions of this paper are summarized as follows:

(1) Based on the analysis of the viscoelasticity of epoxy resin (ER), the viscoelasticity of ER is attributed to the main inner factor of bias-hysteresis phenomenon. Multiple piecewise function is applied to bias-hysteresis model dealing with creep whose influence on deformation of ER is irregular.
(2) The temperature hysteresis of magnetic induction is mainly determined by the temperature and temperature gradient, which is the main inner factor of scale factor hysteresis. Therefore, temperature, temperature gradient and the coupling of temperature and temperature gradient are used to build the scale factor-hysteresis model.
(3) Four-points rotation calibration experiments for QFA is used to build the rough temperature hysteresis model. Furthermore, the rough temperature hysteresis model is refined through system-level temperature experiments.

Considering the analysis of the relevant literature, few studies involving the temperature hysteresis of QFA are conducted and its impact on navigation accuracy remains unclear. In this paper, a novel compensation method is proposed to address the temperature hysteresis issue. The rest of this paper is organized as follows: In Section 2, the temperature hysteresis of bias and scale factor will be mainly analyzed. Hysteresis of magnetic conduction is mainly influenced by temperature and temperature gradient, which is the inner factor of scale factor hysteresis. Based on this fact, the temperature hysteresis model scale factor is built. Through ER temperature experiments the viscoelasticity of the ER is attributed to the main factor of bias-hysteresis. In Section 3, a four-point rotation calibration test is designed for three cases of QFAs to build the rough hysteresis temperature model in the laboratory. Furthermore, this rough model is refined by this system-level temperature calibration experiment to form an explicit model which then is applied to the flight condition. In Section 4, the summary of this paper is concluded.

2. Analysis of Temperature Hysteresis

2.1. Temperature Model of QFA

The common model of QFA is [13]:

$$E_{out} = \begin{aligned} &K_1 K_0 + K_1 a_i + K_1 K_2 a_i^2 + K_1 K_3 a_i^3 + K_1 K_{ip} a_i a_p \\ &+ K_1 K_{io} a_i a_o + K_1 \delta_o a_p + (-K_1 \delta_p) a_o \end{aligned} \tag{1}$$

where E_{out} is the output of QFA; a_i, a_p, a_o are the inputs along with input reference axis, pendulum reference axis, and output reference axis; K_0 is bias; K_1 is scale factor; K_2 is secondary-order nonlinear coefficient; K_3 is third-order nonlinear coefficient; K_{ip} is the coupling coefficient of input reference axis and pendulum reference axis; K_{io} is the coupling coefficient of input reference axis and output reference axis; δ_o is misalignment angle of output axis; δ_p is misalignment angle of pendulum axis.

Through standardized production and installation, K_3, K_{ip}, K_{io}, δ_o and δ_p can be negligible in the simplified model of QFA. The error of K_2 is closely related to high acceleration, whose uncoupling effect with temperature is weak. Therefore, the simple model of QFA which only includes bias and scale factor is used to address the temperature issue.

The simple temperature model of QFA is usually used as follows [14]:

$$E_{out}(T) = K_1(T) \cdot [K_0(T) + a_i] \tag{2}$$

where $E_{out}(T)$ is the output value of QFA at $T°C$; $K_1(T)$ is the scale factor at $T°C$; $K_0(T)$ is the bias at $T°C$; a_i is the specific force acting along the QFA input axis.

Scale factor and bias are compensated separately through a four-point calibration method at different temperatures. Normally, the scale factor and bias are fitted through simple linear regression. The scale factor and bias are shown below:

$$K_1(T) = \frac{E_{90°}(T) - E_{0°}(T)}{2} = \phi \cdot \Delta T + K_1(T_0) \tag{3}$$

$$K_0(T) = \frac{E_{180°}(T) + E_{270°}(T)}{2K_1(T)} = \omega \cdot \Delta T + K_0(T_0) \tag{4}$$

where ϕ and ω are coefficients of scale factor and bias; $\Delta T = T - T_0$ is the difference of temperature compared with reference temperature.

The compensated output of QFA by using (3) and (4) is:

$$a_i(T) = \frac{E_{out}(T)}{K_1(T)} - K_0(T) \tag{5}$$

2.2. Temperature Hysteresis Model of QFA

QFA consists of the permanent magnet, quartz, ER, iron, and polyester, whose temperature characteristics vary greatly [15]. A mass of research has analyzed the temperature hysteretic behavior of the magnet, which has been proved as the main factor affecting the scale factor performance. Taking a $[Ru_2(O_2CMe)_4]_3[Fe(CN)_6]$ permanent magnet as an example, the temperature hysteresis at different temperatures from 40 mK to 4.8 K is shown in Figure 1 [16]. Almost every temperature-dependence permanent magnet, like Alnico and $Nd_2Fe_{14}B$, is temperature-dependence hysteretic and the shape of the hysteretic curve is similar to the inset of Figure 1 [17]. The Alnico permanent magnet which is used in QFA has the similar hysteresis properties as $[Ru_2(O2CMe)_4]_3[Fe(CN)_6]$ in the temperature cycle from $-195\ °C$ to $400\ °C$ [18,19].

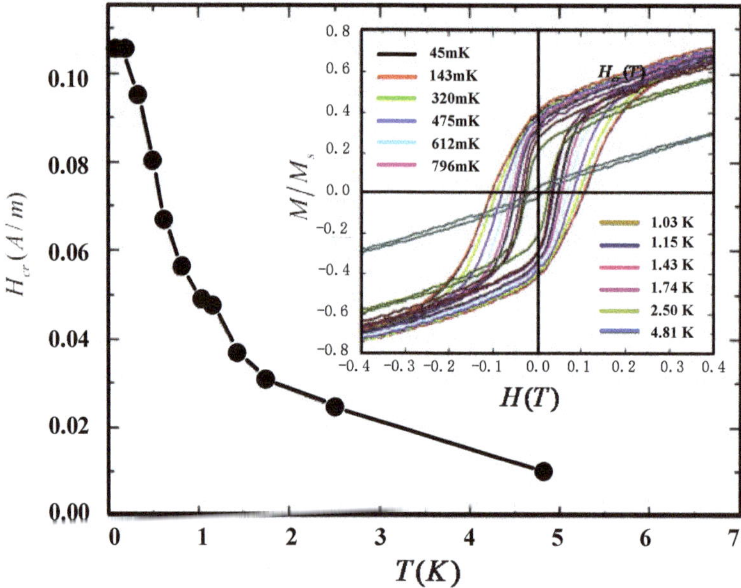

Figure 1. Temperature-dependence hysteresis and coercive force for permanent magnet $[Ru_2(O_2CMe)_4]_3[Fe(CN)_6]$. The main panel shows that the coercive force decreases with temperature decreasing. The coercive force decreases slowly with increasing temperature, but abruptly decays in the vicinity of the transition temperature. The inset shows the magnetic hysteresis loops between 40 mK and 4.8 K. The smallest hysteresis occurs at highest temperature 4.8 K, which then grows with decreasing temperature until saturating below 143 mK.

The QFA scale factor is mainly determined by magnetic induction. Therefore, the temperature hysteretic behavior of magnetic is the key factor of temperature hysteresis of scale factor. Scale factor of QFA can be simply described as a second-order model:

$$K_1(T) = q \cdot (\Delta T)^2 + w \cdot \Delta T + r \tag{6}$$

where q, w, and r are the temperature coefficients of the scale factor.

According to the analysis of the hysteretic phenomenon of scale factor, parameters of scale factor at a specific temperature are not only affected by ΔT, but also affected by gradient of temperature $\frac{\partial T}{\partial t}$ (in this paper, temperature gradient refers to the rate of temperature changing over time). Therefore, the scale factor is remodeled as:

$$K_1(T) = \left(a\frac{\partial T}{dt} + b\right)^2 (\Delta T)^2 + \left(c\frac{\partial T}{dt} + d\right)^2 (\Delta T) + \left(e\frac{\partial T}{dt} + f\right)^2 \tag{7}$$

$$a\frac{\partial T}{\partial t} + b = q \tag{8}$$

$$c\frac{\partial T}{\partial t} + d = w \tag{9}$$

$$e\frac{\partial T}{\partial t} + f = r \tag{10}$$

Linear fitting of scale factor shows the magnitude of coefficients of $\left(\frac{\partial T}{dt}\right)^2 \Delta T$, $\left(\frac{\partial T}{dt}\right)^2 (\Delta T)^2$, and $\frac{\partial T}{dt}(\Delta T)^2$ are 10^{-5} (Using the IMU data in flight experiment). Because the QFA is assembled in the middle of the INS, $\frac{\partial T}{\partial t}$ and ΔT are usually small in aircraft. Therefore, high order terms of Equation (7) are small terms.

After combining similar terms and omitting the high order small terms, Equation (7) can be simplified as below:

$$K_1(T) = \alpha \cdot \left(\frac{\partial T}{dt}\right)^2 + \beta \cdot (\Delta T)^2 + \eta \cdot \frac{\partial T}{dt} \cdot \Delta T + \varepsilon \cdot \frac{\partial T}{dt} + \mu \cdot \Delta T + \gamma \qquad (11)$$

where $\alpha, \beta, \eta, \varepsilon, \mu, \gamma$ are temperature coefficients of scale factor.

Therefore, based on Equation (11), Equation (2) can be remodel as the temperature hysteresis model:

$$E_{out}(T) = \left[\alpha \cdot \left(\frac{\partial T}{dt}\right)^2 + \beta \cdot (\Delta T)^2 + \eta \cdot \frac{\partial T}{dt} \cdot \Delta T + \varepsilon \cdot \frac{\partial T}{dt} + \mu \cdot \Delta T + \gamma\right] \cdot [K_0(T) + a_i] \qquad (12)$$

ER is mainly used as the adhesive to bond the coil with the quartz pendulum whose form relates to bias [19]. Because of the steep temperature gradient and the wide range of temperature in aerial conditions, the thermal influence on ER is typically obvious. Therefore, the deformation of the ER possibly degrades the stability of the pendulum structure, which affects the performance of bias.

ER is of viscoelasticity, which means it has both elastic and viscous properties. Static viscoelasticity and dynamic viscoelasticity are two main characteristics of viscoelasticity. Static viscoelasticity responds to creep effect and hysteretic effect. Dynamic viscoelasticity responds to thermal expansion [20–23]. When ER is affected by the variation of external temperature, internal stress changes. ER stores part of the stress effect and expends the other part, which corresponds to thermal expansion. The stored stress is released when external temperature recovers, which forces the ER to restore to its original condition [24–26]. However, the expended stress causes creep whose deformation is irregular [27]. This phenomenon indicates that the hysteresis of bias has a strong relation with the viscoelasticity of ER. In order to figure out the influence of the static viscoelasticity on QFA, an ER temperature experiment was conducted. The equipment used is shown in Figure 2.

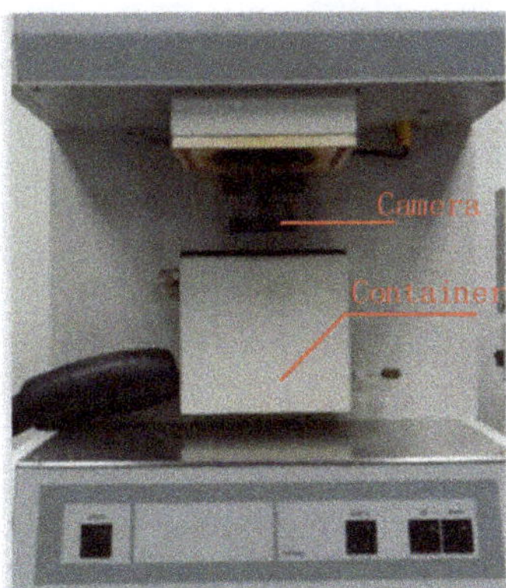

Figure 2. The equipment of creep experiment mainly consists of two parts, a high-resolution camera and a container. Camera can record the micro deformation of object in the container, and the container is a temperature-controlled oven which can provide precise temperature from $-70\,°C$ to $+100\,°C$.

Three blocks of ER of same shape, weight and volume were put in three containers whose inner temperatures were set to $-55\,°C$, $30\,°C$ and $85\,°C$, respectively. Because of the setting of the camera, the measurement of deformation was in negative form. The experiment was conducted for 10 times and the creep results are shown in Table 2. When the structure of ER was relatively stable, the deformation ΔL_v corresponding to creep was recorded.

Table 2. Results of creep deformation of ER.

	$-55\,°\text{C}$	$30\,°\text{C}$	$85\,°\text{C}$
ΔL_{v1} [μm]	-0.377	-0.761	-2.710
ΔL_{v2} [μm]	-0.258	-0.889	-3.804
ΔL_{v3} [μm]	-0.476	-0.575	-2.229
ΔL_{v4} [μm]	-0.477	-0.659	-2.401
ΔL_{v5} [μm]	-0.393	-0.685	-3.373
ΔL_{v6} [μm]	-0.482	-0.982	-2.898
ΔL_{v7} [μm]	-0.579	-0.676	-4.312
ΔL_{v8} [μm]	-0.252	-0.864	-3.915
ΔL_{v9} [μm]	-0.572	-0.961	-2.209
ΔL_{v10} [μm]	-0.281	-0.782	-2.921
Variation	84.88%	42.44%	77.60%

Taking #1 result as benchmark, $Variation = (\Delta L_{v_{\max}} - \Delta L_{v_{\min}})/\Delta L_{v_1} \cdot 100\%$.

The results show that deformation caused by the creep effect related to the temperature and the size of deformation increases while temperature increases. However, deformation at the same temperature varies greatly, which means a precise creep-temperature deformation model is hard to be built.

The experiment results of deformation caused by thermal effect are shown in Table 3. ΔL_c is the severe deformation which caused by thermal effect. The variation of ΔL_c is small, which indicates the relationship between the thermal deformation and the temperature is relative stable. Therefore, thermal-temperature deformation model should exist.

Table 3. Results of thermal deformation of ER.

	$-55\,°\text{C}$	$30\,°\text{C}$	$85\,°\text{C}$
ΔL_{c1} [μm]	-7.696	-7.491	-12.51
ΔL_{c2} [μm]	-7.674	-7.502	-12.56
ΔL_{c3} [μm]	-7.712	-7.489	-13.05
ΔL_{c4} [μm]	-7.703	-7.482	-12.39
ΔL_{c5} [μm]	-7.671	-7.409	-12.18
ΔL_{c6} [μm]	-7.692	-7.511	-12.86
ΔL_{c7} [μm]	-7.710	-7.488	-12.32
ΔL_{c8} [μm]	-7.685	-7.492	-12.54
ΔL_{c9} [μm]	-7.693	-7.508	-12.77
ΔL_{c10} [μm]	-7.749	-7.491	-12.59
Variation	1.01%	1.32%	5.8%

Taking #1 result as benchmark, $Variation = (\Delta L_{c_{\max}} - \Delta L_{c_{\min}})/\Delta L_{c_1} \cdot 100\%$.

The result of #1 is shown in Figure 3. ΔL_c is the severe deformation whose curve is not shown in these figures.

Figure 3a shows that ER shrank 7.696 μm when it deformed severely, then it shrank slowly 0.377 μm in 300 min. Figure 3b shows that ER expanded 7.491 μm when it deformed severely, then it expanded slowly 0.761 μm in 300 min. Figure 3c shows that ER expanded 12.51 μm when it deformed severely, then it expanded slowly 2.71 μm in 300 min. The result shows temperature affects the creep, and the creep speed is proportional to temperature.

The severe deformation of ER is attributed to thermal expansion whose inner factor is dynamic viscoelasticity. The size of slow deformation and the stabilization time is constrained by temperature. This phenomenon verified that creep and hysteresis of ER are influenced by temperature, whose inner factor should be attributed to static viscoelasticity. Because creep of ER results in unexpectable pendulum deformation, $K_0(T)$ is probably inexplicit.

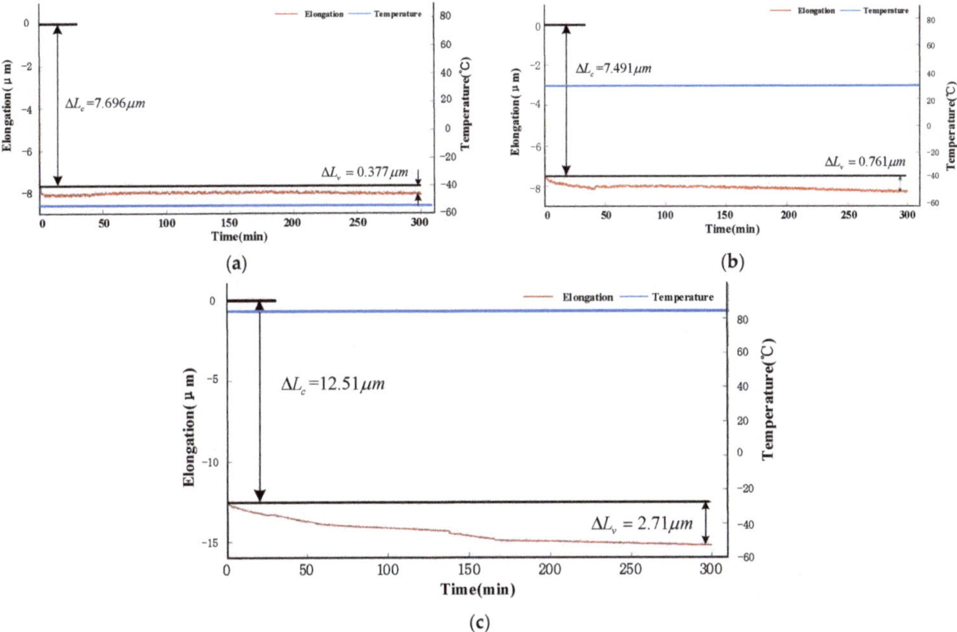

Figure 3. (**a**) is the elongation of ER at −55 °C; (**b**) is the elongation of ER at −35 °C; (**c**) is the elongation of ER at −85 °C. The blue curve represents temperature and the red curve represents the elongation of ER.

3. Experimental Design and Verification

3.1. QFA Temperature Calibration Experiment

The three QFAs are JB-KT8 #1, JB-KT8 #2 and JB-KT8 #3 manufactured by Kaituo Precise Instrument Manufacturing Co., Ltd. (Baoding, China). The three QFAs are assembled in an index head and located in a high-accurate temperature-controlled oven which can provide precise temperature ranging from −75 °C to 120 °C. The temperature accuracy is better than ±0.2 °C. Temperature slew rate is ±1 °C/min to ±5 °C/min. The test equipment is shown in Figure 4.

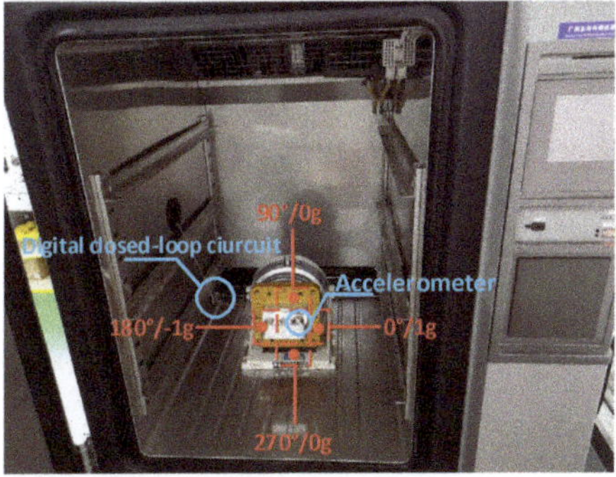

Figure 4. QFA assembled on an index head in temperature-controlled oven, and the index head can be controlled by oven to rotate. The inputs to QFA at 0°, 90°, 180°, 270° are 1 g, 0 g, −1 g, 0 g.

Aerial inertial navigation systems are usually assembled in the central of the aircraft where the variation of temperature is relative stable compared to the temperature near the engine. According to the usual flight data, the gradient of temperature is about 0.2 °C/min in flight. Temperature gradient 1 °C/min can cover all temperature gradient of less than 1 °C/min. Therefore, the use of a 1 °C/min temperature gradient in QFA calibration experiments is reasonable for an aerial environment. The steps of the four-points calibration experiment are as follows:

Step 1: Fix the index head to 0°, then keep the whole device at 22 °C for 120 min;

Step 2: Temperature goes down to −60 °C at the rate of −0.3 °C/min, and it lasts 3 min at each integer temperature point;

Step 3: Temperature rises to 70 °C at the rate of 0.3 °C/min, and it lasts 3 min at each integer temperature point;

Step 4: Temperature goes down to 22 °C at the rate of −0.3 °C/min, and it lasts 3 min at each integer temperature point. Record the output of QFA.

Step 5: Repeat the step 1~step 4 at 90° (0 g), 180° (−1 g), 270° (0 g), and record the outputs of QFA.

Step 6: Repeat the step 1~step 5 with the temperature rate of 0.5 °C/min, 0.7 °C/min, and 0.9 °C/min respectively, and record the outputs of QFA.

Scale factor and bias at $T°C$ can be expressed as:

$$K_1(T) = \frac{A_{90°}(T) - A_{270°}(T)}{2} \tag{13}$$

$$K_0(T) = \frac{A_{0°}(T) + A_{180°}(T)}{2K_1} \tag{14}$$

where $A_{0°}(T)$, $A_{90°}(T)$, $A_{180°}(T)$, $A_{270°}(T)$ are average outputs of QFA at each position at T °C.

Based on Equation (13), scale factors at different temperatures can be calculated. Curves of scale factors vs. temperature are shown in Figure 5. In these figures, the upper curves represent the temperature-falling process, and the lower curve represents the temperature-rising process. Temperature sensors (DS18B20, Maxim Integrated, San Jose, CA, USA) were installed on the surface of QFA. Therefore, the readout of the DS18B20 was not exactly the same as it controlled by the temperature oven. The vertical axes of these three figures are numbers of pulse per gravitational acceleration. The results show that the scale factor of QFA is severely affected by the temperature. The scale factor decreases while temperature rising and the scale factor increases while temperature falling. The scale factor of JB-KT8 #1 changes about 10,010 ppm from about −54 °C to 81 °C and the temperature sensitivity of scale factor is 74 ppm/°C. The scale factor of JB-KT8 #2 changes about 13,600 ppm from about −52 °C to 82 °C and the temperature sensitivity of scale factor is 104 ppm/°C. The scale factor of JB-KT8 #3 changes about 14,200 ppm from about −53 °C to 82 °C, and the temperature sensitivity of scale factor is 105 ppm/°C. (ppm means part per million which is usually used to describe the degree of change. For example, scale factor changes from X to Y, the average of scale factor is A, and the variation of temperature is T The change of scale factor normally defines as $\frac{(X-Y)}{A} \times 10^6$ ppm. The temperature sensitivity of scale factor normally defines as $\frac{(X-Y)}{A \cdot T} \times 10^6$ ppm/°C).

The readout of temperature in this experiment is the surface temperature of QFA. In this temperature model, $\frac{\partial T}{\partial t}$ is a constant, and only ΔT is variable. Therefore, $K_1(T)$ can be fitted by the least squares method.

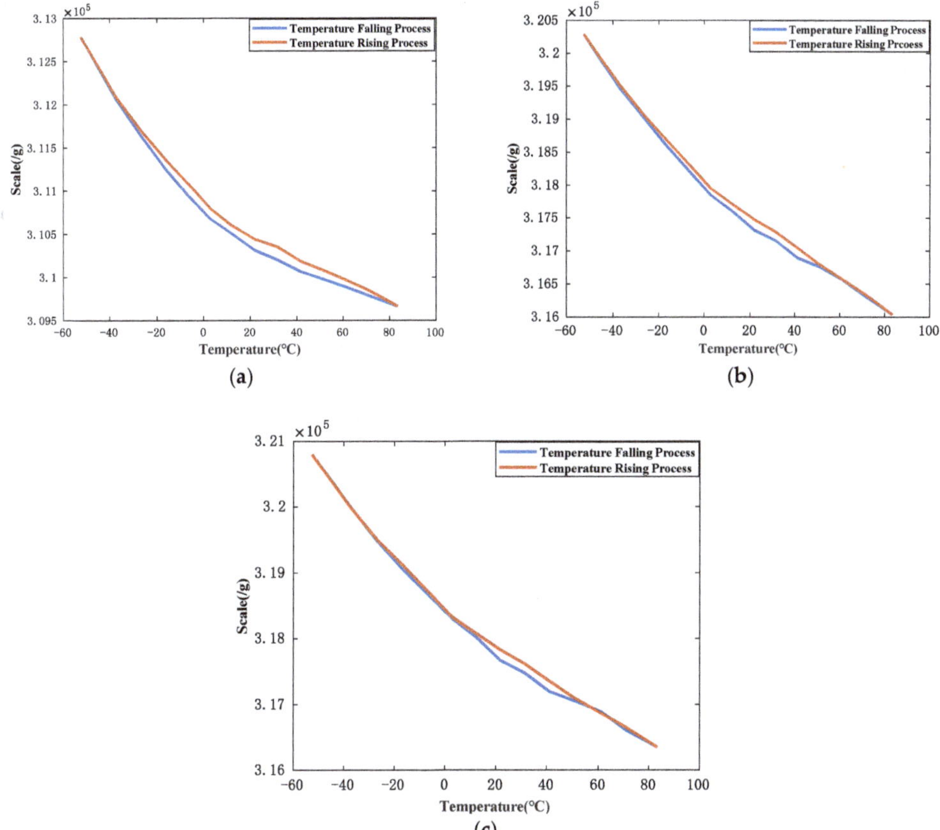

Figure 5. (**a**) is the scale factor of JB-KT8 #1 at different temperature; (**b**) is the scale factor of JB-KT8 #2 at different temperature; (**c**) is the scale factor of JB-KT8 #3 at different temperature.

JB-KT8 #1: The expression of scale factor is shown in Equation (15):

$$K_1(T) = 273 \cdot \left(\tfrac{\partial T}{\partial t}\right)^2 + 0.1008 \cdot (\Delta T)^2 \\ -0.02409 \cdot \left(\tfrac{\partial T}{\partial t}\right) \cdot (\Delta T) - 32.76 \cdot (\Delta T) + 88.14 \cdot \left(\tfrac{\partial T}{\partial t}\right) + 320320 \tag{15}$$

JB-KT8 #2: The expression of scale factor is shown in Equation (16):

$$K_1(T) = 102.76 \cdot \left(\tfrac{\partial T}{\partial t}\right)^2 + 0.08 \cdot (\Delta T)^2 \\ -1.4912 \cdot \left(\tfrac{\partial T}{\partial t}\right) \cdot (\Delta T) - 49.81 \cdot (\Delta T) + 21.01 \cdot \left(\tfrac{\partial T}{\partial t}\right) + 308000 \tag{16}$$

JB-KT8 #3: The expression of scale factor is shown in Equation (17):

$$K_1(T) = 49.12 \cdot \left(\tfrac{\partial T}{\partial t}\right)^2 + 0.2385 \cdot (\Delta T)^2 \\ -0.0972 \cdot \left(\tfrac{\partial T}{\partial t}\right) \cdot (\Delta T) - 41.14 \cdot (\Delta T) + 48.53 \cdot \left(\tfrac{\partial T}{\partial t}\right) + 328000 \tag{17}$$

Based on the temperature hysteresis model, the scale factors of the three QFAs were compensated. One of the best results of compensated scale factor vs. time are shown in Figure 6, and the temperature sensitivity of the compensated scale factor is shown in Table 4.

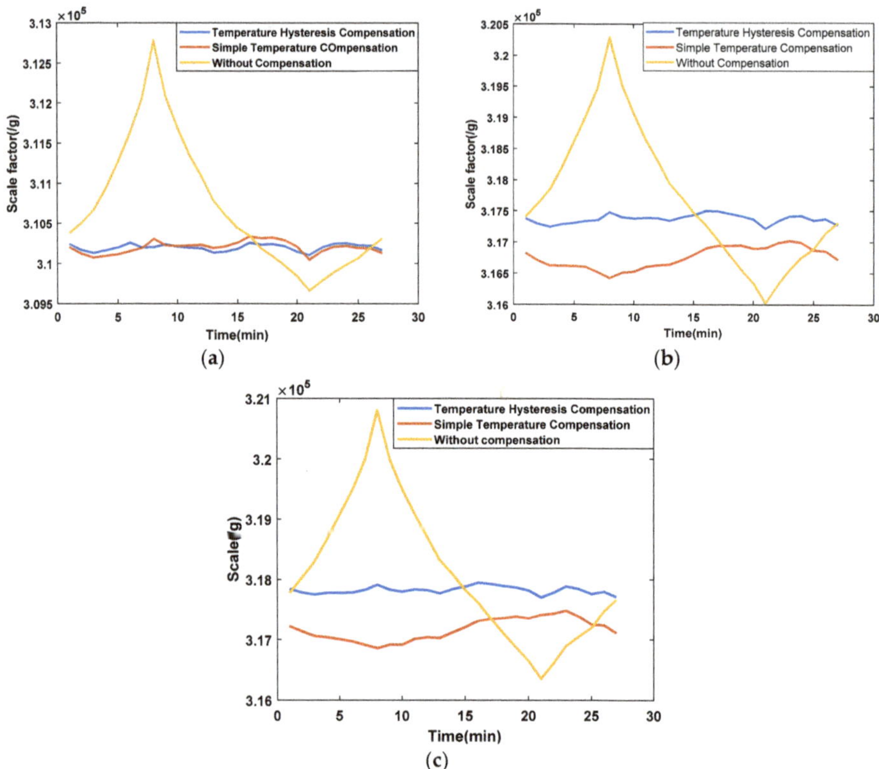

Figure 6. (**a**) is compensation of scale factor of JB-KT8 #1; (**b**) is Compensation of scale factor of JB-KT8 #2; (**c**) is Compensation of scale factor of JB-KT8 #3.

The compensation results in Table 4 indicate that the performance of the proposed temperature compensation model improves at least an order of magnitude compared to the simple model of scale factor. With simple temperature compensation, the temperature sensitivity of scale factor is about 10 ppm/°C. The proposed temperature compensation model can improve temperature sensitivity of scale factor better than 5 ppm/°C.

Table 4. Compensation of temperature sensitivity of scale factors.

	Without Compensation (ppm/°C)	Simple Method (ppm/°C)	Proposed Method (ppm/°C)
JB-KT8 #1	74.14	12.78	2.48
JB-KT8 #2	104.22	14.14	4.23
JB-KT8 #3	105.79	14.46	5.19

Curves of bias vs. temperature are shown in Figure 7. These three figures demonstrate that there is no explicit relation between K_0 and temperature. Therefore, in order to improve the accuracy of temperature compensation of bias, a simple piecewise function is applied. Because the bias-hysteresis model is inexplicit, the detailed model and compensation result for bias are not shown in this paper.

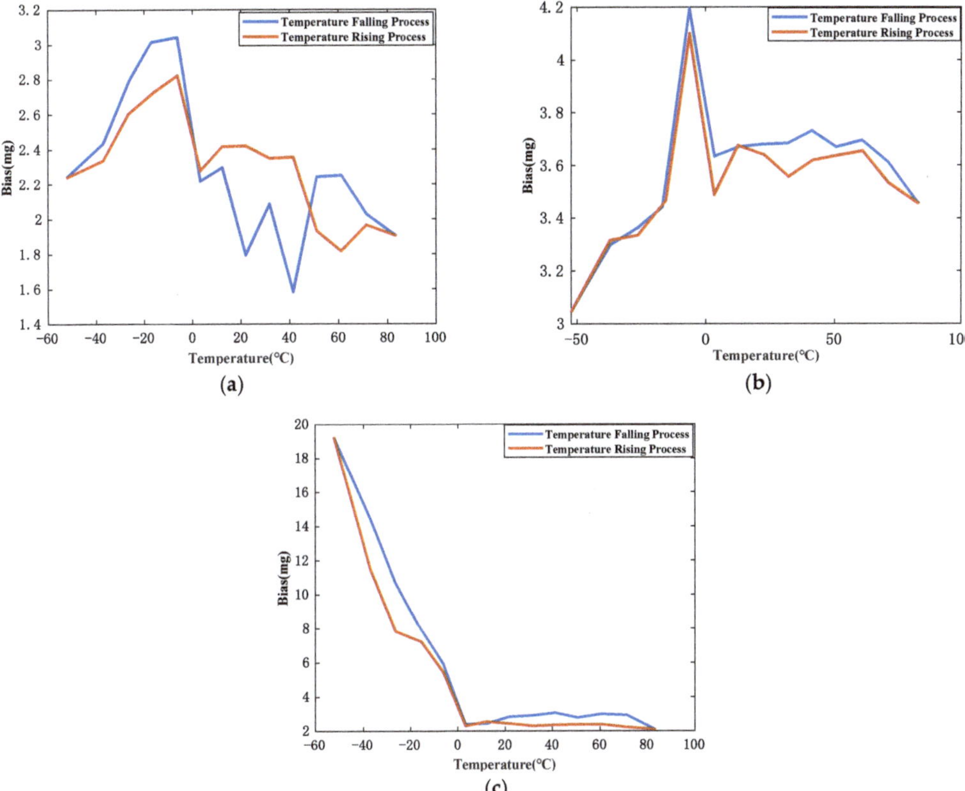

Figure 7. (**a**) is the bias of JB-KT8 #1 at different temperature; (**b**) is the bias of JB-KT8 #2 at different temperature; (**c**) is the bias of JB-KT8 #3 at different temperature.

Before the flight experiment, laboratory test of INS was implemented. The bias stability of gyroscope is 0.01°/h, and the bias stability of QFA (JB-KT8 #1, JB-KT8 #2 and JB-KT8 #3) is 50 μg in stable environment. The performance of tested INS is shown in Table 5. The same INS was used in the flight experiment.

The comparative experiments were conducted in the temperature-controlled oven to verify the efficiency of the temperature hysteresis compensation method. In order to realize the high precision temperature compensation of INS, the temperature compensated output of QFA at different temperature was collected to refine the temperature hysteretic model in system-level.

Table 5. Performance of the tested INS.

	Performance	Quantity
ACC	Bias Stability	50 μg
	Bias Repeatability (one month)	50 μg
	Scale factor-Factor Linearity	30 ppm
	Scale factor-Factor Repeatability (one month)	30 ppm
GYRO	Bias Stability	0.01°/h
	Scale factor Repeatability (one month)	0.01°/h
	Scale factor-Factor Linearity	20 ppm
	Scale factor-Factor Repeatability (one month)	20 ppm

The comparative experiment was employed the simple temperature compensation method (stated in Section 2.1). The testing IMU was placed in the temperature-controlled oven as shown in Figure 8. The temperature is set the same as it is in real flight conditions.

Figure 8. INS in temperature-controlled oven.

The laboratory experiments of INS are conducted for 10 times and one of the best results of the comparative experiments is shown in Figure 9 and Table 6. The blue curve indicates the navigation error compensated by simple compensation and the red curve indicates the navigation error compensated by temperature hysteresis compensation. It can be concluded that the east velocity error (VE error) and the longitude error are restrained obviously, and the north velocity error (VN error) and the latitude error are also slightly reduced.

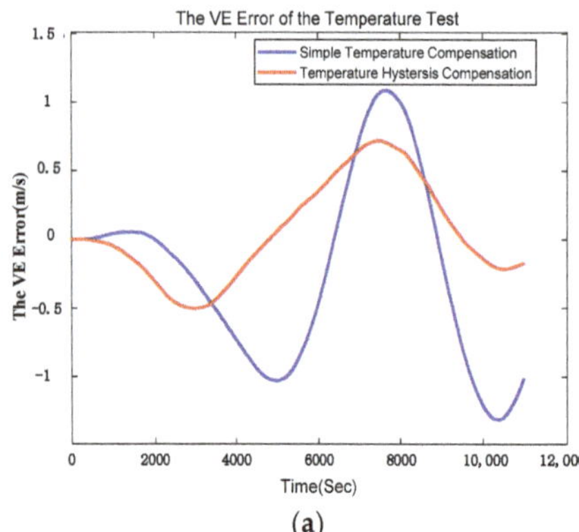

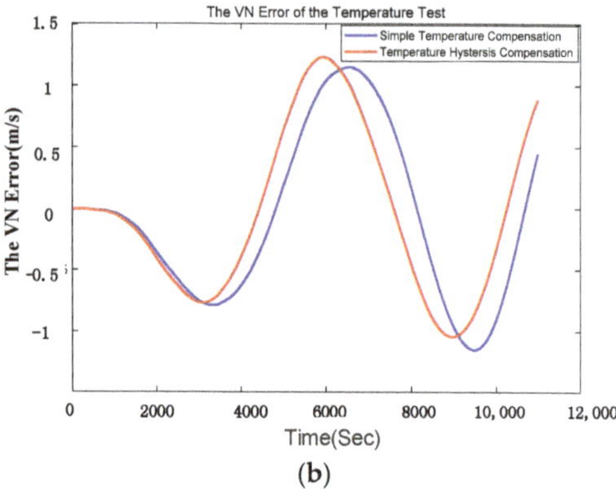

(a) (b)

Figure 9. (**a**) is the comparison of east velocity error in laboratory test; (**b**) is the comparison of north velocity error in laboratory test.

Table 6. Comparison of velocity error in laboratory experiment.

	East Velocity	North Velocity
Simple compensated (RMS) m/s	0.98	1.04
Compensated (RMS) m/s	0.74	0.92

The comparison of east velocity error and the north velocity error in laboratory experiment are shown in Table 6. The errors of east velocity and north velocity are restrained by 24.5% and 11.5%. The position error is shown in Figure 10. It is obvious that the position error decreases after temperature hysteresis compensation.

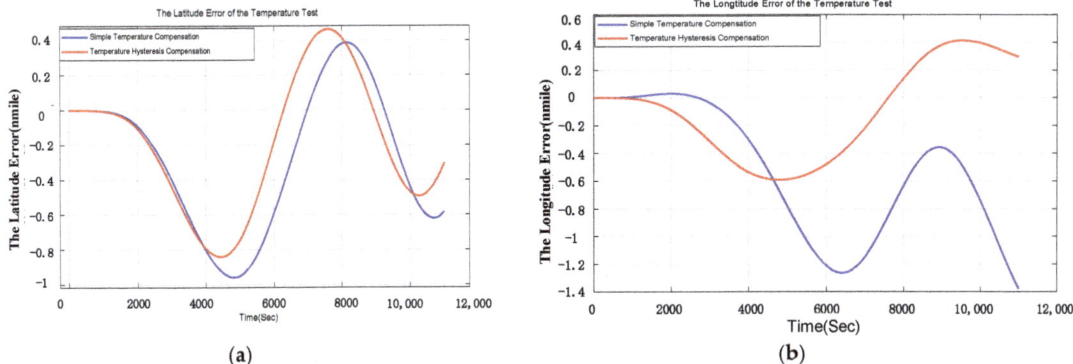

Figure 10. (**a**) is the comparison of longitude error in laboratory test; (**b**) is the comparison of latitude error in laboratory test.

3.2. Temperature Hysteresis Compensation of QFA in Flight Experiment

A flight test was conducted with the GNSS/INS in November 2019. The INS was installed outside the plane. The flight altitude was about 10,000 m, and the flight lasted for around 2.3 h. The GPS was used as a reference to evaluate the performance of the INS. The outputs of INS, including angular rate, acceleration, position, and velocity, are synchronized with the outputs of GPS. The flight path of the aircraft is shown in Figure 11a. There is no temperature control system for INS and the temperature of INS is measured by a temperature sensor assembled inside of IMU. The INS used in the flight experiment is composed of 3 parts: IMU, navigation computer, and power module. IMU is set in a cabin which is isolated from other parts. We used the temperature readout of IMU as the temperature of INS. IMU consists of three accelerometers and three gyroscopes which are assembled on an aluminum alloy fixture whose temperature distributes uniformly. Therefore, the temperature of INS is equivalent to the surface temperature of aluminum alloy fixture whose temperature is almost the same as the surface temperature of the accelerometer. The temperature of IMU is shown in Figure 11b. The flight phase was divided into three parts:

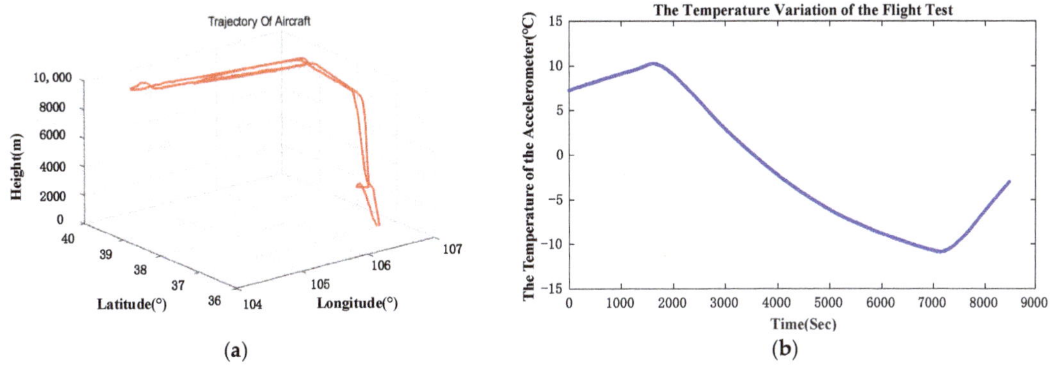

Figure 11. (**a**) is flight path of aircraft; (**b**) is temperature of IMU during the flight.

Start-up and Climbing: It took about 30 min to climb to 10,000 m. The temperature of the INS rose at the beginning period because of the heat generated by the circuits of INS.

Cruise: The aerial vehicle cruised at the altitude of 10,000 m for 1.5 h. The temperature of the INS fell constantly to −11 °C.

Landing: This lasted about 20 min, and the temperature of the INS was rising as the altitude decreased.

The original output of the GNSS/INS was processed offline. The output of the QFA was compensated by the proposed method and simple temperature compensation method respectively. Furthermore, these two groups of processed data were applied to the same pure inertial navigation algorithm to evaluate the compensation effect. The east velocity error and north velocity error are shown in Figure 12.

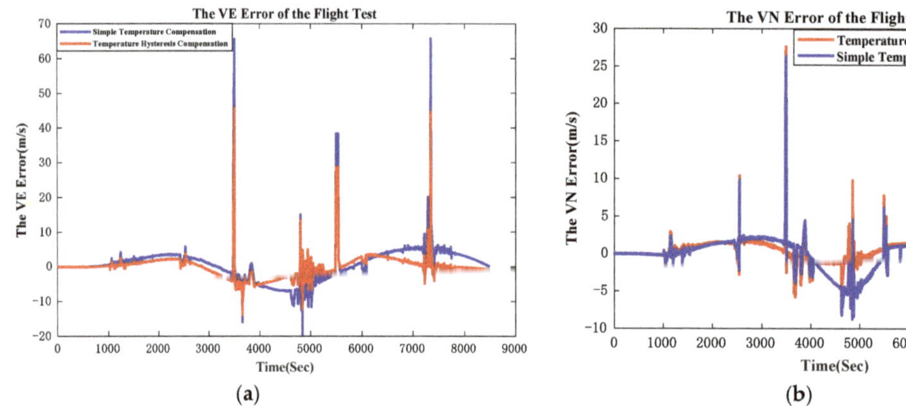

Figure 12. (**a**) is east velocity error; (**b**) is north velocity error. Because the experiment GNSS/INS was assembled inside the cabin and the GPS antenna cannot be installed outside of the cabin, GPS signal was lost when roll angle of the plane was bigger than 30°. GPS signal blocked by structure of cabin at big roll angle may be responsible for GPS loss. The vertexes in the graph are caused by the GPS signal loss.

The comparison of east velocity error and the north velocity error are shown in Table 7. The errors of east velocity and north velocity are restrained by 19.9% and 15.0%.

Table 7. Comparison of velocity error in flight experiment.

	East Velocity	North Velocity
Simple compensated (RMS) m/s	1.36	1.4
Compensated (RMS) m/s	1.09	1.19

Navigation errors are shown in Figure 13. In order to fix the problem of GPS signal loss, GPS data was smoothened through simple linear regression when conducting an off-line navigation solution. The temperature hysteresis compensation result in flight condition shows that the position error (CEP) is restrained from 1.54 nmile/h to 1.29 nmile/h, which means the accuracy of navigation improves at least by 16.2%. In Figure 13, the difference between position errors compensated by two models at early stage is small. This is because the variation of temperature and the gradient of temperature is small, and the simple temperature compensation model works well at this temperature condition. After 6000 s, plane cruised at 10,000 m and then landed. In this period, the temperature varied dramatic, which indicates that the performance of proposed temperature model is better than conventional temperature model.

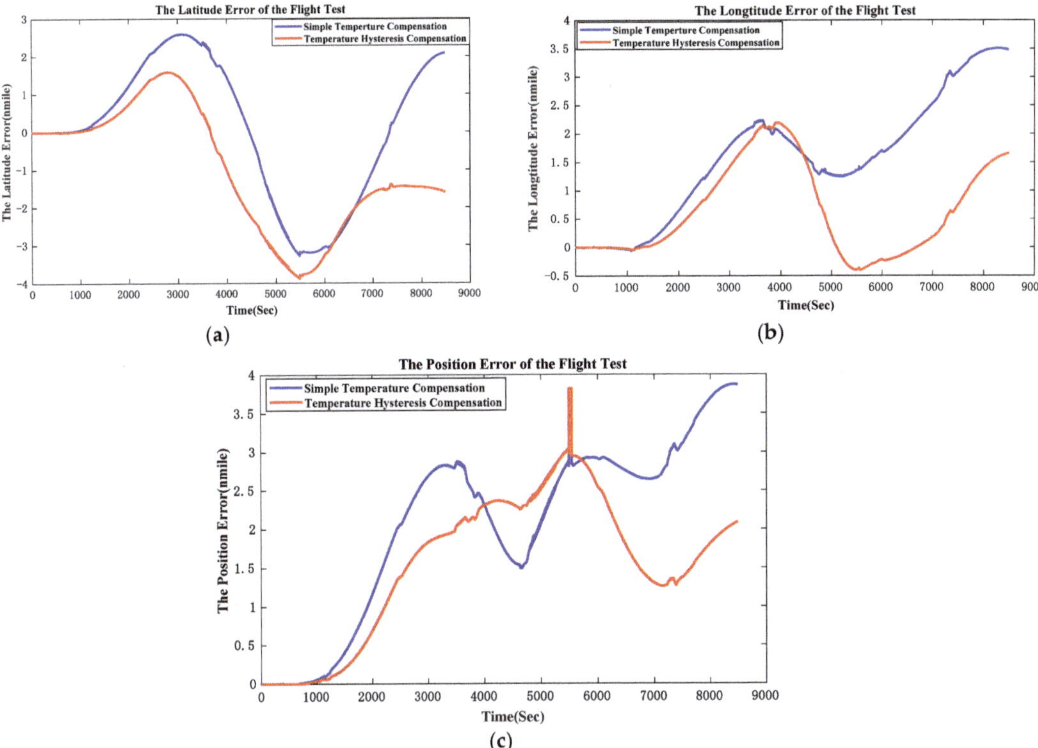

Figure 13. (**a**) is latitude error of the flight test; (**b**) is longitude error of the flight test; (**b**) is position error of the flight test. In (**b**) and (**c**), because the GPS signal was smoothened by simple linear regression, small mutations in position calculation appeared when GPS signal loss.

4. Discussion

According to the analysis above, temperature variation influences the performance of QFA greatly, which degrades the precision of navigation in aerial conditions. Conventional temperature compensation for accelerometers could improve the performance of the INS in aerial conditions. However, the performance of conventional temperature compensation model of QFA cannot meet the demand. Therefore, further research on temperature issues is vital to improve the temperature performance of QFA in severe environments.

In this paper, we propose a temperature hysteresis compensation method for QFA. Through the analysis of the characteristic of ER and magnet, we derived the basic cause of the temperature hysteresis phenomenon which relates to both temperature and the temperature gradient. Based on the analysis above, a rational temperature model is built for accelerometers. Proper QFA temperature compensation experiments were designed to refine the model. The comparative experiments with a simple temperature compensation method in the laboratory indicate that the efficiency of the proposed compensation method is better than the simple one. The result of pure inertial navigation in the flight experiment shows that the navigation accuracy (CEP) improves by 16.2%, from 1.54 nmile/h to 1.29 nmile/h, after temperature hysteresis compensation.

Beyond the proposed temperature compensation method, our future work will focus on optimizing the experiment method and building precise bias-temperature model. The bias has a hysteretic phenomenon, but the bias-hysteresis model remains inexplicit. Using a piecewise function for bias compensation is effective while it still needs to be optimized.

Temperature compensation for QFA is not the only one way to improve its performance. The more effective approach is to optimize QFA itself. For example, the differential structure in the QFA is the focus of our future research. Besides, it should be taken into consideration that the parameters of the QFA temperature model in the INS may be different from the parameters in the single QFA, so precise temperature compensation in the INS level need to be conducted.

Author Contributions: Methodology, X.W.; validation, L.R. and L.S.; formal analysis, X.W.; project administration, C.Z. All authors have read and agreed to the published version of the manuscript.

Funding: This research received no external funding.

Institutional Review Board Statement: Not applicable.

Informed Consent Statement: Not applicable.

Data Availability Statement: Data sharing not applicable.

Conflicts of Interest: The authors declare no conflict of interest.

References

1. Li, Q.; Xu, W.; Chen, X. Investigation on vibration analysis of quartz flexible accelerometer with permutation entropy. In Proceedings of the 11th World Congress on Intelligent Control and Automation, Shenyang, China, 29 June–4 July 2014; pp. 3200–3203.
2. Xu, D.; Chen, Y.-X.; Kang, R. Vibration effects on parameter stability of quartz flexible accelerometer by accelerated test. In Proceedings of the 2013 International Conference on Quality, Reliability, Risk, Maintenance, and Safety Engineering (QR2MSE), Chengdu, China, 15–18 July 2013; pp. 986–990.
3. Tian, D.; Luo, M.; Hou, Z.; Xu, D.; Kang, R. Study on instability mechanism of quartz flexible accelerometer parameters under temperature profile condition. In Proceedings of the Prognostics and System Health Management Conference, Shenzhen, China, 24–25 May 2011.
4. Cao, J.; Wang, M.; Cai, S.; Zhang, K.; Cong, D.; Wu, M. Optimized Design of the SGA-WZ Strapdown Airborne Gravimeter Temperature Control System. *Sensors* **2015**, *15*, 29984–29996. [CrossRef] [PubMed]
5. Zhuo, C.; Du, J.; Tang, H.; Liu, Q. Special Thermal Compensation Experiment and Algorithm Design for Inertial Navigation System. In Proceedings of the 2019 DGON Inertial Sensors and Systems (ISS), Braunschweig, Germany, 10–11 September 2019; pp. 1–16.
6. Niu, X.; Li, Y.; Zhang, H.; Wang, Q.; Ban, Y. Fast Thermal Calibration of Low-Grade Inertial Sensors and Inertial Measurement Units. *Sensors* **2013**, *13*, 12192–12217. [CrossRef] [PubMed]
7. Becker, D.; Nielsen, J.E.; Ayres-Sampaio, D.; Forsberg, R.; Becker, M.; Bastos, L. Drift reduction in strapdown aerial gravimetry using a simple thermal correction. *J. Geod.* **2015**, *11*, 1133–1144. [CrossRef]
8. Araghi, G. Temperature compensation model of MEMS inertial sensors based on neural network. In Proceedings of the Position, Location and Navigation Symposium (PLANS), Monterey, CA, USA, 23–26 April 2018.
9. Gao, J.; Zhang, K.; Chen, F.; Yang, H.-B. Temperature characteristics and error compensation for quartz flexible accelerometer. *Int. J. Autom. Comput.* **2015**, *12*, 540–550. [CrossRef]
10. Chen, F.; Zhang, K. Identification for temperature model and the method for temperature compensation of quartz flexible accelerometer. In Proceedings of the IEEE Advanced Information Technology, Electronic and Automation Control Conference (IAEAC), Chongqing, China, 12–14 October 2015.
11. Liu, D.L.; Zhu, X.B.; Xu, K.L.; Fang, D.M. Temperature Drift Compensation Based on Artificial Fish Swarm Algorithm for Quartz Flexible Accelerometer. *Appl. Mech. Mater.* **2014**, *513–517*, 4030–4034.
12. Zhou, X.; Tan, H. Temperature Drift Compensation Algorithm Based on BP and GA in Quartzes Flexible Accelerometer. *Appl. Mech. Mater.* **2012**, *249–250*, 95–99.
13. Gunhan, Y.; Unsal, D. Polynomial degree determination for temperature dependent error compensation of inertial sensors. In Proceedings of the 2014 IEEE/ION Position, Location and Navigation Symposium—PLANS 2014, Monterey, CA, USA, 5–8 May 2014; pp. 1209–1212.
14. Xie, Q.; Pan, Y.; Ren, C. Temperature Drift Compensation of Quartz Flexible Accelerometer Based on Wavelet Network. *Meas. Control* **2009**, *12*, 15–17.
15. IEEE Standard Specification Format Guide and Test Procedure for Linear, Single-Axis, Digital, Torque-Balance Accelerometer. *IEEE Stand. 1293TM* **1978**. [CrossRef]
16. Chen, C.; Wu, M.; Cao, J.; Wang, Q. Analysis of stress on beam of small-range high-precision quartz flexible accelerometer. In Proceedings of the 2015 IEEE International Conference on Information and Automation, Lijiang, China, 8–10 August 2015; pp. 2943–2946.
17. Haque, F.Z.; Del Barco, E.; Fishman, R.S.; Miller, J.S. Low temperature hysteretic behavior of the interpenetrating 3-D network structured $[Ru_2(O_2CMe)_4]_3[Fe(CN)_6]$ magnet. *Polyhedron* **2013**, *64*, 73–76. [CrossRef]

18. Parilov, A.; Lileev, A.; Lileeva, Y.; Reissner, M.; Steiner, W. Temperature dependence of the magnetic hysteresis in Nd–Fe–B alloy film magnets. *J. Magn. Magn. Mater.* **2004**, *272*, E511–E512. [CrossRef]
19. Zhilin, G.P.; Laptei, D.A.; Saunin, V.N. Temperature magnetic hysteresis of precopitation-hardened alloys. *Sov. Phys. J.* **1976**, *19*, 1259–1264. [CrossRef]
20. Chen, Y.; Deng, F.; Xu, D.; Kang, R. Research on the degradation mechanisms and finite element modeling of accelerometers. In Proceedings of the Prognostics and System Health Management Conference, Shenzhen, China, 24–25 May 2011.
21. Sixdenier, F.; Messal, O.; Hilal, A.; Martin, C.; Raulet, M.-A.; Scorretti, R. Temperature-Dependent Extension of a Static Hysteresis Model. *IEEE Trans. Magn.* **2015**, *52*, 1–4. [CrossRef]
22. Gromala, P.J.; Prisacaru, A.; Jeronimo, M.; Lee, H.; Sun, Y.; Han, B. Non-linear Viscoelastic Modeling of Epoxy Based Molding Compound for Large Deformations Encountered in Power Modules. In Proceedings of the 2017 IEEE 67th Electronic Components and Technology Conference (ECTC), Orlando, FL, USA, 30 May–2 June 2017; pp. 834–840.
23. Ning, X.; Xiang, Z.; Liu, P.; Feng, H.; Peng, Z. Space charge behavior of epoxy resin at high temperature. In Proceedings of the 2013 Annual Report Conference on Electrical Insulation and Dielectric Phenomena, Shenzhen, China, 20–23 October 2013; pp. 222–225.
24. Zhao, Y.; Zheng, T.; Yang, K.; Wang, X.; He, Y. Study on Residual Stress of Epoxy Resin under different cooling methods. In Proceedings of the 2018 IEEE International Conference on High Voltage Engineering and Application (ICHVE), Athens, Greece, 10–13 September 2018; pp. 1–4.
25. Shao, Z.; Wu, K.; Chen, X.; Guo, J. Effects of nanoparticles on trap depth in epoxy resin under different temperature. In Proceedings of the IEEE Conference on Electrical Insulation and Dielectric Phenomena (CEIDP), Toronto, ON, Canada, 16–19 October 2016.
26. Thurn, J.; Hermel-Davidock, T. Thermal stress hysteresis and stress relaxation in an epoxy film. *J. Mater. Sci.* **2007**, *42*, 5686–5691. [CrossRef]
27. Chen, B.; Dillard, D.; Dillard, J.G. Effect of the T-stress in the adhesively bonded joints on the locus of failure in rubber toughened epoxy system. In Proceedings of the 3rd International Conference on Adhesive Joining and Coating Technology in Electronics Manufacturing 1998 (Cat. No.98EX180), Binghamton, NY, USA, 30 September 1998.

Article

Multifrequency Vector Measurement System for Reliable Vehicle Magnetic Profile Assessment

Zbigniew Marszalek * and Krzysztof Duda

Department of Measurement and Electronics, AGH University of Science and Technology,
30 Mickiewicz Avenue, 30-059 Krakow, Poland; kduda@agh.edu.pl
* Correspondence: antic@agh.edu.pl; Tel.: +48-12-617-5049

Received: 3 July 2020; Accepted: 29 August 2020; Published: 31 August 2020

Abstract: This paper describes the design and the performance of simultaneous, multifrequency impedance measurement system for four inductive-loop (IL) sensors which have been developed for vehicle parameters measurement based on vehicle magnetic profile (VMP) analysis. Simultaneous impedance measurement on several excitation frequencies increases the VMP measurement reliability because typical electromagnetic interferences (EMI) are narrowband, and should not simultaneously affect, in the same way, all measurement bands that are spread in the frequency, i.e., it is expected that at least one measurement band is disturbance-free. The system consists of two standard and two slim IL sensors, specially designed and installed, the analogue front-end, and an industrial computer with digital-to-analogue and analogue-to-digital converters accessed via field-programmable gate array (FPGA). The impedance of the IL sensors is obtained by vector measurement of voltages from auto-balancing bridge (ABB) front-end. Complex voltages are demodulated from excitation frequencies with FIR filters designed with the flat-top windows. The system is capable of delivering VMPs in real-time mode, and also storing voltages for off-line postprocessing and analysis. Field distributions and sensitivities of slim and standard IL sensors are also discussed. Field test confirmed assumed increased reliability of VMP measurement for proposed simultaneous multifrequency operational mode.

Keywords: auto-balancing bridge method; FIR filter; FPGA; impedance; inductive-loop sensor; multifrequency; vehicle magnetic profile; vector voltmeter; signal processing

---

1. Introduction

Over the last several decades, inductive-loop (IL) sensors have become increasingly important as reliable vehicle detectors, for proper operation of traffic control systems, as well as for increase of the responsiveness of intelligent transportation systems (ITS) [1]. However, a multifrequency impedance measurement is a new uprising research filed in the IL sensor technology.

Currently, there are many technologies available for measurement of vehicle parameters in traffic, such as vehicle speed, weight, length, its number of axles, the distance between the axles, vehicle class, and so on [2]. The most common sensors used for traffic measurement are inductive, hydraulic, magnetic, pneumatic, reluctance, resistive, optical fiber, capacitive, laser, piezoelectric, quartz, tensometric, acoustic, ultrasonic, microwave and infrared, light curtains, and cameras with image processing algorithms [3–5].

When it is necessary to detect the vehicle axles, and to measure their arrangement in the vehicle body, in principle, above group limits down to sensors providing signals associated with vehicle wheels. It can be accomplished by a proper set of load sensors, i.e., piezoelectric, quartz, tensometric, fiber [6], and at least one standard, i.e., 1 m by 2 m, IL sensor for grouping axles of a given vehicle [7]. Systems equipped with load sensors provide accurate measurements of the axles number and the distance between them. Load sensors usually dedicated to weigh-in-motion (WIM) applications [8], have a nominal two years working life and thus are not a long-life neither low-cost solution. Load sensors differ in price, installation method, and operating principle, but they all have the same drawback as their active element is exposed to the force exerted by the tires of the passing vehicle. This main drawback not only limits the load sensor lifespan, but can also lead to the misclassification of vehicles with lifted axles [9]. In addition to load sensors, WIM systems must also have at least one IL sensor for vehicle detection.

The measurement system consists of an IL sensor, connecting cables, controller cabinet, and detector electronics. The IL sensor is made from a several loops of isolated wire wound permanently mounted in the pavement of a road. The IL sensor is powered by the detector electronics at frequencies typically ranging from 10 to 200 kHz [10]. The IL sensor is and inductive element of a tuned electrical circuit. A vehicle passing over the IL sensor affects its inductance. This change of inductance activates the detector electronics, which further initialize the controller unit. Direct measurement of the inductance change is not possible because signal conditioning systems are based on oscillating circuit [10]. The inductance change causes phase and frequency shifts in the output signal [11].

Advanced IL system can provide an inductance waveform when a vehicle passes over the sensor. This waveform, known as vehicle signature, or vehicle magnetic profile (VMP), can be used in vehicle re-identification, and matching vehicles between an upstream and downstream location [12,13].

The systems presented in [10–13] contain structurally different IL sensors, but they all share common characteristics. The IL sensor works together with a circuit, where the inductance change is measured indirectly by frequency change.

The frequency change of oscillating circuit, i.e., the magnetic profile, contains only the information associated with the inductance of IL sensor. Due to the operational principle of a system based on any resonant circuit, the information associated with the real part of IL impedance, i.e., the resistance, is lost [14]. The loss of resistance waveform leads to limitations in the development of signal processing algorithms based on electrical knowledge.

A different approach to the vehicle magnetic profile assessment is used in the AC-bridge-based system [7], where the impedance components (real and imaginary) in the output are irreversibly mixed together and the mixing depends on the set of synchronous demodulator phase angle.

There are axles detection algorithms based on both resistance and inductance waveform of IL sensor impedance components [9]. However, these algorithms cannot work on the inductance waveform delivered by the system based on oscillating circuit technology. For this reason, a method for obtaining both resistance and inductance waveform from one IL sensor was carried out [9,15].

A vehicle generates strong electromagnetic interference (EMI) in the IL sensors field, with the spectrum overlapping the operational band of impedance measurement unit (IMU), which essentially deteriorate the performance of the IMU with the impedance components separation [16]. In the proposed measurement system, the IL sensor impedance is acquired on several frequencies simultaneously [17–20], which reduces the disturbing influence of a vehicle engine EMI. The impedance waveforms with detected disturbances are then rejected before further processing.

We propose a new approach to impedance waveform measurement. The novelty of our solution is threefold: (1) Applied IL sensors are arranged in original quad configuration consisting of two slim loops and two standard loops. (2) The designed impedance measurement system works simultaneously in three separate frequency bands. (3) A new application of auto-balancing bridge (ABB) method [21] is presented.

Fully functional measurement system was built and validated via extensive field tests. The IL sensors were mounted in the real road near a parking lot. Early signal processing was realized using National Instrument PXI industrial system. Obtained results confirmed overall performance of the proposed measurement system.

The paper is organized as follows. Section 2 presents the proposed IL quad sensors arrangement, sensor model, and sensitivity description of the slim and standard loop. Section 3 describes the multifrequency IMU followed by system hardware implementation and signal processing stage. Section 4 describes the experimental set-up and the experimental results. Section 5 concludes the paper.

2. Applied IL Quad Sensors

2.1. Arrangement and Dimensions of IL Sensors

Vehicles, as metal objects, passing through the alternating magnetic field of IL cause changes to the IL impedance both in the real and imaginary part. These changes, on the level of a few percent of IL nominal impedance value, provide useful information for algorithms measuring vehicle parameters in traffic.

Changes in the monitored values of impedance represented as a function of time during the vehicle passing over the IL are called the vehicle magnetic signature [1] or VMP.

The proposed quad IL arrangement, presented in Figure 1, consists of two standard loops—IL1 and IL3—and two slim loops—IL2 and IL4. Standard loops work as typical dual-loop vehicle detectors [5]. Slim loops are dedicated for axle detection [9] and distance measurement between them.

Dimensions of loops are defined in Figure 1 and given in Table 1. The distance between standard loops, IL1 and IL3, is 0.5 m, and the distance between slim loops, IL2 and IL4, is 1.4 m. This arrangement of compact loops takes 2.8 m in a drive direction. Due to the purpose of the slim loops, they are longer than standard loops (compare dimension B and D). The distances, F, between the front edges of the standard and slim loops are the same.

Table 1. Loops dimensions.

Dimension	A	B	C	D	E	F
Value (cm)	100	200	10	320	20	150

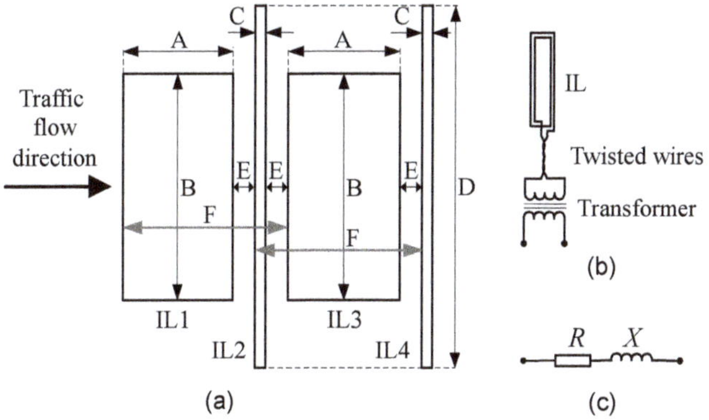

Figure 1. Proposed inductive-loop (IL): (**a**) sensors arrangement on lane. Dimensions are specified in Table 1. (**b**) Hardware components and (**c**) serial model.

Standard loops, IL1 and IL3, are made with 4 turns of wire, and slim loops, IL2 and IL4, are made with 8 turns of the same wire with a cross section of 2.5 mm^2.

Figure 2 presents quad IL mounted in the road. They have the following advantages; the ability to measure vehicle speed [5], length, the direction of movement, and distance between axles as well as the number of axles even if they are lifted. Suspension height as well as front and rear overhang of the vehicle are also possible to estimate.

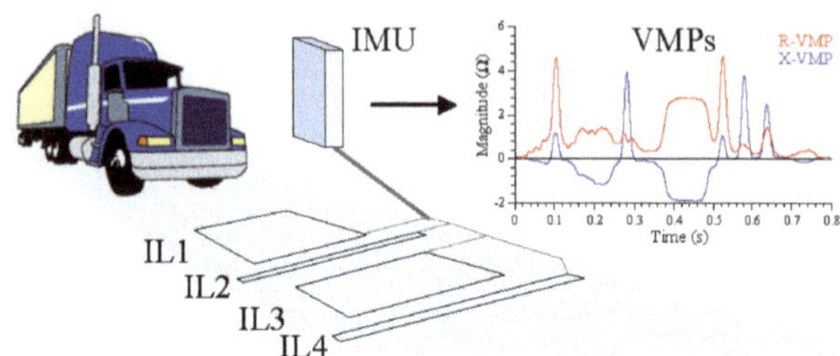

Figure 2. Inductive loop (IL) sensors arrangement on lane, impedance measurement unit (IMU), and vehicle magnetic profiles (VMPs) example from IL2 for a single frequency.

2.2. Galvanic Isolation and IL Sensor Model

The distance between the arrangement of IL sensors and cabinet for IMU may vary from tens to a hundred meters. In measurement practice, a single IL sensor consists of a loop installed on the lane, long twisted connection wires, and a transformer for galvanic separation, see Figure 1b. Due to the long underground installed connection wires, galvanic separation is absolutely necessary to protect the IMU. Although the sensor consists of several elements arranged over a large area, a serial equivalent circuit model of impedance is used to describe the sensor, as shown in Figure 1c.

The transformer for galvanic separation is made with a ferrite core and its transmission ratio is approximately 1:7. Therefore, the nominal impedance of IL with the transformer is higher than the loop impedance. Examples of measured impedances, using E4980A instrument [22], are shown in Table 2.

Table 2. Nominal IL impedance measured at 10 kHz.

Impedance $Z = R + jX$	Slim Loop (Ω)	Standard Loop (Ω)
loop only	0.8 + j15	0.6 + j7
loop with transformer	61 + j748	51 + j373

2.3. The Principle of IL Sensor Operation

The flow of AC current through the IL causes a magnetic field around the winding. According to Maxwell's equations, eddy currents in a conductive metal object present in the field change the global magnetic field distribution. There are dynamic interactions of eddy currents field and inductive loop magnetic field. If the IL impedance is monitored, the effect of these interactions can be seen as impedance changes.

2.4. Field Distribution and Sensitivity of IL

This section describes the sensitivity of the IL in relation to the magnetic field distribution generated by the loop. A sensitivity indicates how much the sensor's output changes with the change of the input measured quantity. Therefore, to characterize the sensitivity of the IL, which is related to its magnetic field distribution, it is necessary to visualize the distribution of its field in the active space that is penetrated by vehicles. The distribution of the field can be calculated using the Biot–Savart law and presented on the plane in selected cross-sections [1].

The wheel rim is the closest part of the car that is extended towards the IL sensor. The distance between the rim and the loop wires depends on the height of the tire (e.g., 8 cm) and the mounting depth of IL in the road (e.g., 2 cm). Therefore, as an example, the magnetic field induction distribution in the air, over the IL, in a horizontal section in a plane parallel to the sensor, located at a distance of 10 cm was computed. For comparison, the constant dimensions of spatial cross-sections for slim and standard IL sensors, and the same current flow of 1 A-turn were adopted.

The field distribution, in a horizontal plane, 10 cm above the standard loop, is shown in Figure 3a,b, which also shows the field distribution in the vertical cross plane. In Figure 3b, the isoline $B = 0.5\ \mu T$ is marked, with the magnetic induction value 4 times smaller than the maximum magnetic induction in the considered area.

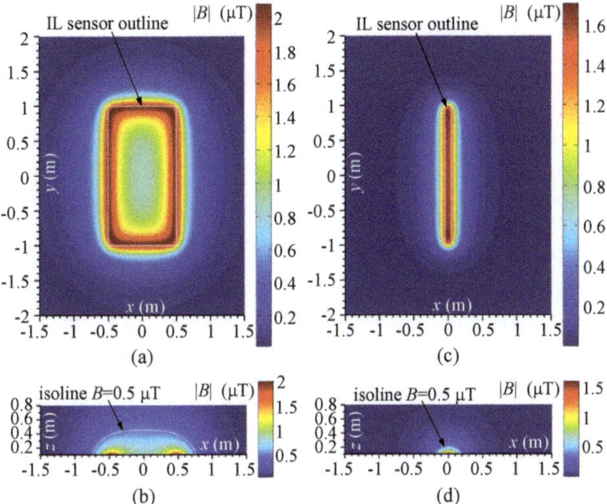

Figure 3. Magnetic field module distribution, $|B|$ (μT), of the standard-loop: (**a**) 10 cm above the plane of the IL wires and (**b**) in vertical cross section. The slim-loop (**c**) 10 cm above the plane of the IL wires and (**d**) in vertical cross section.

Field distributions for the slim loop were calculated analogically and are shown in Figure 3c,d. By comparing and analyzing IL sensors field distributions, one can come to the conclusion that standard loops generate a spread field, which simultaneously includes many vehicle chassis components. The induction values, not less than 0.5 μT, occur where there are highly located vehicle chassis components, e.g., trucks. This explains why standard loops are suitable for detecting entire vehicle bodies, even those with high suspension.

The different situation is in the case of the slim IL sensors because their field is spatially less spread. Values higher than 0.5 μT, see Figure 3d, are focused close to the loop. Thus, the slim-loop sees less than the standard-loop. It is stated that the slim-loops also have a lower range of their magnetic field. Thanks to this feature, slim-loops are dedicated to wheel rims detection, which allows getting information about the number and arrangement of axles in the vehicle body.

The field distribution reveals the sensors features, which have an effect on the resultant sensitivity of the loops considered. The standard-loop is sensitive to objects in a much larger space above the IL than the slim-loop, which exhibits scanning properties of details protruding from large objects such as metal parts of vehicle wheels, rocker arms, bumpers, etc.

The presented field distributions also indicate that the measurement system will not work properly only with slim-loop. It must be equipped with standard-loops providing strong information about the vehicle body when the slim-loop is between the axles and its field does not reach to high suspensions elements. The slim IL sensor used for vehicle axle detection requires more sensitive, and more sophisticated IMU.

3. Multifrequency Impedance Measurement System

This section describes the multifrequency, 4-channel impedance measurement system able to estimate VMPs simultaneously in three different measurement bands for each channel. The system features auto-balancing operation, i.e., the need of time-consuming balancing during the absence of the vehicle in IL field is eliminated. The VMP amplitude resolution was increased by oversampling and averaging [23]. A greater number of samples also improves noise properties of the frequency estimator. The variance of amplitude estimation of a sinusoid disturbed by additive white Gaussian noise is inverse proportional to the number of samples, and the variance of frequency estimation is inverse proportional to the third power of the number of samples [24,25]. The power of quantization noise do not depend on sampling frequency and noise spreads, in the frequency domain, form 0 to the half of the sampling frequency. Thus, by increasing the sampling frequency the noise level is lowered, and the signal to noise ratio in the sub-band of interest is increased [26]. The highest possible sampling frequency enabled in the PXI system, that is, 400 kHz, was used. During amplitude and phase demodulation signals are averaged by FIR filter with the impulse response lasting 0.1 s.

3.1. Analogue Section

The measurement system, used for the complex impedance of the device under test (DUT) acquisition, consists of a sinusoidal excitation source, a vector voltmeter, and a vector ammeter. The auto-balancing bridge (ABB) method is commonly used in modern single-frequency impedance measurement instruments [21]. The ABB circuits used for low-frequency impedance measurement (below 100 kHz) typically exploit a basic current-to-voltage (I–V) converter and an operational amplifier with a negative feedback, as depicted in Figure 4. The DUT consists of IL sensor, connecting wires, and transformer.

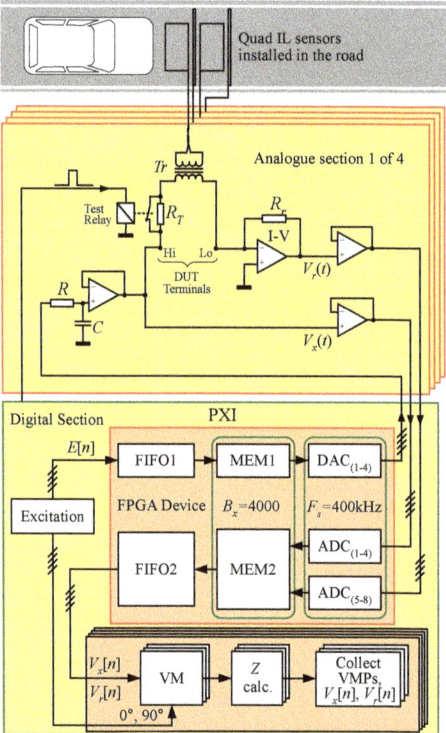

Figure 4. The hardware design of the field-programmable gate array (FPGA)-based four-channel system for multifrequency impedance measurement of IL sensors, where VM is a vector measurement.

The digital-to-analogue converter (DAC$_1$) is used to digitally generate the voltage excitation signal $V_x(t)$, that is next filtered by an analog low-pass filter, and then excites the current flow through the DUT and the I–V converter. The same current, as the one flowing through the DUT, is derived by the operational amplifier, of the I–V converter, in the negative feedback loop. The potential at the Lo terminal of the DUT is automatically balanced to zero because the feedback current flowing through the R_r is equal to the input current [21,27].

The vector voltages, $V_x[n]$ and $V_r[n]$, are calculated in Section 3.2. As the R_r value is known, in static condition, to calculate complex impedance $Z[n]$ of the DUT, we use:

$$Z[n] = R[n] + jX[n] = R_r \frac{V_x[n]}{V_r[n]} \tag{1}$$

where n is the number of the sample.

It should be noted that $V_x[n]$ and $V_r[n]$ are discrete-time versions of continuous-time counterparts $V_x(t)$ and $V_r(t)$. Different digital signal processing algorithms can be used in quadrature demodulation process for obtaining complex-valued voltages $V_x[n]$ and $V_r[n]$ used in (1).

For example, in [27] a simple vector voltmeter algorithm for $V_x[n]$ and $V_r[n]$ calculation for the individual frequencies and in static conditions was used.

In this work, signals $V_x[n]$ and $V_r[n]$ are demodulated simultaneously on all excitation frequencies by complex-valued FIR filters, as explained later in the paper.

The correct system operation on all channels, and on all measurement bands simultaneously is validated by a non-inductive test resistor, $R_T = 1\ \Omega$ installed in series with the transformer Tr. Test resistor is normally bypassed via test relay. The validation of the system operation is based on the short switching the relay and observation of all outputs. Changes visible in the real part of measured impedance, increment equal to one ohm, and also no changes in the imaginary part, confirm the correct overall operation of the system.

3.2. Digital Section

The PXI system with the NI-RIO field-programmable gate array (FPGA) circuit is used for implementation of Section 3.2. The FPGA module contains also digital-to-analog converters DAC$_{(1-4)}$, analogue-to-digital converters ADC$_{(1-8)}$, and data memory for signals MEM1 and MEM2 with a fixed size of B_x = 4000 samples per signal block. The data can be written to memory MEM1 and MEM2, and read from this memory, by FIFO queues via DMA channels. The DACs and ADCs work synchronously with the sampling frequency of F_s = 400 kHz. FPGA transmits 100 blocks of signals per second to the host. The system has been implemented in LabVIEW.

As shown in Figure 4, FPGA supports digital-to-analog and analog-to-digital conversion, whereas the main system CPU is used for excitation generation and impedance parameters calculation and acquisition. Considering the available hardware resources, it is expected that the demodulation can fully be implemented in FPGA, what will be the subject of further work.

The basic configuration of the system hardware is described in Table 3.

Table 3. Configuration of the digital section.

Symbol	Details
PXI	NI-PXIe-1082 Chassis, NI-PXIe-8133 Embedded Controller with 8 Intel Core i7 CPU Q820 1.73GHz and 2GB RAM.
FPGA	NI PXI-7853R (R series) - NI-RIO FPGA Device with 8 ADC and DAC, 750 kS/s, 16 bit, $\pm$10 V.
Op-Amps	AD845
R_r	560 Ω, $\pm$1 %
R_T	1 Ω, $\pm$1 %
Tr	Ratio 1:7

3.3. Excitation

The excitation voltage $E[n]$ is a K-sine discrete-time signal that can be defined in general form by

$$E[n] = \sum_{k=1}^{K} A_k \sin(\omega_k n + \varphi_k) \tag{2}$$

where $\omega_k = 2\pi F_k / F_s$ is a normalized pulsation in radians of a discrete-time signal, F_s is the sampling frequency in hertz, A_k is the amplitude, F_k is the frequency in hertz, φ_k is the phase angle in radians, the lower subscript k refers to the kt frequency component of the excitation signal, and n stands for the sample number.

Amplitudes A_k may be adjusted to provide expected DUT current for given frequency [17]. The phase angles φ_k have been properly selected to minimize the crest factor [18].

To avoid discontinuities in the excitation signal (2), excitation packets include an integer number of sinusoidal cycles of each frequency and are converted to analog signals without any transitions effects, i.e., obtained continues-time signals are smooth. As long as the block of excitation signal B_x has the length of 4000 samples and the sampling frequency F_s is 400 kHz, we obtain the minimum frequency spacing $f_d = F_s / B_x$ ($f_d = 100$ Hz) between working frequencies.

3.4. Signal Demodulation

Discrete-time measurement signals $V_x[n]$ and $V_r[n]$ are bandpass signals with amplitude and phase modulation in the form

$$V_x[n] = \sum_{k=1}^{k} A_k^x[n] \sin(\omega_k n + \varphi_k^x[n]), \tag{3}$$

where the upper subscript x refers to the signal $V_x[n]$. With the symbol x replaced by r in (3) we get the model valid for $V_r[n]$. Signals $V_x[n]$ and $V_r[n]$ are demodulated simultaneously on all excitation pulsations ω_k by means of the flat-top bandpass Hilbert transformers (FTBPHT) implemented as FIR filters [28]. FTBPHT was chosen due to exceptional accuracy of amplitude and phase demodulation, and high rejection of unwanted spectral components, e.g., other carrier frequencies. Complex-valued impulse response $h_k[n]$ of FTBPHT FIR filter is obtained by modulating the flat-top window $w_M[n]$ to the desired excitation pulsation ω_k

$$h_k[n] = w_M[n]e^{j\omega_k n}, \tag{4}$$

where the flat-top window is the following cosine window [29],

$$w_M[n] = \begin{cases} \sum_{m=0}^{M} c_M[m] \cos(m\frac{\pi}{N}n), & n = -\frac{N-1}{2}, ..., \frac{N-1}{2} \\ 0, & \text{otherwise} \end{cases} \tag{5}$$

where M is the window order, $c_M[m]$ are coefficients of an M-order window, and N is odd window length. We used the 3rd order flat-top window with coefficients $c_3[0] = 1$, $c_3[1] = 1.9918$, $c_3[2] = 1.7038$, $c_3[3] = 0.71199$, and length $N = 40{,}001$ samples.

Figure 5 presents frequency responses of three FTBPHTs configured for excitation signal containing three sinusoidal components with frequencies $f_1 = 8$ kHz, $f_2 = 12$ kHz, and $f_3 = 16.2$ kHz.

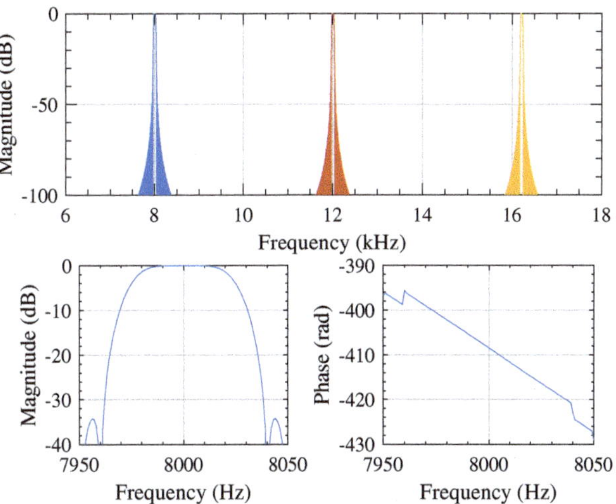

Figure 5. Frequency magnitude responses of the flat-top bandpass Hilbert transformers (FTBPHTs) configured for three working frequencies f_1 = 8 kHz, f_2 = 12 kHz, and f_3 = 16.2 kHz (top), and zoomed magnitude and phase response for the FTBPHT working with f_1 = 8 kHz (bottom). The length of each FTBPHT filter is 40,001 samples. The shapes of magnitude and phase responses of the remaining two filters in passbands are approximately the same.

The decaying of the side lobes of the filters is very fast thus for a specific excitation component high robustness against other frequency components, including remaining excitations, is ensured. The passband of each FTBPHT is flat what gives some margin for excitation frequency fluctuation. For the signal $V_x[n]$ (3), the output of the kth FTBPHT is

$$v_k^x[n] = A_k^x[n]e^{j\varphi_k^x[n]},$$ (6)

and similarly for the signal $V_r[n]$.

3.5. Implementation of Vector Measurement

The output (6) of the kth FTBPHT is the convolution sum of measurement signals $V_x[n]$, $V_r[n]$ (3) and the complex-valued filter impulse response $h_k[n]$ (4)

$$v_k^x[n] = \sum_{m=0}^{N-1} h_k[m]V_x[n-m]$$ (7)

and similarly for $v_k^r[n]$, with indexes x replaced with r in (7).

Measurement signals $V_x[n]$ and $V_r[n]$ are heavily oversampled narrowband signals, and the filter outputs $v_k^x[n]$ and $v_k^r[n]$ may be computed every L samples instead of every single sample. By substituting $n = lL, l = 0, 1, 2, \ldots$ in (7) the filter $h_k[n]$ is shifted by L samples along the filtered signal which L times reduces computational effort. The impedance value over time for kth excitation component is computed by

$$Z_k[n] = R_k[n] + jX_k[n] = R_r\frac{v_k^x[n]}{v_k^r[n]}.$$ (8)

4. Results

The proposed four-channel system for simultaneous, multifrequency impedance measurement of IL sensors was built using FPGA and tested. In every single channel, three unique excitation frequencies have been applied. Taking into account the physical properties of the measurement system and its environment, and also the previous test and simulation results [16], the frequency range from 6 to 17.2 kHz with spacing of 4 kHz was adopted. This range of frequencies ensures acceptable measurement sensitivity and disturbance immunity. The list of excitation frequencies is presented in Table 4.

Table 4. The list of excitation frequencies applied in the system.

Frequency Value in kHz in a Given Channel	f_1	f_2	f_3
#1: for the first standard IL1 sensor	8	12	16.2
#3: for the second standard IL3 sensor	9	13	17.2
#2: for the first slim IL2 sensor	6	10	14.2
#4: for the second slim IL4 sensor	7	11	15.2

where: f_1, f_2, f_3—denote excitation frequencies

The sampling frequency of the ADC and DAC converters was set to F_s = 400 kHz. For signal demodulation the $h_k[n]$ filter length N = 40,001 coefficients, and the downsampling parameter of L = 400 samples was set. In this set-up configuration, the system outputs 1000 samples of impedance per second for a single excitation frequency, which in overall is $1000 \times 4 \times 3$ samples for 4 IL sensors each working on 3 excitation frequencies.

Computed VMPs ($R_k[n]$, $X_k[n]$) are adjusted to begin with the value equal zero for presenting in common plots.

Figure 6 shows the system operation validation results for the first standard IL1 sensor channel during the short switching the relay (see Figure 4). Changes visible in the real part of measured impedance signals (R), increment equal to 1 Ω, and also negligible changes in the imaginary part (X), confirm the system correct operation.

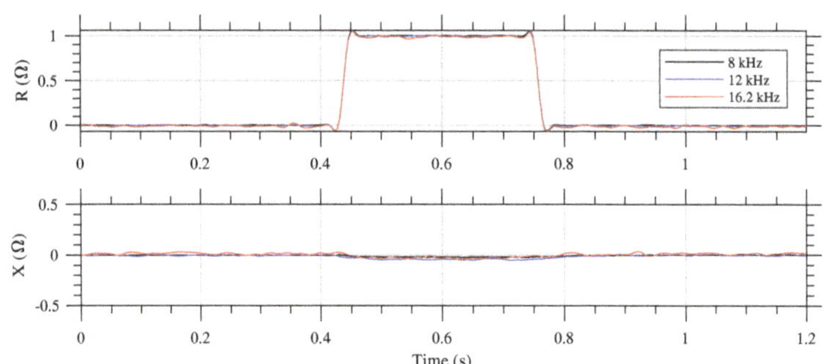

Figure 6. The test-relay system operation validation results for the first standard IL1 sensor channel. On the top, there is the R component, bottom X, and the legend describes the frequencies, see Table 4.

Vehicles that passed through the sensor stand were also photographed. Figure 7 shows vehicles for which VMPs are further presented and discussed. Figures 8–10 present exemplary VMPs as estimated by the FTBPHTs shown in Figure 5 for vehicles depicted in Figure 7.

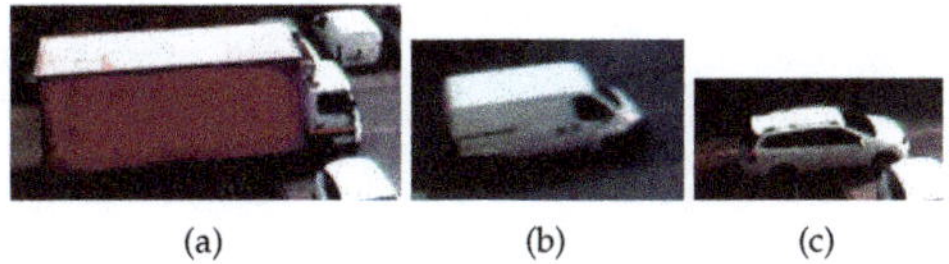

Figure 7. Vehicles for which VMPs were registered and analyzed: (**a**) a truck, (**b**) a delivery car, and (**c**) a passenger car.

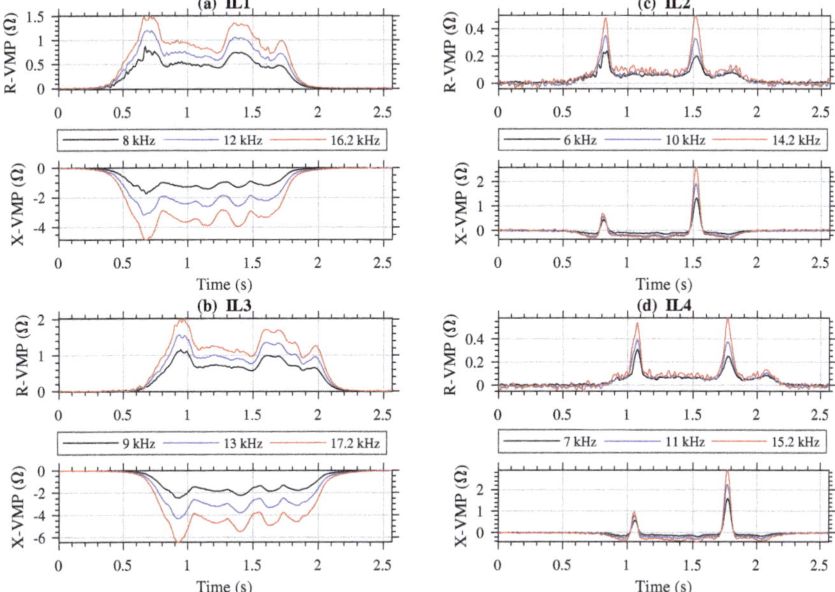

Figure 8. VMPs of a truck shown in Figure 7a; R-VMP and X-VMP denote the real and the imaginary impedance components; stimulation frequencies are listed in the legend; (**a**) the first standard IL1 sensor, (**b**) the second standard IL3 sensor (**c**) the first slim IL2 sensor, and (**d**) the second slim IL4 sensor.

Figure 8 shows VMPs of a truck presented in Figure 7a. Figure 8a presents the VMPs obtained from the first standard IL1 sensor (see Figures 1a and 2), Figure 8b presents the VMPs obtained from the second standard IL3 sensor, Figure 8c presents the VMPs obtained from the first slim IL2 sensor, and Figure 8d presents the VMPs obtained from the second slim IL4 sensor, for frequencies listed in Table 4.

In the case of a truck, we conclude that (1) all R-VMPs components are positive, (2) all X-VMPs from the standard IL1 and IL3 sensors are negative, and (3) X-VMP components of slim IL2 and IL4 sensors have a characteristic spikes that comes from ferromagnetic elements in wheels, that is steel belts. Those spikes allow the measurement of vehicle axles arrangement in vehicle body and also the distance between axles.

In addition, the absolute maximum values of VMPs from the slim IL2, and IL4 are much smaller than from the standard IL1 and IL3. Due to the fact that the IL sensors are arranged one after the other on the lane, VMPs are shifted in time. Distances between IL sensors are known, therefore measurement of vehicle speed [5], and vehicle length is possible.

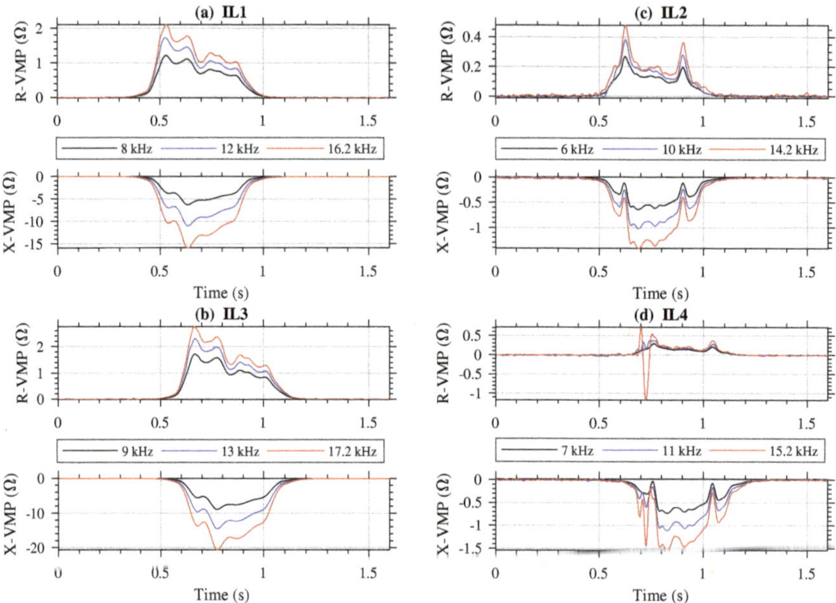

Figure 9. VMPs of a delivery car shown in Figure 7b, further description the same as in Figure 8.

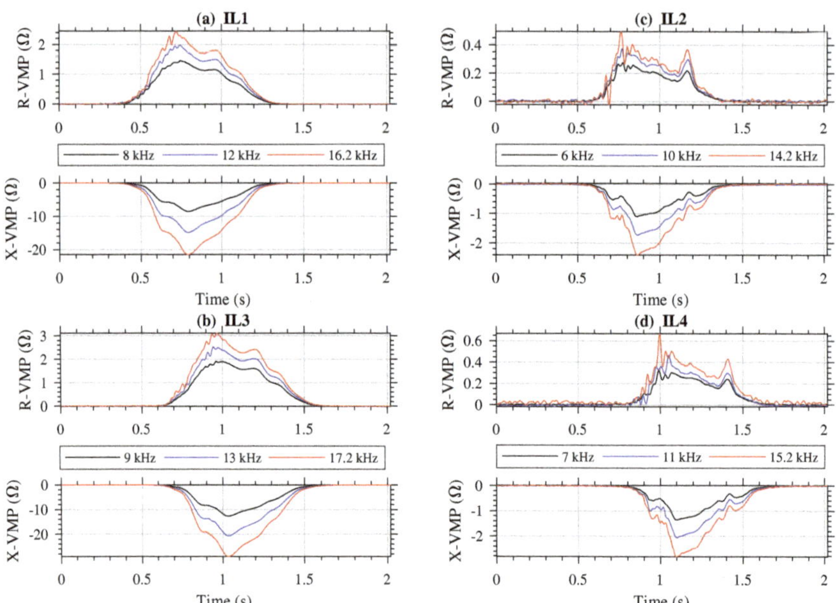

Figure 10. VMPs of a passenger car shown in Figure 7c, further description the same as in Figure 8.

The VMPs of a delivery vehicle shown in Figure 7b are presented in Figure 9. It can be seen that high truck-specific spikes do not occur. Instead, there are visible subtle local maxima in VMPs from IL2 and IL4 sensors, see Figure 9c,d. Delivery vehicles generally have lower suspension than truck vehicles. Low vehicle suspension causes intense eddy currents in flat metal elements, and weakens the effect from the steel belt of a tire. R-VMP and X-VMP obtained at frequency f_3 = 15.2 kHz in channel 4 to which the IL4 sensor is connected are disturbed during interval from 0.7 to 0.8 s (see Figure 9d). A subtle local spike that comes from the vehicle axle is deformed and cannot be used in axle detection process.

The VMPs of a truck, see Figure 8, are not disturbed. The results in all channels coincide well.

The case in Figure 9d illustrates the main advantage of the proposed simultaneous multifrequency estimation. It is observed that the measurement for the excitation frequency f_3 is heavily disturbed; however, the measurement for the excitation frequency f_1 is disturbance-free. Thus, by introducing some redundancy we increased measurement reliability.

Analyzing Figure 10, that is, VMPs obtained for a passenger car presented in Figure 7c, we conclude that (1) VMPs from the standard IL1 and IL3 are different from those of the delivery vehicle, (2) VMPs from the slim IL2 and IL4 have even less visible spikes coming from the vehicle axles, (3) VMPs from the slim IL2 at $f_3 = 14.2$ kHz are disturbed in 0.7–0.8 s time interval, and (4) VMPs from the slim IL4 at $f_2 = 11$ kHz and $f_3 = 15.2$ kHz are also disturbed during the 0.95–1.05 s time interval. Axle detection in low-suspension passenger vehicles is the most difficult case. If there is at least one undisturbed X-VMP and one undisturbed R-VMP from the slim IL sensor, then axle detection is feasible. By using a 3-frequencies system for impedance measurement, there is increased probability of obtaining undisturbed one R-VMP and X-VMP.

Further data processing and data fusion of VMPs are expected to improve the axle counting and also the measurement of other vehicle parameters, however it is outside the scope of this work.

5. Conclusions

In the paper, the proposed VMP measurement system was thoroughly presented. Full system realization and operation were described. The properties of the standard and slim IL sensors and the way of their pavement installation were examined. It was practically verified that the simultaneous operation with multifrequency excitations improves the overall reliability of VMP measurement. Simultaneous measurement of VMPs at different frequencies, and processing only disturbance-free VMPs, reduces measurement system uncertainty as compared to the system working at only one frequency.

Another open problem is the possibility of retrieving the signal of interest from several possibly disturbed channels working on different frequencies.

The advantages of the described VMPs acquisition method over another methods [5,7,14–16] are as follows.

(1) The described method does not lose information on the real part of the IL sensor impedance and provides the R-VMP. The R-VMP and the X-VMP signals may be used in the future to develop algorithms for visualizing the vehicle undercarriage structure. It should be noted that the most commonly exploited LC-generator-based method is unable to provide the R-VMP, only the X-VMP associated with the inductance of IL sensor is acquired. On the other hand, the AC-bridge-based system provides only one VMP where the real and imaginary components of the IL sensor impedance are irreversible mixed together.

(2) The ABB method does not require time-consuming balancing and significantly simplifies the design of the analogue part which makes the system more reliable. Other methods usually exploit AC-bridge or other circuit, which requires time-consuming balancing and is generally a cumbersome problem.

(3) The described VMPs acquisition method is multifrequency and short-time. Several VMPs at selected frequencies are measured simultaneously during the short time of vehicle passage through the measurement stand. Simultaneous multifrequency measurement diminishes negative effects of a vehicle engine EMI disturbance, which, in turn, reduces the number of incorrectly measured vehicles.

(4) The described system is also a multisensor method, therefore the VMPs signal fusion is possible to be applied for more accurate or sophisticated vehicle parameters measurement.

Author Contributions: Conceptualization, Z.M.; methodology, Z.M.; software, Z.M.; validation, Z.M. and K.D.; formal analysis, Z.M. and K.D.; investigation, Z.M.; resources, Z.M.; data curation, Z.M.; writing—original draft preparation, Z.M. and K.D.; writing—review and editing, Z.M. and K.D.; visualization, Z.M.; supervision, Z.M.; project administration, Z.M.; funding acquisition, Z.M. All authors have read and agreed to the published version of the manuscript.

Funding: This work was supported by AGH University of Science and Technology contract no 11.11.120.774.

Conflicts of Interest: The authors declare no conflicts of interest.

Abbreviations

The following abbreviations are used in this manuscript.

IL	Inductive Loop
ABB	Auto Balancing Bridge
IMU	Impedance Measurement Unit
VMP	Vehicle Magnetic Profile
R-VMP	Resistance VMP
X-VMP	Reactance VMP
FTBPHT	Flat Top Band Pass Hilbert Transformers

References

1. Mocholi-Salcedo, A.; Arroyo-Nunez, J.H.; Milian-Sanchez, V.M.; Palomo-Anaya, M.J.; Arroyo-Nunez, A. Magnetic Field Generated by the Loops Used in Traffic Control Systems. *IEEE Trans. Intell. Transp. Syst.* **2017**, *18*, 2126–2136. [CrossRef]
2. Klein, L.A. *Sensor Technologies and Data Requirements for ITS*; Artech House: Massachusetts, MA, USA, 2001.
3. Sivaraman, S.; Trivedi, M.M. Looking at vehicles on the road: A survey of vision-based vehicle detection, tracking, and behavior analysis. *IEEE Trans. Intell. Transp. Syst.* **2013**, *14*, 1773–1795. [CrossRef]
4. Mohammed Ali, S.S.; George, B.; Vanajakshi, L. An Efficient Multiple-Loop Sensor Configuration Applicable for Undisciplined Traffic. *IEEE Trans. Intell. Transp. Syst.* **2013**, *14*, 1151–1161. [CrossRef]
5. Belenguer, F.M.; Salcedo, A.M.; Ibañez, A.G.; Sánchez, V.M. Advantages offered by the double magnetic loops versus the conventional single ones. *PLoS ONE* **2019**, *14*, e0211626. [CrossRef]
6. Al-Tarawneh, M.; Huang, Y.; Lu, P.; Tolliver, D. Vehicle Classification System Using In-Pavement Fiber Bragg Grating Sensors. *IEEE Sensors J.* **2018**, *18*, 2807–2815. [CrossRef]
7. Gajda, J.; Piwowar, P.; Sroka, R.; Stencel, M.; Zeglen, T. Application of inductive loops as wheel detectors. *Transp. Res. Part Emerg. Technol.* **2012**, *21*, 57–66. [CrossRef]
8. Gajda, J.; Burnos, P.; Sroka, R. Accuracy Assessment of Weigh-in-Motion Systems for Vehicle's Direct Enforcement. *IEEE Intell. Transp. Syst. Mag.* **2018**, *10*, 88–94. [CrossRef]
9. Marszalek, Z.; Zeglen, T.; Sroka, R.; Gajda, J. Inductive Loop Axle Detector Based on Resistance and Reactance Vehicle Magnetic Profiles. *Sensors* **2018**, *18*, 2376. [CrossRef] [PubMed]
10. Ki, Y.K.; Baik, D.K. Model for accurate speed measurement using double-loop detectors. *IEEE Trans. Veh. Technol.* **2006**, *55*, 1094–1101. [CrossRef]
11. Liu, H.; Tok, Y.C.A.; Ritchie, S.G. Development of a real-time on-road emissions estimation and monitoring system. In Proceedings of the 14th International IEEE Conference on Intelligent Transportation Systems (ITSC), Washington, DC, USA, 5–7 October 2011; pp. 1821–1826. [CrossRef]
12. Ndoye, M.; Totten, V.F.; Krogmeier, J.V.; Bullock, D.M. Sensing and Signal Processing for Vehicle Reidentification and Travel Time Estimation. *IEEE Trans. Intell. Transp. Syst.* **2011**, *12*, 119–131. [CrossRef]
13. Guilbert, D.; Le Bastard, C.; Ieng, S.S.; Wang, Y. Re-identification by Inductive Loop Detector: Experimentation on target origin–destination matrix. In Proceedigs of the 2013 IEEE Intelligent Vehicles Symposium (IV), Gold Coast, QLD, Australia, 23–26 June 2013; pp. 1421–1427.
14. Marszalek, Z.; Sroka, R.; Zeglen, T. Inductive loop for vehicle axle detection from first concepts to the system based on changes in the sensor impedance components. In Proceedings of the 20th International Conference on Methods and Models in Automation and Robotics (MMAR), Miedzyzdroje, Poland, 24–27 August 2015. [CrossRef]

15. Marszalek, Z. Maxwell-Wien bridge with vector voltmeter system for measurement small and rapid changes in inductive-loop sensor impedance components. *Measurement* **2018**, *121*, 57–61. [CrossRef]
16. Marszalek, Z.; Sroka, R.; Zeglen, T. Multi-frequency conditioning system of the inductive loop sensor—Simulation investigations. In Proceedings of the 22nd International Conference on Methods and Models in Automation and Robotics (MMAR), Miedzyzdroje, Poland, 28–31 August 2017; pp. 889–893. [CrossRef]
17. Yin, W.; Dickinson, S.J.; Peyton, A.J. A multi-frequency impedance analysing instrument for eddy current testing. *Meas. Sci. Technol.* **2006**, *17*, 393–402. [CrossRef]
18. Sanchez, B.; Bragos, R. Multifrequency simultaneous bioimpedance measurements using multitone burst signals for dynamic tissue characterization. *J. Phys. Conf. Ser.* **2010**, *224*, 012004. [CrossRef]
19. Sanchez, B.; Fernandez, X.; Reig, S.; Bragos, R. An FPGA-based frequency response analyzer for multisine and stepped sine measurements on stationary and time-varying impedance. *Meas. Sci. Technol.* **2014**, *25*, 015501. [CrossRef]
20. Koster, D.; Du, G.; Battistel, A.; Mantia, F.L. Dynamic impedance spectroscopy using dynamic multi-frequency analysis: A theoretical and experimental investigation. *Electrochim. Acta* **2017**, *246*, 553–563. [CrossRef]
21. Keysight. Impedance Measurement Handbook. Available online: https://www.cmc.ca/wp-content/uploads/2019/07/Keysight-Technologies-impedance-measurement-handbook.pdf (accessed on 31 August 2020).
22. E4980A Precision LCR Meter, 20 Hz to 2 MHz. Available online: https://www.keysight.com/en/pd-715495-pn-E4980A/precision-lcr-meter-20-hz-to-2-mhz?&cc=HK&lc=eng (accessed on 31 August 2020).
23. Temes, G.; Candy, J. A tutorial discussion of the oversampling method for A/D and D/A conversion. In Proceedings of the Circuits and Systems (ISCAS), New Orleans, LA, USA, 1–3 May 1990; pp. 910–913.
24. Kay, S. *Fundamentals of Statistical Signal Processing: Estimation Theory*; Prentice-Hall, Englewood Cliffs: New Jersey, NJ, USA, 1993.
25. Yao, Y.X.; Pandit, S.M. Cramer-Rao lower bounds for a damped sinusoidal process. *IEEE Trans. Signal Process.* **1995**, *43*, 878–885. [CrossRef]
26. Devices, A. *A Technical Tutorial on Digital Signal Synthesis*; Analog Devices: Norwood, MA, USA, 1999.
27. Marszalek, Z.; Duda, K.; Turcza, P. Design and Experimental Validation of Multi-Frequency Impedance Measurement System. In Proceedings of the 2018 International Conference on Signals and Electronic Systems (ICSES), Kraków, Poland, 10–12 September 2018. [CrossRef]
28. Duda, K.; Turcza, P.; Zielinski, T.; Marszalek, Z. Flat-Top Bandpass FIR Hilbert Transformers. In Proceedings of the 2018 International Conference on Signals and Electronic Systems (ICSES), Kraków, Poland, 10–12 September 2018. [CrossRef]
29. Duda, K.; Zieliński, T.P.; Barczentewicz, S.H. Perfectly flat-top and equiripple flat-top cosine windows. *IEEE Trans. Instrum. Meas.* **2016**, *65*, 1558–1567. [CrossRef]

© 2020 by the authors. Licensee MDPI, Basel, Switzerland. This article is an open access article distributed under the terms and conditions of the Creative Commons Attribution (CC BY) license (http://creativecommons.org/licenses/by/4.0/).

Article

Compact Current Reference Circuits with Low Temperature Drift and High Compliance Voltage

Sara Pettinato, Andrea Orsini and Stefano Salvatori *

Engineering Department, Università degli Studi Niccolò Cusano, via don Carlo Gnocchi 3, 00166 Rome, Italy; sara.pettinato@unicusano.it (S.P.); andrea.orsini@unicusano.it (A.O.)
* Correspondence: stefano.salvatori@unicusano.it

Received: 7 July 2020; Accepted: 25 July 2020; Published: 28 July 2020

Abstract: Highly accurate and stable current references are especially required for resistive-sensor conditioning. The solutions typically adopted in using resistors and op-amps/transistors display performance mainly limited by resistors accuracy and active components non-linearities. In this work, excellent characteristics of LT199x selectable gain amplifiers are exploited to precisely divide an input current. Supplied with a 100 μA reference IC, the divider is able to exactly source either a ~1 μA or a ~0.1 μA current. Moreover, the proposed solution allows to generate a different value for the output current by modifying only some connections without requiring the use of additional components. Experimental results show that the compliance voltage of the generator is close to the power supply limits, with an equivalent output resistance of about 100 GΩ, while the thermal coefficient is less than 10 ppm/°C between 10 and 40 °C. Circuit architecture also guarantees physical separation of current carrying electrodes from voltage sensing ones, thus simplifying front-end sensor-interface circuitry. Emulating a resistive-sensor in the 10 kΩ–100 MΩ range, an excellent linearity is found with a relative error within ±0.1% after a preliminary calibration procedure. Further advantage is that compliance voltage can be opposite in sign of that obtained with a passive component; therefore, the system is also suitable for conditioning active sensors.

Keywords: selectable gain amplifier; resistive-sensor; current divider; current reference

---

1. Introduction

Transistor-based current sources and current mirrors represent fundamental solutions in analog Integrated Circuits (IC) design [1], as well as in discrete circuits, whenever a biasing current source or a reference current is needed [2]. Current sources applications range from biasing and stabilization of circuits to reference or linearizing systems for conditioning of resistive-sensors widely used in industrial applications to measure different quantities (temperature, pressure, strain, and gas concentration, to name a few), as well as to set a test condition or simply regulate a signal for actuation. In these contexts, different solutions are available on the market [3]. Current generators capable of providing stable currents in the order of nA are also required for instrument calibration (i.e., picoammeters) and for materials characterization, both in experiments and in production tests [4]. As an example, in the widely studied field of nano-photonics and quantum photonics, it is of fundamental importance to have a stable low-level current source to precisely characterize the electroluminescence of fabricated devices [5,6].

Regarding resistive sensors, the basic signal conditioning circuit for allowing a measurable output voltage is the Wheatstone Bridge (WB), the output of which may in turn be converted into time intervals by means of time width modulators or into frequency by means of oscillators. In the literature, there are several low-cost solutions used to convert the sensor resistance variation into a time interval, such as Resistance to Period Converter (RPC) or Resistance to Pulsewidth Modulation (RPM) circuits [7–10]. Although WB is widely used, it shows the inconvenient of nonlinear dependence from the sensor impedance. Several linearization techniques have been reported, as the Current Mode WB [11], switches driven integration/deintegration method [12], and the double differential potential subtractor [13]. The latter only requires a constant and stable current flowing through resistive-sensor, as well as resistance matching. This method is well suited for conditioning resistive-sensors when relatively high resistances are concerned, i.e., the estimated error for a sensor resistance range of 15 kΩ–1.1 MΩ may be lower than ±1% [14–17]. In all the above-mentioned cases, it appears essential to provide a stable current in the range of 0.1 μA–10 μA flowing through resistive sensor. It is therefore clear how particular resistive-sensors are and how much they need a dedicated front-end capable of measuring resistance variations over wide ranges. As described, several solutions for resistive-sensors interfacing have been proposed. To evaluate the performance of these solutions, in some cases resistive-sensors were used [18]; more often, experimental evaluations were conducted using commercial resistors to simulate the sensor over a wide resistance range [19–22].

Even if high-performance dedicated monolithic current source ICs are commercially available [23], for particular applications it is necessary to design a specific circuit able to meet some given requirements [24]. At the circuit board level, the simplest ways to build such current source is to apply a linear voltage ramp to a differentiation capacitor or to use precision both voltage references and resistor in order to increase the system accuracy and ensure it over time [25,26]. However, they are mainly realized starting from op-amps acting as Voltage-to-Current (V-I) converters (i e , transconductance amplifiers) [27]. V-I converters accept a voltage as input to source or sink a current from a load. They would also act as a current-reference if the input is a voltage reference. Since V-I converters output is a current, they need a load to work and, within their voltage compliance value, they work properly independently on the load nature.

Op-amp based Howland current-pump solution [28] is the most widely used circuit for the implementation of accurate V-I converters. For instance, they have been proposed as fully-integrated solutions for piezoresistive and resistive-sensors supply [29], as well as for electrotactile stimulation in sensory substitution systems for blind people [30]. Moreover, many Electrical Impedance Tomography (EIT) applications apply Howland circuits as current sources [31–33]. Howland circuit topology being similar to that of a difference amplifier (DA), low-power, high-precision integrated DAs have been recommended for current sources implementation [34]. Although several commercially available devices integrate both low noise op-amps and precision thin film resistors (with a fairly low temperature coefficient and an extremely good matching ratio), DA-based solutions require at least an external precision resistor to establish the desired V-I conversion factor. Conversely, a compact single-chip solution based on commercially available Selectable-Gain Amplifiers (SGA) requiring no additional external component has been proposed for Howland circuit implementation [35]. Due to the SGA internal thin film resistors excellent matching ratio and the low temperature coefficient, the designed circuit shows outstanding performance in terms of linearity, output resistance, and temperature dependence, with the possibility of implementing high-precision current sources operating in the μA–mA range. Moreover, its effectiveness for biasing and conditioning of platinum resistor-based temperature sensors has been verified [36].

In this paper, exploiting the aforementioned excellent performance of SGA devices, the realization and characterization of a precision current divider are illustrated. Experimental results highlight that circuit accuracy is mainly limited to that of the REF200 precision current reference [23] used to supply the single-chip current divider. Moreover, with a thermal coefficient lower than 10 ppm/°C in the range 10–40 °C, the circuit represents a good choice as on-board calibration system for compact pA-meters [26,37–39]. In addition, the adopted circuit architecture both guarantees a voltage compliance close to power supply rails and the separation of current carrying electrodes of the sensor and the sensing point for voltage reading. The system, tested on the field by emulating a resistive sensor with a sample of commercial resistors, shows excellent linearity over more than four decades of load resistance values.

Taking also into account its compactness, the proposed system would represent a valid solution as front-end electronics in sensor-interfacing to AD conversion and processing circuitry.

2. Circuit Description

In this section, design consideration as well as simulated characteristics of two different current references are illustrated. Describing the system requirements, the next section will finally show the experimental results obtained on a prototype based on the design choices indicated hereafter.

2.1. SGA-Based Current Divider Circuit Analysis

Exploiting the tightly-matched on-chip resistor ratios, selectable-gain amplifiers LT1991/5/6 (by Analog Devices) can be configured into precision current dividers by strapping their pins [40]. With absolute values of tens of kΩ, LT1991 [41] integrates resistor with 1/1, 1/3, and 1/9 ratios, whereas 1/9, 1/27, and 1/81 are available for the LT1996 [42]. The latter well adapts for the realization of current dividers down to a hundredth by proper parallel connection of feedback resistors. Figure 1 shows an example of current divider based on the LT1996, able to source or sink an output current of about 1 μA with an input signal generated by the precision 100 μA current-reference REF200 chip [23]. Indeed, LT1996 pins have been connected for a 450 kΩ/109 resistance between op-amp output and its inverting input. By means of the 450 kΩ resistance between non-inverting input and output, the circuit acts as high precision 1:109 current divider.

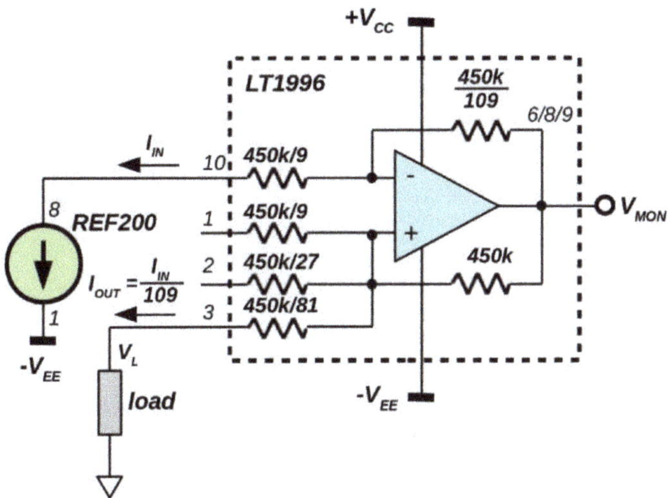

Figure 1. Precision current-divider based on an LT1996 selectable-gain amplifier. A REF200 current reference (by Texas Instruments) is used to have a source current of about 1 μA into the load.

Due to a resistor matching within ±0.05% [42], as worst case, a ±0.2% of accuracy is expected for the 1:109 ratio. For the proposed circuit, the positive compliance voltage of the output current is limited to about V_{CC} − 1.2 V, which is the common mode voltage limit of the op-amp inputs. Conversely, a negative compliance voltage of about $-V_{EE}$ + 8 V is obtained, due to both the minimum 3 V supply for the REF200 and the 5 V voltage drop on the 50 kΩ resistor (at pin 10 of LT1996). Due to the dual power supply, negative values of voltage compliance allow to source also an active load.

It is worth to observe that the circuit is well suited to the conditioning of a high-resistance sensor used as load. Indeed, decoupled from the sensor itself, the voltage amplitude on the load can be measured at the low-impedance op-amp output node having $V_L = V_{MON} - V_{OFF}$, with V_{OFF}, around 0.42 V, the V_{MON} value measured short-circuiting the load during a preliminary calibration procedure.

Equivalent current source non-idealities are mainly related to the input bias I_B and offset I_{OS} currents, as well as the offset voltage V_{OS} of the op-amp. By inspection of the schematic of Figure 1, we obtain:

$$I_{OUT} = \frac{I_{IN}}{109} + \frac{V_{OS}}{450 \times 10^3} + \frac{I_{B(-)}}{109} - I_{B(+)} = \frac{I_{IN}}{109} + \frac{V_{OS}}{450 \times 10^3} - \frac{108}{109}I_B - \frac{110}{109}I_{OS} \qquad (1)$$

where $I_{B(+)} = I_B + I_{OS}$ and $I_{B(-)} = I_B - I_{OS}$ are the bias currents absorbed by non-inverting and inverting inputs, respectively.

Assuming for I_B, I_{OS} and V_{OS} the values declared in [42], typical 2.7 nA and maximum 5.2 nA absolute errors are then estimated for I_{OUT}. Hence, current divider circuit accuracy would be within ±0.55%. In addition, a thermal coefficient (TC) of about 68 ppm/°C is also evaluated by Equation (1) taking into account the 7.5 nA maximum value for the bias current in the −40/+85 °C range, i.e., 60 pA/°C, and the $\Delta V_{OS}/\Delta T = 1$ μV/°C (corresponding to 2.2 pA/°C for I_{OUT}) values reported for the LT1996 [42].

Equivalent output resistance $\Delta V_L/\Delta I_{OUT}$ of implemented current source should be evaluated by means of op-amp parameter dependence on common mode voltage V_{CM} amplitude. A change in V_{CM} will alter the operating point of op-amp input-stage giving rise to a change at the output and reflected at the input in the form of an offset error $\Delta V_{OS} = \Delta V_{CM} / CMRR$, where CMRR is the common-mode rejection-ratio of the op-amp. For the circuit illustrated in Figure 1, neglecting the voltage drop at the 5.56 kΩ (pin 3) resistor connecting load to the op-amp non-inverting input, $V_{CM} \approx V_L$, hence:

$$R_{OUT} = \frac{\Delta V_L}{\Delta I_{OUT}} \approx CMRR \frac{\Delta V_{OS}}{\Delta I_{OUT}} \sim CMRR \times 450k\Omega \qquad (2)$$

where the last equality has been found differentiating Equation (1) Assuming for CMRR the typical value of 120 dB for the LT1996 chip, a 450 GΩ value for R_{OUT} is estimated. Although the absolute tolerance of LT1996 internal resistors is fairly poor (±30%), R_{OUT} would range between 300 GΩ and 600 GΩ, a remarkable high value for the proposed reference-current generator.

In order to get a better insight into system characteristics, a circuit simulation has been performed by means of the equivalent SPICE model for LT1996. In order to evaluate main performance of LT1996-based current divider, an ideal 100 μA constant current generator has been used as REF200. Simulation result reported in Figure 2 outlines excellent performance in terms of the output resistance with a value as high as 700 GΩ. From Equation (1), it is worth noting that this value corresponds to a CMRR around 124 dB, a quantity declared for LT1996 for a voltage gain of 81 which is actually the ratio of the resistors seen at op-amp output.

A sink current is simply obtained by reversing the REF200 connections indicated in Figure 1 and obviously changing its supply voltage to V_{CC}. In this case both the accuracy and the equivalent output resistance assume values equal to those estimated by previous analysis and performed by means of Equation (1) and simulations. Conversely, positive and negative compliance voltages assume almost complementary values to those aforementioned for the source current reference of Figure 1, i.e., $-V_{EE} + 1$ V $< V_L < V_{CC} - 8$ V. In this case, too, it is possible to provide a current supply for an active load.

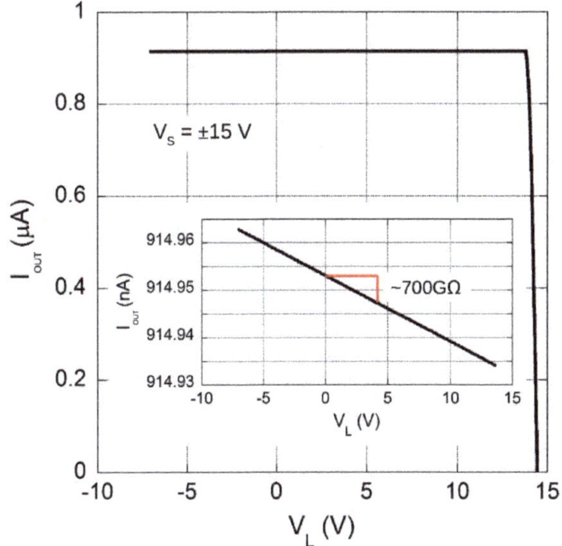

Figure 2. Simulated I_{OUT} as a function of the load voltage V_L for the circuit of Figure 1 using the SPICE model of LT1996 chip, whereas a 100 µA constant current was used of I_{IN}. Simulation has been performed for the allowed $V_L > -7$ V voltage range (see text). The inset reports the zoomed I-V characteristics in order to evaluate the equivalent output resistance.

Simulating either source or sink output currents also allow to evaluate the amplitude of expected values of I_{OUT} and I_B (@ $V_L = 0$ V). Indeed, from Equation (1), considering the two cases $I_{OUT} = I_{SOURCE}$ (with I_{IN} flowing as indicated in Figure 1) and $I_{OUT} = I_{SINK}$ (with I_{IN} sourcing pin 10 of LT1996), and neglecting voltage V_{OS} and current I_{OS} offset contributions we have:

$$I_{OUT} = \frac{I_{SOURCE} - I_{SINK}}{2} = 917.431nA \tag{3}$$

$$I_{B(+)} = \frac{I_{SOURCE} + I_{SINK}}{2} = 2.48nA \tag{4}$$

The former is actually the expected $I_{IN}/109$ value. Hence, Equation (3) allows the value of output current to be evaluated by eliminating the contribution due leakage currents of Equation (1) (and estimated with Equation (4)). This idea will be verified with experimental characterization carried out on the assembled prototype.

Finally, Table 1 summarizes main characteristics of the proposed circuit taking into account also REF200 features. It should be noted that other current-division values can be obtained simply by modifying the connections between LT1996 pins. For example, referring to the schematic of Figure 1, connecting pin 10 to 6 and using 8 as input, a 1:10 ratio is obtained.

2.2. 92 nA Current Reference Analysis

The presence of unused tightly-matched resistors at the op-amp non-inverting input of LT1996 suggests to achieve an additional division of the output current for the circuit of Figure 1. In particular, a 1:4 ratio is obtained by using resistors either at pins 1–2 or at pins 2–3, whereas a 1:10 factor is gained by means of the resistors connected at pin 1 and pin 3. In order to assure a current division independent of the V_L load voltage, hence a high voltage compliance for the generator, used pins must work at the same V_L. Figure 3 shows the schematic of the implemented circuit. By virtual connection of pins 1 and 3 of U2 performed by op-amp U3, output current is further divided by a factor of 10. Neglecting bias current and offset induced errors by both the two ICs, the load current is here equal to $I_{IN}/1090$, therefore 91.74 nA as nominal value. It is worth noting that in this solution V_L can be measured at U3 output (V_{MON} pin in Figure 3). Unlike the above-mentioned circuit illustrated in Figure 1, in this case, the error is ideally represented only by the input-offset voltage of U3 and, again, can be evaluated by short-circuiting the load during a preliminary calibration procedure.

Table 1. Main characteristics evaluated for the circuit of Figure 1.

	Min	Typ	Max	Unit
Current Value (nominal)		917.4		nA
Current Accuracy		±0.54%	±1.55%	
Temperature Drift			95	ppm/°C
Output Impedance	500	700	900	GΩ
Voltage Compliance (source)	$-V_{EE} + 8$ V		$V_{CC} - 1.2$ V	
Voltage Compliance (sink)	$-V_{EE} + 1.2$ V		$V_{CC} - 8$ V	
Supply Voltage			±20	V

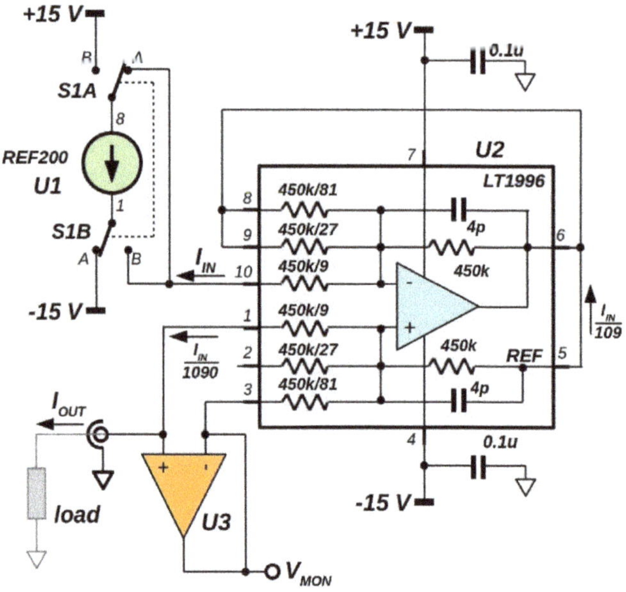

Figure 3. Schematic of the implemented current source. LT1996 acts as precision 1:109 current divider, whereas U3 allows a further 1:10 division of the output current. For a 100 µA source, a ~92 nA either source or sink current flows into the load by means of the S1 double-pole, double-throw switch.

Although REF200 integrates two independent 100 µA references, in order to maintain the same accuracy, a double-pole, double-throw switch has been inserted allowing either sourcing or sinking current into the load by using the same generator. As aforementioned, in accordance with Equation (3), in this way, the $(I_{SOURCE} - I_{SINK})/2$ amplitude can be evaluated independently of the error induced by op-amps bias currents. This solution would reveal effective for calibration of a measuring instrument as underlined by the good accuracy found for the prototype as described in the next section.

As for the circuit shown in Figure 1, also in the new schematic the virtual connection that U3 makes on the output current divider should guarantee a constant output signal over a wide range of V_L between the supply rails. In particular, for a source reference, S1 in position A, the highest value for V_L coincides with the smallest common mode limit between U2 and U3. Conversely, as found for the aforesaid circuit, minimum value for V_L is limited around $V_{EE} + 8$ V. For a sink generator, S1 in position B, the voltage compliance will be in the range between $-V_{LIM}$ and $V_{CC} - 8$ V, with $-V_{LIM}$ the highest negative common mode limit between U2 and U3. In both cases, for a purely passive load, the proposed circuit would assure a maximum load voltage fairly close to the supply rails, and dictated by common mode limits of either U2 or U3.

In order not to degrade the expected excellent performance of the current generator of Figure 1 highlighted in the previous section, U3 must be chosen within op-amp families with ultra-low offset voltage and bias current. In particular, by inspection of the circuit illustrated in Figure 3, the absolute leakage current on I_{OUT} is now expressed as

$$I_{leak} \approx I_{B(U3)} + I_{OS(U3)} + 0.1 \times \left[I_{B(U2)} + I_{OS(U2)}\right] + 1.8 \times 10^{-5} \times V_{OS(U3)} + 2.2 \times 10^{-7} \times V_{OS(U2)} \quad (5)$$

with terms having the same meaning as those of Equation (1). As expected, contributions introduced by LT1996 are now reduced by a factor of 10. On the contrary, the output current is directly unbalanced by the current $I_{B(+)}$ of U3, and its offset voltage gives a leakage quantity 81 times that induced by U2.

As shown by experimental results obtained for the realized prototype and illustrated afterwards, the dependence of op-amps input bias current cannot be neglected. However, as a rough estimation of TC, previous equation can be used to evaluate how thermal drift of op-amp offset voltages affect that of the output current:

$$TC(I_{OUT}) \approx \frac{1.8 \times 10^{-5} \times TC\left[V_{OS(U3)}\right] + 2.2 \times 10^{-7} \times TC\left[V_{OS(U2)}\right]}{I_{OUT\_NOM}} \quad (6)$$

where $I_{OUT\_NOM} = 91.74$ nA represents the nominal value for I_{OUT}. The contribution from U3 being 100 times larger than that of LT1996, Equation (6) stresses the need to use for U3 a device having a very low offset voltage drift, too.

For the proper choice of the device, circuit simulations have been performed for different op-amps and results are summarized in Table 2. The main characteristics of the op-amps used for U3 are shown in the first columns of the same table. Circuit simulations allowed to evaluate the output current amplitudes for either source or sink references, both for $V_L = 0$ V. The R_{OUT} values have been calculated as shown in Figure 2 evaluating the slope of the I_{OUT}-V_L characteristic outside any saturation condition. Finally, column I_B shows the leakage current values, calculated with Equation (4), for I_{SOURCE} and I_{SINK} at $V_L = 0$ V, while for all cases, Equation (3) gave the expected value of 91.743 nA.

Table 2. Main characteristics and simulation results for some ICs chosen for U3.

U3	I_B (nA)	V_{OS} (µV)	TC_{OS} (µV/°C)	CMRR (dB)	I_{OUT_SOURCE} (nA)	I_{OUT_SINK} (nA)	R_{OUT} (GΩ)	I_B (nA)
OP07	4	75	1.3	120	91.495	−91.991	38.7	−0.248
OP27	80	100	0.6	120	91.495	−91.991	5.7	−0.248
OP1177	3.8	61	2.2	126	93.488	−89.998	3.1	1.745
OPA189	0.3	3	0.005	168	91.430	−92.056	270	−0.313
OP191	65	500	1.1	90	120.01	−63.475	10.7	28.27
LTC1022	0.15	1000	3	92	91.495	−91.991	22	−0.248
LTC2054	0.003	10	0.1	130	91.496	−91.990	258	−0.247

Despite to superior performances obtained with LTC2054, it should be noted that the last three rows of Table 2 refer to low noise and precision devices that work in a more limited range of the supply voltage than the LT1996. Following the aforesaid requirements, by contrast OPA189 [43] was used for the final prototype. This device, with both common-mode voltage range and voltage supply comparable to that of the LT1996, should show the best performance thanks to the extremely high CMRR, as well as its very low bias current and the ultra-low absolute value and thermal coefficient of the offset voltage. Compared to the current divider alone, simulated output resistance of the circuit with OPA189 drops to 270 GΩ, a value however high enough to have an error in the order of only

0.04 nA for a 10 V compliance voltage value. It is worth observing that 270 GΩ corresponds to a CMRR of 134 dB which is actually the low-frequency value declared for OPA189.

3. Prototype Characterization

A current-reference circuit was assembled following the schematic illustrated in Figure 3. In order to reduce electromagnetic interference, the circuit was encapsulated in a metal box and a triaxial connector was used for the output with outer shield connected to line-earth. In Figure 4, a picture of the realized prototype is reported. At $V_L = 0$ V, source and sink currents were (91.060 ± 0.002) nA and (−92.183 ± 0.002) nA, respectively, as measured by a Keithley 6517A electrometer. Hence, by Equation (3), an average value of (91.622 ± 0.004) nA is evaluated, in very good agreement with the ideal 91.743 nA, with an inaccuracy of −0.13% within that of REF200 (±0.25% typical). In addition, the absolute error of (0.562 ± 0.004) nA, estimated with Equation (4), is in perfect agreement with the amplitude of the current offset input for the OPA189 (up to ± 600 pA) although the other contributions expressed in Equation (5) cannot be ruled out. Disconnecting REF200 from pin 10 of LT1996, a (99.895 ± 0.09) μA value was measured from its output by means of the 6517A electrometer. Therefore, the evaluated 1:(1090.3 ± 0.9) factor underlines the excellent accuracy of on-chip resistor ratios of LT1996.

Figure 4. Picture of the prototype used for circuit characterization. The circuit was placed in a metal box. A triaxial cable with external shield connected to the earth-line was used for output connection.

Output load regulation has been evaluated by measuring the output current amplitude as a function of the allowed range of output voltage V_L generated by the source unit integrated in the Keithley 6517A. As expected, when REF200 biasing decreases below ~3 V, a sharp change in the output current value is observed. To highlight circuit performance outside any saturation condition, Figure 5 shows the output currents in a region around the 92 nA absolute values. By the data shown in the figure, an equivalent output resistance of about 100 GΩ, hence a 0.2 nA error in a wide range for V_L, is evaluated. This value is almost three times lower than the estimated one reported in Table 2.

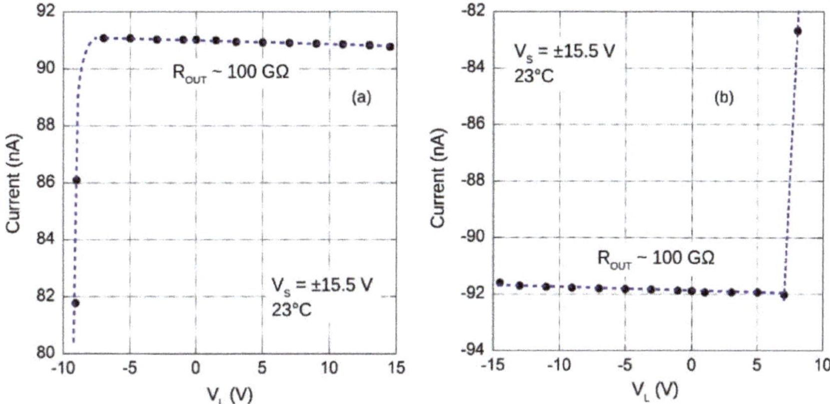

Figure 5. Output current as a function of the load voltage for both source (**a**) and sink (**b**) currents, with S1 of Figure 3 in A and B position, respectively.

As expected by simulation and theoretical analysis, with the applied ±15.5 V supply voltage, a high voltage compliance is verified: $-7.5\ \text{V} < V_L < 14.5\ \text{V}$ and $-14.5\ \text{V} < V_L < 7.3\ \text{V}$ for the source (Figure 5a) and sink (Figure 5b) reference, respectively.

A climate chamber was used for temperature characterization of the circuit. The output current amplitude in the source configuration was acquired in the range 0–125 °C, although LT1996 performances are guaranteed up to only 85 °C. Experimental results are reported in Figure 6. A sharp decrease in the current value above 80 °C is clearly observed. However, as depicted in Figure 7, it is worth noting that in the range 0–60 °C the temperature drift, calculated by dividing the min-max current difference (156 pA) in the temperature range shown in the figure, is slightly lower than 30 ppm/°C and comparable to that of REF200. In addition, in the wide range around the ambient temperature (25 ± 15) °C, current is practically stable and no temperature drift is clearly observed.

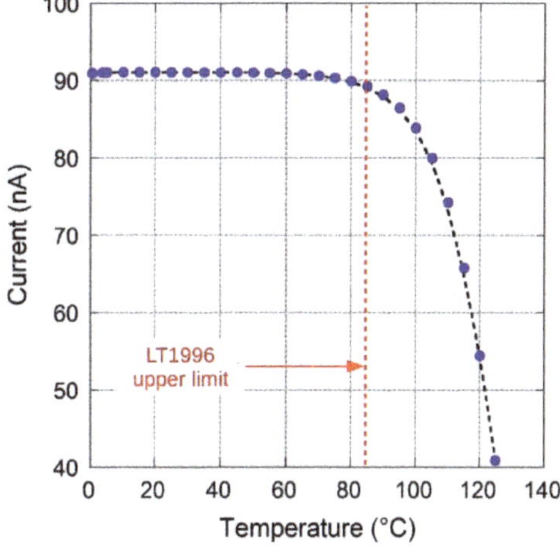

Figure 6. Circuit output source current I_{SOURCE} values as a function of temperature.

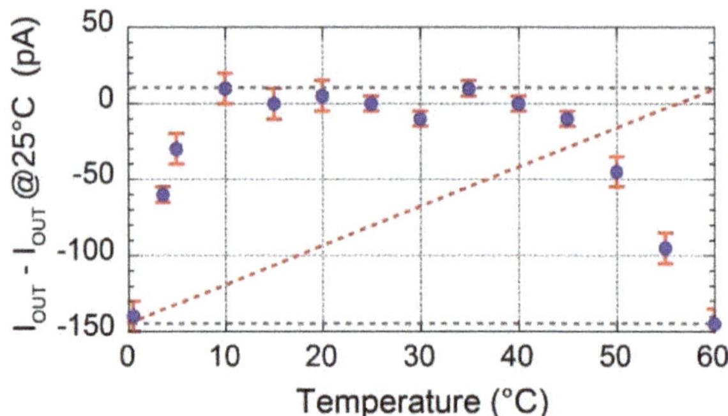

Figure 7. Currents have been calculated subtracting the value at 25 °C to evaluate the temperature drift in the 0–60 °C range.

To get an insight into the role that op-amps have on the temperature characteristics of the circuit, the $| I_{OUT}$ (T) − 91.622 Na|quantity has been calculated. Here, I_{OUT} (T) represents data reported in Figure 6, whereas subtracting value is the above-mentioned current estimated by Equation (3) at ambient temperature. Hence, according to Equation (5), calculated values represent an estimate of leakage current due to both bias currents and offset voltages of op-amps. Plot of data, reported in Figure 8, shows that error contributions by op-amps is quite constant up to 40 °C. Conversely, an exponential behavior is clearly observed for $T > 70$ °C. In particular, current amplitude double every 9 °C (red dotted line) which is typical for reverse-biased pn-junction, hence attributed to diodes inserted for ESD protection of op-amp input stage. This increased bias current would be a significant problem for high temperature applications. However, the low thermal drift depicted in Figure 7 highlights excellent performance of the prototype in the relatively wide 0–60 °C range and underlines the good performance of chosen devices.

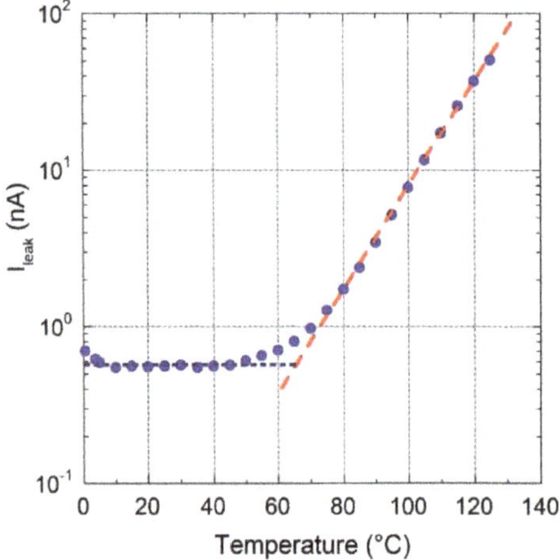

Figure 8. Current values calculated subtracting the 91.622 nA value estimated at ambient temperature to highlight leakage contribution due to reverse-biased op-amp input diodes (red dotted line).

To evaluate LT1996 role on circuit performance, a further characterization was carried out on the same prototype but disconnecting U3. In this case, following the schematic depicted in Figure 1, the output current from pin 3 of LT1996 was measured. Experimental results are reported in Figure 9a in the range 0–125 °C. Here output current amplitude shows a sharp decrease for $T > 85$ °C, whereas it is almost constant, within ±300 pA, at lower temperatures. Hence, up to the maximum temperature of 85 °C declared for LT1996, a very low temperature coefficient (TC) of about 10 ppm/°C is found. From Equation (1), with data reported in [42] and neglecting the contribution due to resistors' matching TC, the worst case for I_{OUT} temperature dependence can be evaluated as:

$$\frac{\Delta I_{OUT}}{\Delta T} \approx 2.2 \times 10^{-6} \left| \frac{\Delta V_{OS}}{\Delta T} \right| + \left| \frac{\Delta I_B}{\Delta T} \right| + \left| \frac{\Delta I_{OS}}{\Delta T} \right| \approx 40 pA/°C \tag{7}$$

i.e., a TC ≈ 44 ppm/°C. Taking into account also the REF200 thermal drift, as worst case a TC around 90 ppm/°C should be obtained. The lower value for TC experimentally estimated for the realized prototype highlights that a partial compensation between different contributions would exist.

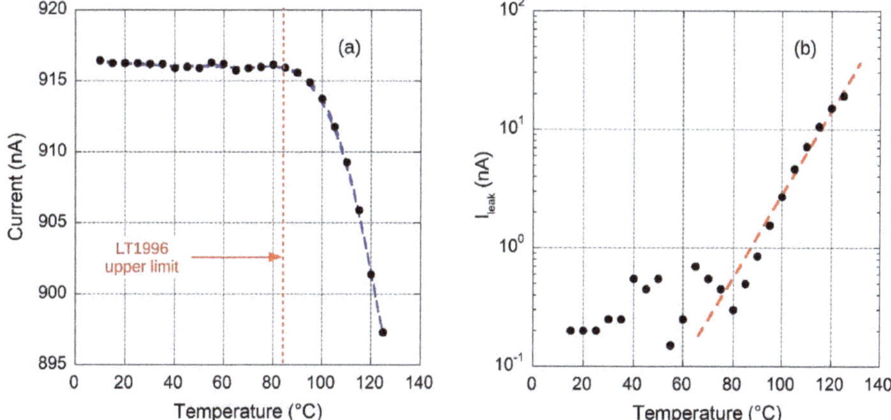

Figure 9. (**a**) Output of ~1 μA current reference circuit as a function of temperature. In (**b**), the value measured at 25 °C was subtracted to data reported in (**a**) to highlight leakage contribution due to reverse-biased op-amp input diodes (red dotted line).

The increase in op-amp bias current is also observed at the high temperature regime. Figure 9b shows the plot of data obtained subtracting the 916.6 nA average value measured at 25 °C to those of Figure 9a. In this case too, an exponential behaviour is clearly observed for $T > 85$ °C, with amplitude doubling every 9 °C (red dotted line). It is worth to note that I_{leak} values are about one order of magnitude lower than those found with OPA189 inserted, highlighting again the good characteristics of the implemented LT1996-based current divider.

In order to better evaluate the temperature dependence degradation induced by OPA189, Figure 10 shows relative errors r_ε for the two circuits calculated as:

$$r_\varepsilon = \frac{I_{OUT}(T) - I_{OUT}(25 °C)}{I_{OUT}(25°C)} \tag{8}$$

where $I_{OUT}(T)$ are the values acquired at temperature T, and $I_{OUT}(25$ °C$)$ the one acquired at ambient temperature.

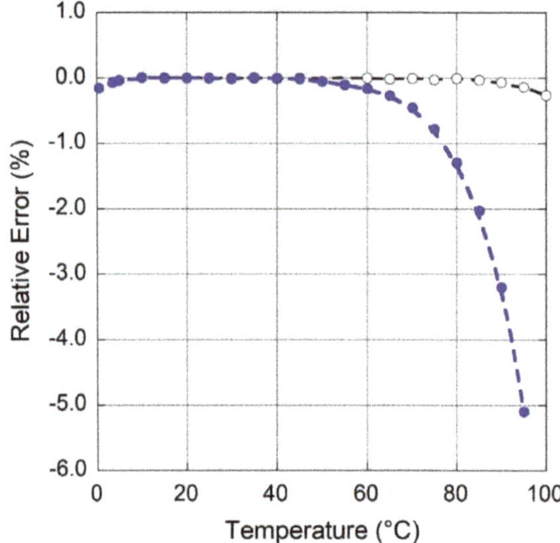

Figure 10. Relative error versus temperature for the circuits following the schematics of Figure 1 (open dots) and Figure 3 (full dots).

For the 916 nA current source, open dots, a maximum error of about 0.3% up to 100 °C is found. Conversely, the 91 nA current source, where OPA189 is included, displays a relative error greater than 5% in the same temperature range. However, an error of about 2% is found in the temperature lower than 85 °C allowed for LT1996. It is worth to observe that this error is comparable to $I_{OS} + I_B = \pm 2.6\,nA$ declared for OPA189 [43] pointing out that most of the errors come from a change of the bias current at non-inverting input of U3 since the error induced by $\Delta V_{OS}/\Delta T$ of the chip is negligible.

To evaluate the long term stability of the output signal, a preliminary characterization has been performed. Raw data acquired continuously in 16 h are reported in Figure 11 together with a smoothed curve (red line). A ±30 pA peak-to-peak noise amplitude of raw data remains over the investigated time interval and the standard deviation is equal to 9.78 pA. The smoothed curve shows peak-peak fluctuations of about ±15 pA, too large to be attributed to laboratory ambient temperature fluctuations but likely depending on noise sources coupled to the circuitry.

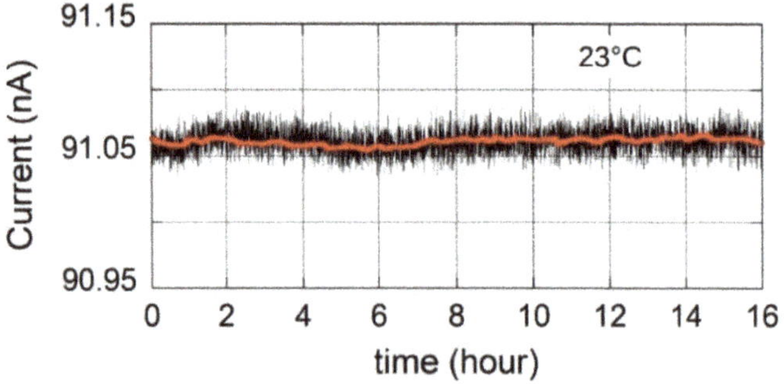

Figure 11. 16 h of continuous acquisition of the output current.

To evaluate the system noise performance, the Keithley 6517A electrometer has been set to acquire 8192 subsequent samples in fast mode: integration time equal to 200 µs and sampling frequency around 79 Hz. Compared to data illustrated in Figure 11, where the instrument was set for an integration time of 20 ms, output current signal displays a higher noise content, with a root mean square value of about 0.2 nA. Amplitude spectral density of the output current was estimated by means of FFT algorithm implemented in MATLAB® (R2019b). Results reported in Figure 12 (blue curve) state that an almost constant distribution is found in the investigated range with a value around 30 pA/√Hz. The same connection set-up was adopted to acquire a 92 nA signal generated by a Keithley 6221 precision current source, too. Results, reported in the same figure (red curve), highlight a less-noisy signal with values around 2 pA/√Hz for f < 10 Hz and as low as 0.6 pA/√Hz at higher frequencies. It is worth to observe that worst case estimation of noise signal induced by ICs as declared by manufactures [23,42,43] should be settle around 1 pA/√Hz level (see green dotted line in the Figure 12). Then, obtained results underline that the noisy-nature of the circuit could be tentatively attributed to a poor-shielding of the used circuit case.

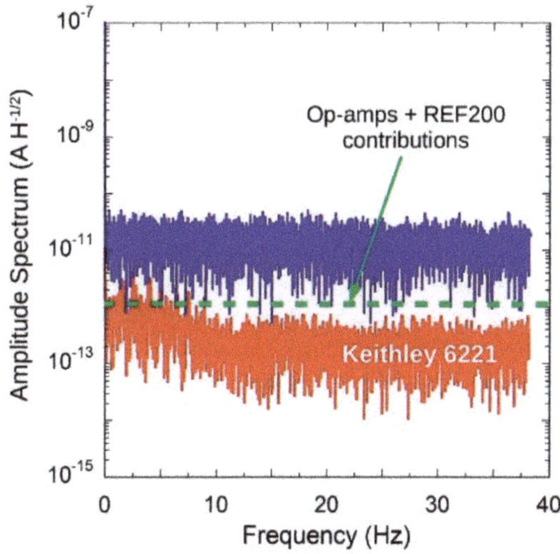

Figure 12. Amplitude spectrum of noise (blue curve) compared to that obtained in the same condition by a Keithley 6221 precision current source (red curve). Green dotted line represents the expected spectral noise level due to ICs used in the circuit.

Although a more detailed investigation of very long-term stability, as well as noise analysis performed by means of a spectrum analyzer, is required, experimental results shown here point out outstanding performance on the characterized prototype in terms of low thermal drift and high stability over time.

As already mentioned, the proposed circuit is suitable for conditioning resistive-sensors, which may have resistances ranging from a few kΩ to hundreds of MΩ. Finally, in this work an experimental test of the proposed circuit, for an output current of about 92 nA, was performed simulating a sensor with different commercial sample resistors. Each resistance value was measured with the Keithley 6517A high resistance meter. Conversely, an Agilent 34401A digital multimeter was used to acquire output voltage V_{MON} values corresponding to the particular resistor connected at the input of current-reference prototype. Figure 13 summarizes obtained results for both I_{SOURCE} (circles) and I_{SINK} (rhombus symbols) currents. Axis on the right refers to the resistance values calculated by dividing the voltages by the 91.74 nA nominal value.

Experimental results demonstrate excellent performance in terms of linearity over more than four decades of resistance variation, from 10 kΩ to 122 MΩ. Best fit of experimental data (dotted line) gives slopes of (90.99 ± 0.02) nA and (92.02 ± 0.03) nA, in good agreement to the values preliminary measured by a Keithley 6517A at $V_L = 0$.

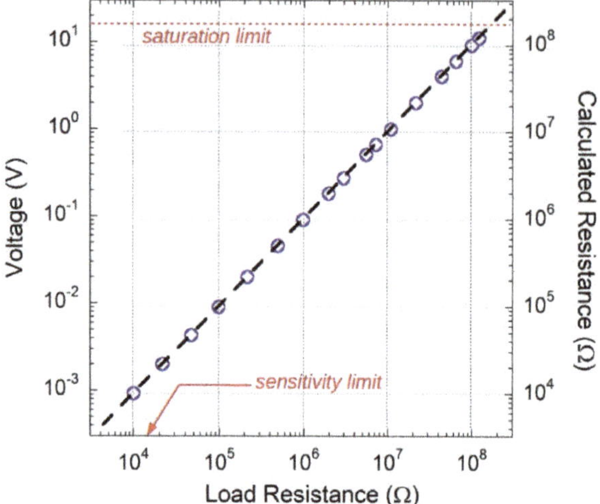

Figure 13. Proposed circuit exhibits an excellent linearity of the output voltage for load resistance values within the 10 kΩ–100 MΩ range.

The Figure 14a shows the relative error values of the resistors calculated using the nominal value of 91.743 nA, compared to the values measured with the 6517A Ohmmeter. It should be noted that the error is around −0.2% for voltages measured with I_{SOURCE} (red circles) and −0.75% for those at I_{SINK} (blue circles). The green squares, on the other hand, refer to the error calculated by considering the "average" voltage ($V_{SOURCE} + |V_{SINK}|$)/2, with which the offset voltage and bias current contributions of the U3 buffer are eliminated. In this case the error is about −0.6%, in agreement with the accuracy of the REF200 reference. It is worth noting that if the resistance value error is calculated taking into account the measured values of I_{SOURCE} and I_{SINK} at $V_L = 0$ (Figure 14b), resistance values within ±0.1% error are obtained. This means that the system allows the resistive load to be accurately measured by a simple preliminary calibration operation with an ammeter connected in parallel to the load.

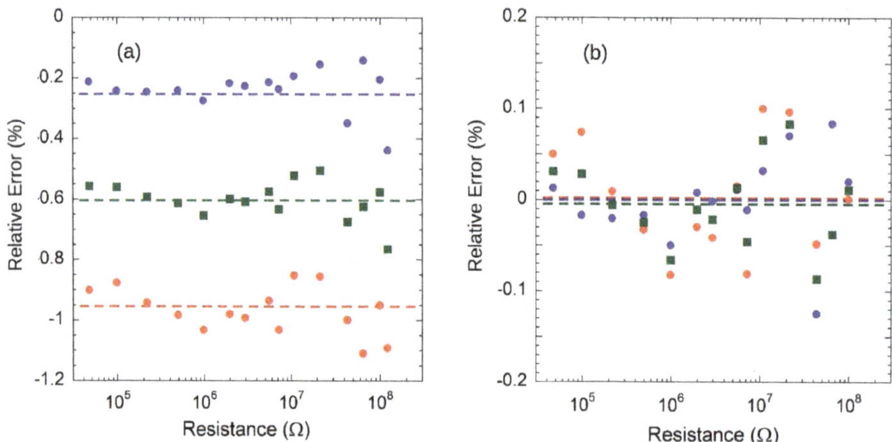

Figure 14. (a) Relative error of calculated resistance values for both I_{SOURCE} (red circles), I_{SINK} (blue circles) and "mean" (green squares) currents (see text). (b) Relative error if measured currents are considered (symbols have the same meaning of (a)). Dotted lines represent mean values of errors.

4. Conclusions

Current reference circuits are widely used in different applications, from biasing and stabilization of circuits, to resistive-sensors conditioning. In this work simple, high precision current references for grounded load have been described. The design and characterization of the realized prototype have been illustrated in terms of equivalent output resistance, thermal coefficient, as well as long-term stability. Main experimental results are summarized in Table 3. In particular, due to its excellent features, LT1996 selectable gain amplifier demonstrates particularly effective for the realization of a high-precision current divider. Two different ratios have been verified in the present work, but it is worth remarking that an appropriate choice of the LT1996 pin connections allows to obtain others I_{OUT}/I_{IN} values according to a specific requirement.

Table 3. Measured electrical characteristics of realized prototype for the two circuits depicted in Figure 1 (1:109 ratio) and in Figure 3 (1:1090 ratio). Supply voltage ±15.5 V, T = 25 °C (unless otherwise specified).

	Min	Typ	Max	Unit
Supply Voltage ($\pm V_S$)	8		20	V
Current Value (source)		91.06		nA
Current Value (sink)		−92.183		nA
Initial accuracy	−0.74		+0.48	%
Temperature drift				ppm/°C
10–40 °C			10	
0–70 °C		30		
Output Resistance		100		GΩ
Voltage Compliance (source)	−7.5		14.5	V
Voltage Compliance (sink)	−14.5		7.3	V
Noise Current (f < 3 Hz)		60		pA$_{P-P}$
		10		pA$_{rms}$
Noise Current BW = 0.1 Hz to 40 Hz		30		pA/$\sqrt{Hz}$
Current Value (source)		916.6		nA
Current Value (sink)		−915.8		nA
Initial accuracy	−0.09		−0.2	%
Temperature drift 0–85 °C		10		ppm/°C
Voltage Compliance (source)	−7.5		14.5	V
Voltage Compliance (sink)	−14.5		7.5	V

Supplied by a REF200 100 µA current reference, realized prototype exhibits good performance for either ~1 µA or ~0.1 µA source/sink current generator. Measured current values at ambient temperature are close to those expected by performed current division, with an absolute inaccuracy lower than 1%, mainly attributed to that of the REF200 IC. Voltage compliance extends at about 1 V of the supply rails (±20 V maximum) assuring good performance even for wide change of load-resistance. For a voltage supply of ±15.5 V, experimental results show that op-amps operate outside any saturation condition between −7.5 V and +14.5 V, and between −14.5 V and +7.5 V for source and sink reference, respectively. In this regard, the circuit also allows active load supply. Moreover, it is worth remarking that circuit architecture allows sensor-voltage sensing separated by current carrying sensor-electrodes, simplifying acquisition circuitry interfacing.

Realized circuit also demonstrates extremely good performance in terms of equivalent output resistance and low thermal coefficient (see Table 3). Experimental results show a relatively low noise amplitude of the generated current and a good stability has been verified up to 16 h. The noise performance of the system was also evaluated allowing to estimate an approximately constant distribution, with a value of about 30 pA/$\sqrt{\text{Hz}}$ up to about 40 Hz to be mainly attributed to a poor shielding of the circuit.

The extremely low components count, up to three ICs and a few by-pass capacitors, means compactness and cost effectiveness of the proposed solution which would find effective application for resistive-sensor biasing as well as conditioning. In this regard, an experimental test was ultimately performed by simulating a sensor with a sample of commercial resistors. In the range from 10 kΩ to 122 MΩ, the system shows excellent performance in terms of linearity. In addition, the relative error values of the resistances have been calculated: errors are around -0.2% and -0.75% for the voltages measured with I_{SOURCE} and I_{SINK}, respectively. The error is around -0.6% when the average voltage value $(V_{SOURCE} + |V_{SINK}|)/2$ is considered, which compensates for the offset voltage and bias current of ICs. The error value is in good agreement with the accuracy of the REF200 reference. Moreover, thanks to a preliminary calibration, the system is appropriate to measure resistive loads accurately. In fact, the error reduces to $\pm0.1\%$ if calculated considering the measured values of I_{SOURCE} and I_{SINK} at $V_L = 0$.

Author Contributions: S.P. and S.S. conceived the ideas and innovations; S.S. and S.P. realized the prototype; S.P. and S.S. performed the simulations and the measurements; all the authors analyzed the data; A.O. contributed a lot to the principle verification, results and discussion of the work; S.P. wrote the paper. All authors reviewed and commented on the manuscript; S.S. provided supervision and guidance in this work. All authors have read and agreed to the published version of the manuscript.

Funding: This research received no external funding.

Conflicts of Interest: The authors declare no conflict of interest.

References

1. Gray, P.R.; Hurst, P.J.; Lewis, S.H.; Meyer, R.G. *Analysis and Design of Analog Integrated Circuits*, 5th ed.; John Wiley & Sons, Inc.: New York, NY, USA, 2009; Ch. 4.
2. Murnane, M. Current Sources: Options and Circuits, AN-968, Analog Devices, Inc. 2008. Available online: https://www.analog.com/media/en/technical-documentation/application-notes/AN-968.pdf (accessed on 5 July 2020).
3. Keithley Instruments Inc. Ultra-sensitive Current Sources. Available online: https://www.tek.com/keithley-low-level-sensitive-and-specialty-instruments/keithley-ultra-sensitive-current-sources-seri (accessed on 5 July 2020).
4. TDK-Lambda Inc. Uses for Constant Current Power in Industrial Applications. Available online: https://www.emea.lambda.tdk.com/uk/KB/Uses-for-constant-current-power-in-Industrial-Applications.pdf (accessed on 5 July 2020).
5. van Treeck, D.; Ledig, J.; Scholz, G.; Lähnemann, J.; Musolino, M.; Tahraoui, A.; Brandt, O.; Waag, A.; Riechert, H.; Geelhaar, L. Electroluminescence and current–voltage measurements of single-(In,Ga)N/GaN-nanowire light-emitting diodes in a nanowire ensemble. *Beilstein J. Nanotechnol.* **2019**, *10*, 1177–1187. [CrossRef]
6. Palacios-Berraquero, C.; Barbone, M.; Kara, D.M.; Chen, X.; Goykhman, I.; Yoon, D.; Ott, A.K.; Beitner, J.; Watanabe, K.; Taniguchi, T.; et al. Atomically thin quantum light-emitting diodes. *Nat. Commun.* **2016**, *7*, 1–6. [CrossRef] [PubMed]
7. Flammini, A.; Marioli, D.; Taroni, A. A low-cost interface to high-value resistive sensors varying over a wide range. *IEEE Trans. Instrum. Meas.* **2004**, *53*, 1052–1056. [CrossRef]
8. Depari, A.; Falasconi, M.; Flammini, A.; Marioli, D.; Rosa, S.; Sberveglieri, G.; Taroni, A. A new low-cost electronic system to manage resistive sensors for gas detection. *IEEE Sens. J.* **2007**, *7*, 1073–1077. [CrossRef]
9. Hoon, K.; Chung, W.S.; Kim, H.J.; Son, S.H. A resistance deviation-to-pulsewidth converter for resistive sensors. *IEEE Trans. Instrum. Meas.* **2009**, *58*, 397–400. [CrossRef]

10. Kokolanski, Z.; Gavrovski, C.; Dimcev, V.; Makraduli, M. Simple interface for resistive sensors based on pulse width modulation. *IEEE Trans. Instrum. Meas.* **2013**, *62*, 2983–2992. [CrossRef]

11. Azhari, S.J.; Kaabi, H. AZKA cell, the current-mode alternative of Wheatstone bridge. *IEEE Trans. Circuits Syst. I Fundam. Theory Appl.* **2000**, *47*, 1277–1284. [CrossRef]

12. Nagarajan Ramanathan, P.; George, B.; Kumar, V.J. A linearizing digitizer for wheatstone bridge based signal conditioning of resistive sensors. *IEEE Sens. J.* **2017**, *17*, 1696–1705. [CrossRef]

13. Anderson, K.F. The new current loop: An instrumentation and measurement circuit topology. *IEEE Trans. Instrum. Meas.* **1997**, *46*, 1061–1067. [CrossRef]

14. Kishore, K.; Malik, S.; Baghini, M.S.; Akbar, M.S. A Dual-Differential Subtractor-Based Auto-Nulling Signal Conditioning Circuit for Wide-Range Resistive Sensors. *IEEE Sens. J.* **2020**, *20*, 3047–3056. [CrossRef]

15. Wu, L.; Song, F.; Fang, X.; Guo, Z.-X.; Liang, S. A practical vacuum sensor based on a ZnO nanowire array. *Nanotechnology* **2010**, *21*, 475502. [CrossRef] [PubMed]

16. Polese, D.; Mattoccia, A.; Giorgi, F.; Pazzini, L.; Ferrone, A.; Di Giamberardino, L.; Maiolo, L.; Pecora, A.; Convertino, A.; Fortunato, G.; et al. Layered Double Hydroxides intercalated with Chlorine used as low temperature gas sensors. *Procedia Eng.* **2015**, *120*, 1175–1178. [CrossRef]

17. De Iacovo, A.; Venettacci, C.; Bruno, S.; Colace, L. Lead Sulphide Colloidal Quantum Dots for Sensing Applications. In Proceedings of the 7th International Conference on Photonics, Optics and Laser Technology (PHOTOPTICS 2019), Prague, Czech Republic, 25–27 February 2019; pp. 235–240, ISBN 978-989-758-364-3. [CrossRef]

18. De Marcellis, A.; Ferri, G.; Mantenuto, P. A novel 6-decades fully-analog uncalibrated Wheatstone bridge-based resistive sensor interface. *Sens. Actuators B Chem.* **2013**, *189*, 130–140. [CrossRef]

19. De Marcellis, A.; Depari, A.; Ferri, G.; Flammini, A.; Sisinni, E. A CMOS integrated low-voltage low-power time-controlled interface for chemical resistive sensors. *Sens. Actuators B Chem.* **2013**, *179*, 313–318. [CrossRef]

20. Depari, A.; Flammini, A.; Marioli, D.; Sisinni, E.; De Marcellis, A.; Ferri, G.; Stornelli, V. A new and fast-readout interface for resistive chemical sensors. *IEEE Trans. Instrum. Meas.* **2010**, *59*, 1276–1283. [CrossRef]

21. Depari, A.; Flammini, A.; Sisinni, E.; de Marcellis, A.; Ferri, G.; Mantenuto, P. Fast, versatile, and low-cost interface circuit for electrochemical and resistive gas sensor. *IEEE Sens. J.* **2014**, *14*, 315–323. [CrossRef]

22. Mantenuto, P.; Ferri, G.; De Marcellis, A. Uncalibrated automatic bridge-based CMOS integrated interfaces for wide-range resistive sensors portable applications. *Microelectron. J.* **2014**, *45*, 589–596. [CrossRef]

23. Texas Instruments. REF200, "Dual, 100-µA Current Sink/Source". Available online: http://www.ti.com/product/REF200 (accessed on 5 July 2020).

24. Harrisson, L.T. *Current Sources & Voltage References*; Newnes-Elesevier: Oxford, UK, 2005.

25. Callegaro, L.; D'Elia, V.; Trinchera, B. A current source for picoammeter calibration. *IEEE Trans. Instrum. Meas.* **2007**, *56*, 1198–1201. [CrossRef]

26. D'Antonio, E.; Salvatori, S.; Oliva, P.; Patanè, F.; Girolami, M.; Trucchi, D.M. High Precision Integrator for CVD-Diamond Detectors for Dosimetric Applications. In Proceedings of the 2018 IEEE International Symposium on Medical Measurements and Applications (MeMeA), Rome, Italy, 11–13 June 2018; pp. 1–6. [CrossRef]

27. Pandiev, I.M. Analysis and design of voltage-controlled current sources for a grounded load. *Int. J. Circ. Theor. Appl.* **2015**, *43*, 756–775. [CrossRef]

28. Sheingold, D.H. Impedance and Admittance Transformations Using Operational Amplifiers. *Lightning Empiricist* **1964**, *12*, 1–8.

29. Pelcia, M.M.; dos Reis Filho, C.A. Fully Integrated Programmable Howland Current Source for Sensors Excitation. In Proceedings of the Fourth IEEE International Caracas Conference on Devices, Circuits and Systems 2002, Oranjestad, Aruba, Dutch Caribbean, 17–19 April 2002. C028–CI-4. [CrossRef]

30. Poletto, C.J.; Van Doren, C.L. A High Voltage, Constant Current Stimulator for Electrocutaneous Stimulation through Small Electrodes. *IEEE Trans. Biomed. Eng.* **1999**, *46*, 929–936. [CrossRef] [PubMed]

31. Li, Z.; Xu, Z.; Ren, C.; Wang, W.; Zhao, D.; Zhang, H. Study of Voltage Control Current Source in Electrical Impedance Tomography System. In Proceedings of the 4th International Conference on Bioinforrnatics and Biomedical Engineering (iCBBE) 2010, Chengdu, China, 18–20 June 2010; pp. 1–4. [CrossRef]

32. Lee, J.W.; Oh, T.I.; Paek, S.M.; Lee, J.S.; Woo, E.J. Precision Constant Current Source for Electrical Impedance Tomography. In Proceedings of the 25th Annual International Conference of the IEEE EMBS, Cancun, Mexico, 17–21 September 2003; pp. 1066–1069. [CrossRef]

33. Al-Obaidi, A.A.; Meribout, M. A New Enhanced Howland Voltage Controlled Current Source for EIT Applications. In Proceedings of the IEEE GCC Conference and Exhibition (GCC), Dubai, UAE, 19–22 February 2011; pp. 327–330. [CrossRef]
34. Zhao, N.; Malik, R.; Liao, W. Difference Amplifier Forms Heart of Precision Current Sources. *Analog Dialogue* **2009**, *43*, 22–24.
35. Salvatori, S. Gain-selectable IC yields voltage-to-current converter. *EDN Eur.* **2015**, *7*, 30–32.
36. Salvatori, S.; Rossi, M.C.; Girolami, M. High-precision voltage-to-current converters based on single-chip gain-selectable amplifiers. *Analog Integr. Circuits Signal Process.* **2019**, *99*, 491–495. [CrossRef]
37. Salvatori, S.; Masarone, N.; Di Nucci, G.; Conte, G. Compact front-end electronics for low-level current sensor measurements. *Electron. Lett.* **2006**, *42*, 682–684. [CrossRef]
38. Pettinato, S.; Orsini, A.; Girolami, M.; Trucchi, D.M.; Rossi, M.C.; Salvatori, S. A High-Precision Gated Integrator for Repetitive Pulsed Signals Acquisition. *Electronics* **2019**, *8*, 1231. [CrossRef]
39. Pettinato, S.; Orsini, A.; Rossi, M.C.; Tagnani, D.; Girolami, M.; Salvatori, S. A Compact Gated Integrator for Conditioning Pulsed Analog Signals. In *Applications in Electronics Pervading Industry, Environment and Society*; Saponara, S., De Gloria, A., Eds.; Springer International Publishing: Cham, Switzerland, 2020; pp. 33–39. [CrossRef]
40. Brisebois, G.; Munson, J. Versatile Op Amps Need No Resistors. *Linear Technol. J.* **2004**, *14*, 1–4.
41. Analog Devices. LT1991, Precision, 100µA Gain Selectable Amplifier. Available online: https://www.analog.com/media/en/technical-documentation/data-sheets/1991fh.pdf (accessed on 5 July 2020).
42. Analog Devices. LT1996, Precision, 100µA Gain Selectable Amplifier. Available online: https://www.analog.com/media/en/technical-documentation/data-sheets/1996f.pdf (accessed on 5 July 2020).
43. Texas Instruments. OPA189 Single, 14MHz, MUX-Friendly, Low-Noise, Zero-Drift, RRO, CMOS Precision Operational Amplifier. Available online: https://www.ti.com/lit/gpn/opa189 (accessed on 5 July 2020).

© 2020 by the authors. Licensee MDPI, Basel, Switzerland. This article is an open access article distributed under the terms and conditions of the Creative Commons Attribution (CC BY) license (http://creativecommons.org/licenses/by/4.0/).

MDPI

St. Alban-Anlage 66

4052 Basel

Switzerland

Tel. +41 61 683 77 34

Fax +41 61 302 89 18

www.mdpi.com

Sensors Editorial Office

E-mail: sensors@mdpi.com

www.mdpi.com/journal/sensors